METHODS IN MOLECULAR BIOLOGY

Series Editor
John M. Walker
School of Life and Medical Sciences
University of Hertfordshire
Hatfield, Hertfordshire, UK

For further volumes:
http://www.springer.com/series/7651

For over 35 years, biological scientists have come to rely on the research protocols and methodologies in the critically acclaimed *Methods in Molecular Biology* series. The series was the first to introduce the step-by-step protocols approach that has become the standard in all biomedical protocol publishing. Each protocol is provided in readily-reproducible step-by-step fashion, opening with an introductory overview, a list of the materials and reagents needed to complete the experiment, and followed by a detailed procedure that is supported with a helpful notes section offering tips and tricks of the trade as well as troubleshooting advice. These hallmark features were introduced by series editor Dr. John Walker and constitute the key ingredient in each and every volume of the *Methods in Molecular Biology* series. Tested and trusted, comprehensive and reliable, all protocols from the series are indexed in PubMed.

In Vitro Differentiation of T-Cells

Methods and Protocols

Edited by

Shin Kaneko

Shin Kaneko Laboratory, Department of Cell Growth and Differentiation, Center for iPS Cell Research and Application, Kyoto University, Kyoto, Japan

Humana Press

Editor
Shin Kaneko
Shin Kaneko Laboratory, Department
of Cell Growth and Differentiation
Center for iPS Cell Research and Application, Kyoto University
Kyoto, Japan

ISSN 1064-3745 ISSN 1940-6029 (electronic)
Methods in Molecular Biology
ISBN 978-1-4939-9727-5 ISBN 978-1-4939-9728-2 (eBook)
https://doi.org/10.1007/978-1-4939-9728-2

This Humana imprint is published by the registered company Springer Science+Business Media, LLC, part of Springer Nature.
The registered company address is: 233 Spring Street, New York, NY 10013, U.S.A.

Preface

The dream of in vitro replication of cell differentiation occurring within our bodies has become a reality with the emergence of pluripotent stem cells, which have paved the way to clinical application with the advent of induced pluripotent stem (iPS) cells. On this occasion, I am honored to be able to publish this protocol book, *In Vitro Differentiation of T-Cells: Methods and Protocols.*

T-cells play a pivotal role in host defense from pathogens; however, sometimes, they cause a serious disease. The in vitro T-cell differentiation system is of vital importance to the pathological analysis, drug development, and cell therapy of human T-cells.

In this protocol book, T-cell induction methods from human pluripotent stem cells, such as human embryonic stem cells (Chapter 6) and human iPS cells (Chapters 6–10), are mainly compiled. Among the various reported methods, this book cites in vitro T-cell differentiation protocols through co-culture with murine stromal cells (Chapter 7), feeder cell-free condition (Chapter 8), T-cell receptor gene-transduced iPS cells (Chapter 9), and chimeric antigen receptor gene-transduced iPS cells (Chapter 10). It also contains a protocol for establishing T-cell-derived iPSCs, which offers an extremely useful tool for research on in vitro T-cell differentiation induction (Chapter 5). Furthermore, this book also briefs the readers on a protocol for in vitro differentiation for T-cell-related immune cells, such as human NK cells (Chapter 12) and human NKT cells (Chapter 13). They both are near to clinical application and are considered a helpful reference for clinical development.

In recent years, when medical practice of using peripheral blood T-cells as drugs has become a reality in the form of CART and TCR-T cells, we cannot dismiss the possibility of clinical applications of in vitro differentiated T-cells. In addition to the protocols for in vitro cell differentiation, this book also compiles useful techniques to develop and evaluate cellular medicines. Specifically, this book contains the details on in vitro T-cell differentiation from pluripotent stem cells of mouse (Chapter 14) and monkey (Chapter 11) that are useful in basic research for predicting their effectiveness and safety in vivo. Furthermore, to assist the characterization of in vitro and in vivo differentiated T-cells, the following protocols have been included: the basics of flow cytometry analysis (Chapters 1 and 2), gene transfer to peripheral blood T-cells and effector function analysis (Chapter 3), single T-cell transcriptome analysis (Chapter 16), in vivo research on blood immune cells using immunodeficient mice (Chapter 19), and in vivo assays for effectiveness and safety of iPSC-derived T-cells through suicide gene switch (Chapter 20).

In addition, this volume collects recently reported interesting technologies, including "rejuvenating" technology for human peripheral T-cells without pluripotent stem cell technology (Chapter 4); characterization techniques for thymic epithelial stem cells located in the thymus, which is the main field of T-cell differentiation (Chapter 15); and methods for the structural analysis of antigens and their specific T-cell receptors (Chapter 17).

I would like to extend my cordial gratitude to the authors and the editors who contributed to the creation of this protocol book despite the fact that my inexperience as an editor caused them so much trouble.

It is my sincere wish that this book would be of help to the researchers who are working on the differentiation of T-cells from stem cells.

Kyoto, Japan ***Shin Kaneko***

Contents

Contributors

MAKOTO ANDO • *Department of Microbiology and Immunology, Keio University School of Medicine, Shinjuku-ku, Tokyo, Japan*

VICTOR APPAY • *Centre d'Immunologie et des Maladies Infectieuses (CIMI-Paris), Sorbonne Université, INSERM, Paris, France; International Research Center of Medical Sciences (IRCMS), Kumamoto University, Kumamoto, Japan*

CURTIS CAI • *Kirby Institute for Infection and Immunity, The University of New South Wales, Sydney, NSW, Australia; School of Medical Sciences, The University of New South Wales, Sydney, NSW, Australia*

XIAONIAO CHEN • *Department of Ophthalmology, Chinese PLA General Hospital, Beijing, China; Department of Ophthalmology, Harvard Medical School, Boston, MA, USA*

SHUNSUKE CHIKUMA • *Department of Microbiology and Immunology, Keio University School of Medicine, Shinjuku-ku, Tokyo, Japan*

ANA DAVILA • *Graduate School of Medicine, Osaka University, Osaka, Japan*

AUDA A. ELTAHLA • *Kirby Institute for Infection and Immunity, The University of New South Wales, Sydney, NSW, Australia; School of Medical Sciences, The University of New South Wales, Sydney, NSW, Australia*

TINHINANE FALI • *Centre d'Immunologie et des Maladies Infectieuses (CIMI-Paris), Sorbonne Université, INSERM, Paris, France*

JACQUELINE FLYNN • *Rheumatology Research Group, Centre for Inflammatory Diseases, School of Clinical Sciences at Monash Health, Monash University, Melbourne, VIC, Australia; School of Health and Biomedical Sciences, RMIT University, Melbourne, VIC, Australia; Burnet Institute, Melbourne, VIC, Australia*

PAUL GORRY • *School of Health and Biomedical Sciences, RMIT University, Melbourne, VIC, Australia*

YOKO HAMAZAKI • *Department of Immunology and Cell Biology, Graduate School of Medicine, Kyoto University, Kyoto, Japan; Laboratory of Immunobiology, Graduate School of Medicine, Center for iPS Cell Research and Application (CiRA), Kyoto University, Kyoto, Japan*

NAOTO HIRANO • *Tumor Immunotherapy Program, Princess Margaret Cancer Centre, University Health Network, Toronto, ON, Canada*

YASUMICHI HITOSHI • *Thyas Co. Ltd., Kyoto, Japan*

YUKI IMURA • *Department of Microbiology and Immunology, Keio University School of Medicine, Shinjuku-ku, Tokyo, Japan; Sohyaku, Innovative Research Division, Mitsubishi Tanabe Pharma Corporation, Aoba-ku, Yokohama, Kanagawa, Japan*

SHOICHI IRIGUCHI • *Shin Kaneko Laboratory, Department of Cell Growth and Differentiation, Center for iPS Research and Application, Kyoto University, Kyoto, Japan*

SHIN KANEKO • *Shin Kaneko Laboratory, Department of Cell Growth and Differentiation, Center for iPS Cell Research and Application (CiRA), Kyoto University, Kyoto, Japan*

KAZUTAKA KATOH • *Research Institute for Microbial Diseases, Osaka University, Osaka, Japan; Immunology Frontier Research Center, Osaka University, Osaka, Japan*

DAN S. KAUFMAN • *Division of Regenerative Medicine, Department of Medicine, University of California, San Diego, San Diego, CA, USA*

YOHEI KAWAI • *Shin Kaneko Laboratory, Department of Cell Growth and Differentiation, Center for iPS Research and Application, Kyoto University, Kyoto, Japan*
SHUICHI KITAYAMA • *Shin Kaneko Laboratory, Department of Cell Growth and Differentiation, Center for iPS cell Research and Application (CiRA), Kyoto University, Kyoto, Japan*
TAISUKE KONDO • *Department of Microbiology and Immunology, Keio University School of Medicine, Shinjuku-ku, Tokyo, Japan*
CAMILLE K'ROS • *Centre d'Immunologie et des Maladies Infectieuses (CIMI-Paris), Sorbonne Université, INSERM, Paris, France*
FENGYANG LEI • *Department of Ophthalmology, Harvard Medical School, Boston, MA, USA*
SONGLING LI • *Research Institute for Microbial Diseases, Osaka University, Osaka, Japan*
TIANYI LIU • *Key Laboratory of Cancer Center, Chinese PLA General Hospital, Beijing, China*
FABIO LUCIANI • *Kirby Institute for Infection and Immunity, The University of New South Wales, Sydney, NSW, Australia; School of Medical Sciences, The University of New South Wales, Sydney, NSW, Australia*
MASAKI MARUTA • *Department of Hematology, Clinical Immunology, and Infectious Diseases, Ehime University Graduate School of Medicine, Toon, Ehime, Japan*
HIROAKI MASHIMA • *Department of Gastroenterological and Transplant Surgery, Graduate School of Biomedical and Health Sciences, Hiroshima University, Hiroshima, Japan*
ATSUTAKA MINAGAWA • *Shin Kaneko Laboratory, Department of Cell Growth and Differentiation, Center for iPS Cell Research and Application, Kyoto University, Kyoto, Japan*
YUKIKO MURMANN • *Institute for Stem Cell Biology and Regenerative Medicine, Stanford University School of Medicine, Stanford, CA, USA*
HIROMITSU NAKAUCHI • *Institute for Stem Cell Biology and Regenerative Medicine, Stanford University School of Medicine, Stanford, CA, USA*
TOSHINOBU NISHIMURA • *Institute for Stem Cell Biology and Regenerative Medicine, Stanford University School of Medicine, Stanford, CA, USA*
TOSHIKI OCHI • *Department of Hematology, Clinical Immunology, and Infectious Diseases, Ehime University Graduate School of Medicine, Toon, Ehime, Japan; Proteo-Science Center, Ehime University Graduate School of Medicine, Toon, Ehime, Japan*
HIDEYUKI OGURO • *Cellular Engineering, The Jackson Laboratory for Genomic Medicine, Farmington, CT, USA*
HIDEKI OHDAN • *Department of Gastroenterological and Transplant Surgery, Graduate School of Biomedical and Health Sciences, Hiroshima University, Hiroshima, Japan*
SIMONE RIZZETTO • *Kirby Institute for Infection and Immunity, The University of New South Wales, Sydney, NSW, Australia; School of Medical Sciences, The University of New South Wales, Sydney, NSW, Australia*
JOHN ROZEWICKI • *Research Institute for Microbial Diseases, Osaka University, Osaka, Japan; Immunology Frontier Research Center, Osaka University, Osaka, Japan*
JEROME SAMIR • *Kirby Institute for Infection and Immunity, The University of New South Wales, Sydney, NSW, Australia; School of Medical Sciences, The University of New South Wales, Sydney, NSW, Australia*
DELPHINE SAUCE • *Centre d'Immunologie et des Maladies Infectieuses (CIMI-Paris), Sorbonne Université, INSERM, Paris, France*
MIHO SEKAI • *Department of Immunology and Cell Biology, Graduate School of Medicine, Kyoto University, Kyoto, Japan; Laboratory of Immunobiology, Graduate School of*

Medicine, Center for iPS Cell Research and Application (CiRA), Kyoto University, Kyoto, Japan

JIANXUN SONG • *Department of Microbial Pathogenesis and Immunology, Texas A&M University, Bryan, TX, USA*

DARON M. STANDLEY • *Research Institute for Microbial Diseases, Osaka University, Osaka, Japan; Immunology Frontier Research Center, Osaka University, Osaka, Japan*

SHUNSUKE TERAGUCHI • *Research Institute for Microbial Diseases, Osaka University, Osaka, Japan; Immunology Frontier Research Center, Osaka University, Osaka, Japan*

YUMI TOKUMITSU • *Division of Cancer Immunotherapy, Exploratory Oncology Research and Clinical Trial Center, National Cancer Center, Kashiwa, Japan*

NORIHIRO UEDA • *Division of Cancer Immunotherapy, Exploratory Oncology Research and Clinical Trial Center, National Cancer Center, Kashiwa, Japan; Division of Immunology, Aichi Cancer Center Research Institute, Nagoya, Japan; Shin Kaneko Laboratory, Department of Cell Growth and Differentiation, Center for iPS cell Research and Application (CiRA), Kyoto University, Kyoto, Japan*

TATSUKI UEDA • *Shin Kaneko Laboratory, Department of Cell Growth and Differentiation, Center for iPS Research and Application, Kyoto University, Kyoto, Japan*

YASUSHI UEMURA • *Division of Cancer Immunotherapy, Exploratory Oncology Research and Clinical Trial Center, National Cancer Center, Kashiwa, Japan; Division of Immunology, Aichi Cancer Center Research Institute, Nagoya, Japan*

WILLEM VAN DER BYL • *Kirby Institute for Infection and Immunity, The University of New South Wales, Sydney, NSW, Australia; School of Medical Sciences, The University of New South Wales, Sydney, NSW, Australia*

FLORIS J. VAN EERDEN • *Research Institute for Microbial Diseases, Osaka University, Osaka, Japan*

JONATHAN VILLANUEVA • *Institute for Stem Cell Biology and Regenerative Medicine, Stanford University School of Medicine, Stanford, CA, USA*

JIANWEI WANG • *Department of Immunology and Cell Biology, Graduate School of Medicine, Kyoto University, Kyoto, Japan; Laboratory of Immunobiology, Graduate School of Medicine, Center for iPS Cell Research and Application (CiRA), Kyoto University, Kyoto, Japan*

LIQIANG WANG • *Department of Ophthalmology, Chinese PLA General Hospital, Beijing, China*

JAN WILAMOWSKI • *Research Institute for Microbial Diseases, Osaka University, Osaka, Japan*

XIAOFANG XIONG • *Department of Microbial Pathogenesis and Immunology, Texas A&M University, Bryan, TX, USA*

ZICHANG XU • *Graduate School of Medicine, Osaka University, Osaka, Japan*

YUTAKA YASUI • *Thyas Co. Ltd., Kyoto, Japan; Shin Kaneko Laboratory, Department of Cell Growth and Differentiation, Center for iPS Cell Research and Application, Kyoto University, Kyoto, Japan*

AKIHIKO YOSHIMURA • *Department of Microbiology and Immunology, Keio University School of Medicine, Shinjuku-ku, Tokyo, Japan*

RONG ZHANG • *Division of Cancer Immunotherapy, Exploratory Oncology Research and Clinical Trial Center, National Cancer Center, Kashiwa, Japan; Division of Immunology, Aichi Cancer Center Research Institute, Nagoya, Japan*

HUANG ZHU • *Division of Regenerative Medicine, Department of Medicine, University of California, San Diego, San Diego, CA, USA*

Chapter 1

Flow Cytometry Analysis to Identify Human CD8$^+$ T Cells

Jacqueline Flynn and Paul Gorry

Abstract

Flow cytometry is a powerful technique allowing multiparameter detection and quantification of single cells or particles including cell size, granularity, cell components (DNA, mRNA), surface receptors, intracellular proteins, and signaling events. The flow cytometer operates via three main systems: the fluidics, optics, and electronics, which work together to analyze the physical and chemical properties of your sample. The first system, the fluidics, transports your sample in a single stream through the instrument, from the sample tube, pass the lasers, and is either sorted for further experiments or discarded into the waste vessel. The second system, the optical system, is composed of a series of lasers; for excitation of your sample and attached fluorescence antibodies as it passes, a series of lenses; and a detector system. The third system is the electronic component, which enables the photocurrent from the detector system to be changed into electronic pulses to be processed by a computer and analyzed by flow cytometry software. Flow cytometry is thus a powerful technique, which is commonly used to determine the expression of cell surface markers and intracellular molecules to define cells into different populations by fluorescently labeled antibodies.

The staining procedure outlined below creates a single-cell suspension for staining with a panel of flow cytometry antibodies, which target different surface markers, to identify an array of cell types. After staining the sample is loaded into the flow cytometer, where the fluorescently labeled cells are excited as they pass by the laser emitting light at various wavelengths which are detected by the flow cytometer. Each fluorescent antibody has its own excitation and emission spectrum allowing the use of multiple antibodies. However, the emission spectrums of different fluorochromes can overlap each other, called spectral overlap. Thus, it is important to have good compensation controls to eliminate any antibody spillover.

The staining methods from this technique can be used for different cell types by changing the surface marker targeted by the flow antibody. It is also important to use knowledge of the density of surface molecule for detection and brightness of fluorochrome to guide antibody selection and also to titrate all antibodies prior to use.

This chapter's protocol has been designed specifically for detection of human CD8$^+$ T cells defining the activation status of the cells by surface marker phenotyping.

Key words CD8$^+$ T cells, Flow cytometry, Activation, Naïve T cells, Memory T cells, CD markers, T cells

Shin Kaneko (ed.), *In Vitro Differentiation of T-Cells: Methods and Protocols*, Methods in Molecular Biology, vol. 2048, https://doi.org/10.1007/978-1-4939-9728-2_1, © Springer Science+Business Media, LLC, part of Springer Nature 2019

1 Introduction

Human $CD8^+$ T cells also known as cytotoxic T lymphocytes (CTLs) play a critical role in immune surveillance and defense, particularly against intracellular pathogens, such as viruses and bacteria, and for tumor surveillance [1]. When a $CD8^+$ T cell recognizes foreign antigen presented by major histocompatibility complex (MHC) class I, it becomes activated. Once activated, the $CD8^+$ T cell engages its ability to be cytotoxic, hence the name CTL commonly given to $CD8^+$ T cells [2].

There are three main ways an activated $CD8^+$ T cell can kill infected or malignant cells: The first is to secrete small proteins called cytokines, commonly interferon gamma (IFN-gamma) and tumor necrosis factor alpha (TNF-alpha) [3]. These cytokines are known for their antiviral and antitumor effects [3]. The second mechanism is via the production and release of cytotoxic granules called perforin and granzymes. Perforin is able to form pores in the membrane of target cells, and granzymes are serine proteases which are able to cleave proteins inside the target cells; these cytotoxic granules often cause apoptosis of the target cell [3, 4]. The third mechanism is via the Fas/Fas ligand interactions which can cause the destruction of the infected cell [4]. The activated $CD8^+$ T cell express the Fas ligand (FasL) which binds to the Fas receptor on the target infected cell. This causes the signaling molecules to activate the caspase cascade, which causes the apoptosis of the target cell.

Similar to other cells, the subpopulations on $CD8^+$ T cells and detection of their effector functions can be defined by multiparameter flow cytometry. These include the analysis of surface marker expression (by phenotyping surface markers), immune function (by intracellular staining of cytokines), and T cell differentiation pathways (intracellular staining of transcription factors).

This chapter focuses on surface marker staining to identify $CD8^+$ T cell subsets and to demonstrate the activation of $CD8^+$ T cells through the use of well-characterized activation markers HLA-DR and CD38. Similar to $CD4^+$ T cells, $CD8^+$ T cells can also be defined into naïve and memory populations using the same set of surface markers. This can distinguish the differentiation status of $CD8^+$ T cell and its effector function, for example, whether it is able to secrete IFN-gamma upon encounter with a foreign antigen presented by MHC class I. The effector function of $CD8^+$ T cells can be defined further through adding flow cytometry-based intracellular cytokine staining to the staining protocol below [5].

2 Materials

1. FACS buffer: 500 mL sterile PBS (Ca^{2+}/Mg^{2+} free), 0.1% BSA, 1 mM EDTA (*see* **Note 1**). Filter sterilize any unsterile components to keep buffer sterile. Keep at 4 °C.
2. FACS FIX: 500 mL sterile PBS, 1% paraformaldehyde (*see* **Note 2**). Keep at 4 °C.
3. Flow cytometry antibodies:

   Panel 1: CCR7 AF-647, CD3 APC Cy7, CD4 FITC, CD8 PE, CD122 PerCP EF710, CD45RO EF450, viability dye EF506, CD95 PE CF594, CD27 PE Cy7, PD-1 BV650, CTLA-1 BV786, and LAG-3 APC-R7.

   Panel 2: CCR7 AF-647, CD3 APC Cy7, CD8 PE, CD122 PerCP EF710, CD45RO EF450, viability dye EF506, CD95 PE CF594, CD27 PE Cy7, HLA-DR BV786, and CD38 BB515.
4. Activation reagents: Soluble CD28 antibody (5 μg/mL) and anti-CD3 antibody (5 μg/mL) diluted in sterile PBS (Ca^{2+}/Mg^{2+} free).
5. Compensation beads: OneComp eBeads.
6. Flow Tubes: 5 mL polypropylene tubes capped (*see* **Note 3**).
7. Range of pipettes (0.3 μL to 1 mL).

3 Methods

Important notes to read prior to starting your surface marker staining which discuss the use of viability markers, compensation controls, experimental controls, and titration of flow antibodies have been added to the end of Subheading 4 (*see* **Notes 4–7**). Useful considerations for flow cytometry panel design have also been added to the end of Subheading 4 (*see* **Notes 8–10**).

Carry out all procedures in a class II biological hazard cabinet to maintain a sterile environment unless described otherwise.

3.1 Coat Plate or Flask with Anti-CD3 and CD28 for Activation of PBMC

1. Make up a solution of anti-CD3 (5 μg/mL) and CD28 (5 μg/mL) and in sterile PBS, and add enough of solution to cover the each well of a 24-well plate or to coat a T25 flask (*see* **Note 11**).
2. Coat 24-well plates or T25 flask with CD28 and anti-CD3 (5 μg/mL) diluted in sterile PBS, either at 4 °C overnight or for 2 h at 37 °C.
3. After coating, rinse wells or flask twice with sterile PBS, and add cells in cell culture media to the wells or flask as outlined in the steps below.

Table 1
Panel 1: flow cytometry CD8$^+$ T cell subset antibody amounts

Flow antibody*	Fluorochrome	Amount per compensation tube	Amount per cell sample	Amount for a ten-sample tube master mix
CCR7	AF-647	2 μL	12 μL	120 μL
CD3	APC Cy7	3 μL	3 μL	30 μL
CD122	PerCP EF710	0.3 μL	1 μL	10 μL
CD45RO	EF450	5 μL	10 μL	100 μL
Viability dye	EF506	N/A	5 μL	50 μL
CD8	PE	0.5 μL	3 μL	30 μL
CD95	PE CF594	0.3 μL	2 μL	20 μL
CD27	PE Cy7	0.5 μL	0.7 μL	7 μL
PD-1	BV650	2 μL	5 μL	50 μL
CTLA-4	BV786	2 μL	5 μL	50 μL
LAG-3	APC-R7	2 μL	5 μL	50 μL

Please note depending upon supplier of flow antibody you use, you may need to titrate the amount of antibody for your experiments. Add the amount per master mix and divide by the number of tubes, for example, for ten tubes as shown in the last column of the table, add 51.7 μL of the master mix per tube. *You can add CD4 FITC to the analysis if needed

3.2 Flow Cytometry Staining

Using thawed peripheral blood mononuclear cells (PBMCs) or freshly isolated PBMCs (from blood using Ficoll-Paque gradient), perform a cell count and resuspend cells in FACS buffer to 10×10^6/mL.

3.2.1 Compensation Tubes

1. Label a flow tube with each antibody and fluorochrome for your compensation control tubes (one tube for each fluorochrome, Table 1 and/or Table 2) (*see* **Note 12**).
2. Vortex compensation beads well.
3. Add one drop of the compensation bead solution to each labeled compensation tube.
4. Add the appropriate amount of antibody to each compensation tube (Table 1 and/or Table 2) (*see* **Note 12**).
5. Stain the beads for 30 min at 4 °C in the dark (in the fridge, covered with foil) (*see* **Note 13**).
6. Add 1 mL of FACS buffer to each compensation tube and spin beads at 400 × *g* for 10 min.
7. Remove supernatant carefully and add 200 μL of FACS FIX.
8. Fix the beads for 30 min at 4 °C in the dark (in the fridge, covered in foil).

Table 2
Panel 2: flow cytometry CD8+ T cell subset antibody amounts

Flow antibody	Fluorochrome	Amount per compensation tube	Amount per cell sample	Amount for a ten-sample tube master mix
CCR7	AF-647	2 μL	12 μL	120 μL
CD3	APC Cy7	3 μL	3 μL	30 μL
CD122	PerCP EF710	0.3 μL	1 μL	10 μL
CD45RO	EF450	5 μL	10 μL	100 μL
Viability Dye	EF506	N/A	5 μL	50 μL
CD8	PE	0.5 μL	3 μL	30 μL
CD95	PE CF594	0.3 μL	2 μL	20 μL
CD27	PE Cy7	0.5 μL	0.7 μL	7 μL
HLA-DR	BV786	2 μL	5 μL	50 μL
CD38	BB515	2 μL	5 μL	50 μL

Please note depending upon supplier of flow antibody you use, you may need to titrate the amount of antibody for your experiments. Add the amount per master mix and divide by the number of tubes, for example, for ten tubes as shown in the last column of the table, add 46.7 μL of the master mix per tube

9. Add 1 mL of FACS buffer to each compensation tube and spin beads at 400 × *g* for 10 min (*see* **Note 11**).
10. Remove supernatant carefully and resuspend in 200 μL of FACS buffer.
11. Store flow tubes at 4 °C covered in foil (in the fridge). The compensation tubes can be stored once fixed for up to 1 week.

3.2.2 CD8+ T Cell Tubes

1. Label flow cytometry tubes with CD8+ activation T cell panel 1 or panel 2, keep one tube labeled as cells alone.
2. Aliquot 200 μL of cell suspension into labeled FACS tubes (approximately one million cells per tube).
3. Set aside the cells alone tube (unstained by antibodies) to allow for correct setup of voltages on the flow cytometer.
4. Make a master mix of titrated flow antibodies (Table 1 and/or Table 2 as appropriate) and to the labeled tubes the appropriate CD8+ activation T cell panel antibody master mix. Resuspend well.
5. Stain cells for 30 min at 4 °C in the dark (in the fridge, covered in foil) (*see* **Note 13**).
6. Add 1 mL of FACS buffer to each flow tube and spin cells at 400 × *g* for 10 min.
7. Remove supernatant carefully and add 200 μL of FACS FIX.

8. Fix cells for 30 min at 4 °C in the dark (in the fridge, covered in foil) (*see* **Note 14**).
9. Add 1 mL of FACS buffer to each flow tube and spin cells at 400 × *g* for 10 min.
10. Remove supernatant carefully and resuspend in 200–400 μL of FACS buffer (*see* **Note 15**).
11. Store flow tubes at 4 °C covered in foil (in the fridge) for up to 1 week.

3.3 Flow Cytometer Data Acquisition

Please note, you will need training to use the flow cytometer to ensure setup and data collection is correct. Contact your flow cytometry department for assistance.

Brief notes for setting up an LSR II flow cytometer:

1. Start the machine and switch on the computer.
2. Ensure adequate sheath fluid levels and that the waste tank is empty (except for 5% bleach for human cells).
3. Prime instrument twice and remove any air bubbles in the system. Remove any air bubbles remaining by gently releasing the relief value at the bubble filter and bleeding the lines into a waste container.
4. Open the software program and setup your templates.
5. Configure the instrument by selecting the lasers and fluorochrome combinations used (an example for the antibodies used in this protocol and configurations for an LSR II flow cytometer is in Table 3).
6. Using your cells alone tube set your voltages for all fluorochromes.
7. Run your compensation control tubes collecting at least 10,000 events.
8. Run your T cell panel tubes uncompensated collecting all events (*see* **Note 16**). Compensation will be done post data acquisition.
9. Export your data and follow the flow cytometry department cleaning and shut down procedure. Turn off the computer and flow cytometer.

3.4 Data Analysis

As multiple fluorochromes are used in this experiment, it is easier to compensate post-acquisition. With a good panel design and good software, this is not too difficult to do post-acquisition compensation, and this also means if you need to change any compensation applied, you can create a new compensation matrix and apply it to your data set. You need to ensure, however, that your compensation is done correctly for each fluorochrome. The compensated populations have the median of each cell population in line with

Table 3
Flow cytometry CD8^{+} T cell antibodies and LSR II instrument configuration

Flow antibody	Fluorochrome	LSR laser	Description
CCR7	AF-647	640 nm	R 670/14
CD3	APC Cy7	640 nm	R 780/60
CD4	FITC	488 nm	B 530/30
CD122	PerCP EF710	488 nm	B 710/50
CD45RO	EF450	405 nm	V 450/50
Viability dye	EF506	405 nm	V 525/50
CCR5	PE	561 nm	YG 582/15
CD95	PE CF594	561 nm	YG 610/20
CXCR4	PE Cy5	561 nm	YG 685/35
CD27	PE Cy7	561 nm	YG 780/60
PD-1	BV650	405 nm	V 660/20
CTLA-4	BV786	405 nm	V 780/60
LAG-3	APC-R7	640 nm	R 730/45
HLA-DR	BV786	405 nm	V 780/60
CD38	BB515	488 nm	B 530/30

Please note several of these antibodies cannot be used together as they are either the same fluorochrome or too closely matched. Refer to the spectral guides and instrument configuration

each other for each fluorochrome used in the experiment (an example is shown in Fig. 1). An example of incorrect compensated populations is shown in Fig. 2. Once compensation is completed and you have saved your compensation matrix, you are ready to start analysis.

Load your FSC files into your software program, and apply the compensation matrix so all files will now be compensated. Below is a strategy of how to set up your gates to define different CD8^{+} T cells via the parameters listed in Table 4 using the panel of CD8^{+} T cells flow cytometry antibodies shown in Table 1 (Fig. 3) [6–9]. Select the statistics you wish to display in your software program, and calculate your parameters of interest, for example, percentage of each CD8^{+} T cell subtype from your clinical samples. Save your experiment and gating template in your software program for future experiments and data analysis (*see* **Note 17**).

To detect the activation status of your CD8^{+} T cell population, you can use activation markers HLA-DR and CD38 (Table 2, Fig. 4b). You will be able to detect a difference between cells activated with CD28 and anti-CD3 and those which aren't

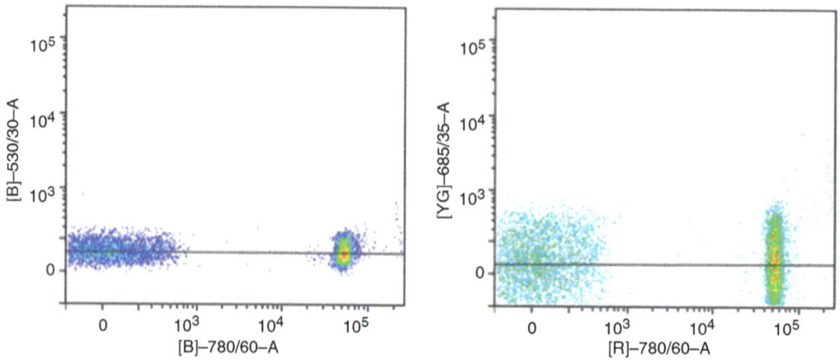

Fig. 1 Compensation example for CD3 APC Cy7. Compensation can be performed post-acquisition using flow cytometry software. Here OneComp eBeads have been stained with CD3 APC Cy7 and analyzed on an LSR II flow cytometer. In the software program, the beads are selected on a FSC-A and SSC-A plot with a gate and then represented as histograms of the negative and positive populations, shown by the markers and percentage. This then creates a series of dot plots showing the median of the negative (or unlabeled beads) and positive populations for each fluorochrome used in the experiment against each other. This figure shows a representation of two APC Cy7 plots (R780/60-A) against two fluorochromes. B530/30-A on the left commonly used for FITC and YG685/35-A on the right commonly used for PE Cy5 dyes. For correct compensation, each plot needs to show the median fluorescence intensity of both the negative and positive populations (shown by the line) for each fluorescence control is aligned to ensure no spillover into other channels occurs. You can adjust the alignment of the median fluorescence intensity of the populations in the analysis software if required by fluorescence subtraction

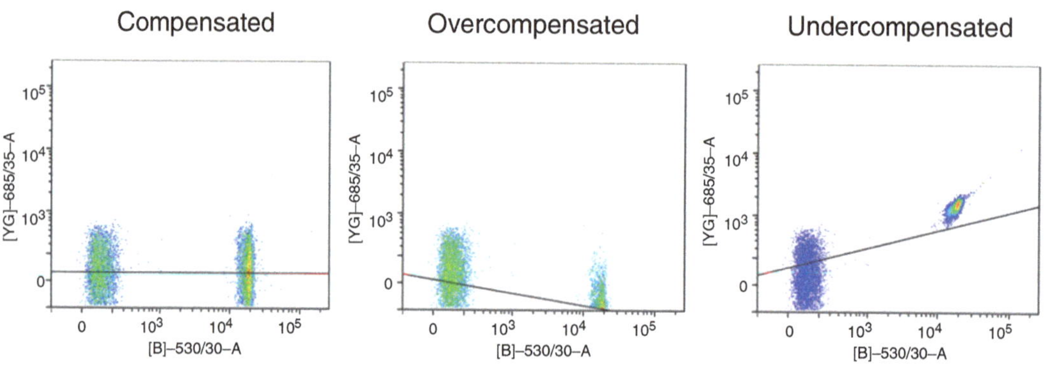

Fig. 2 Correct compensation example. The left plot shows correct compensation of FITC (B 530/30-A) versus Pe-Cy5 (YG 685/35-A) where the median fluorescence intensity (MFI) of the negative and positive populations are equal. The middle plot shows an example of overcompensation, where too much compensation also termed fluorescence subtraction has been applied to the FITC population. Here the Pe-Cy5 MFI of the negative FITC population is greater than the MFI of the positive FITC population. The right dot plot shows an example of under-compensation, where not enough compensation/fluorescence subtraction has been applied. Here the Pe-Cy5 MFI of the positive FITC population is greater than that of the negative FITC population

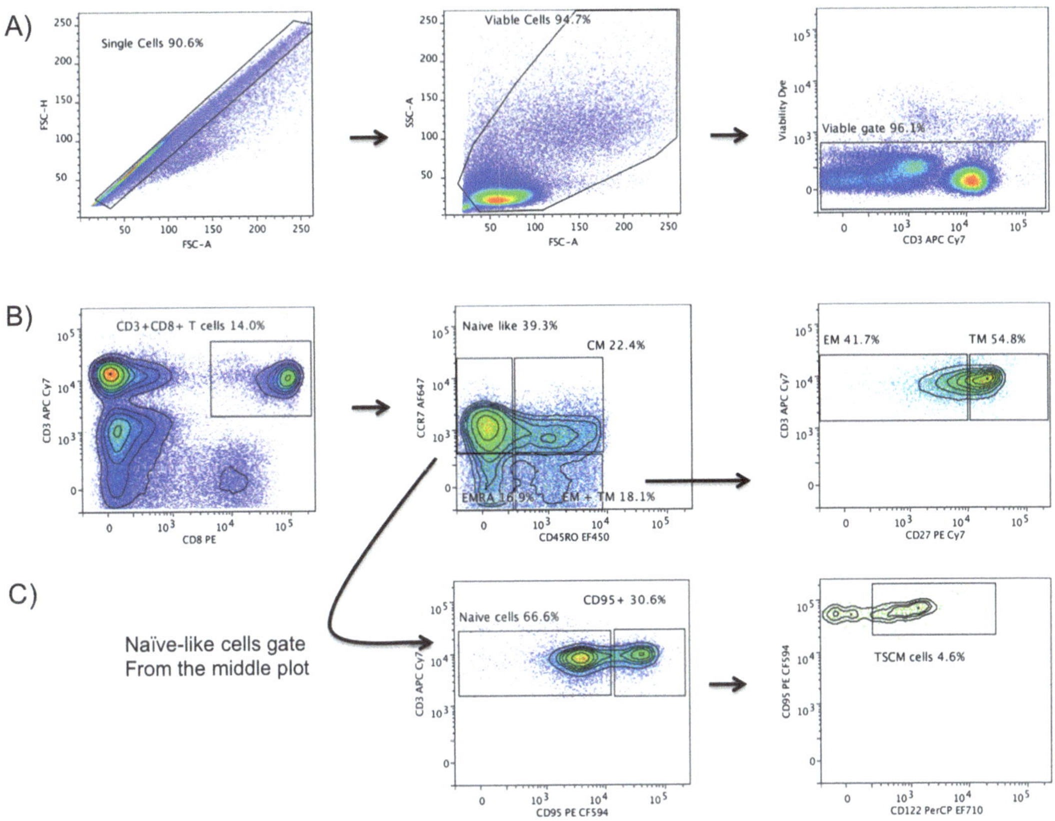

Fig. 3 $CD8^+$ T cell subset gating strategy. To identify $CD4^+$ T cells from peripheral blood mononuclear cells (PBMCs), the cells are first gated using a series of gates to ensure viable single-cell populations. (**a**) Moving from left to right on the top row, these gates include single cells (FSC-A versus FSC-H) and then a general viable cell gate to exclude apoptotic and dying cells (FSC-A versus SSC-A) and distinguish cells based on size and granularity. Next is the viable cell gate using the fixable viability dye EF506, which is shown against CD3 APC-Cy7. Viable cells are able to exclude the dye and are thus negative. (**b**) The second row shows a series of plots to identify the $CD4^+$ T cells. Firstly, the $CD4^+$ T cells are identified as the $CD3^+CD4^+$ double-positive population using CD4 FITC and CD3 APC-Cy7. The middle plot shows the $CD4^+$ T cells being distinguished using CD45RO EF450 to identify naïve-like T cells from memory subsets. This is shown against CCR7 AF-647 which is a chemokine receptor expressed on naïve and the less differentiated central memory cells (CM) and is used for homing to secondary lymphoid organs. Loss of this receptor is associated with a more differentiated memory T cell, and thus it is a useful marker to distinguish $CD4^+$ T cells subsets. The third plot in the second row uses CD27 to enable separation of effector memory (EM) from transition memory (TM) $CD4^+$ T cell subsets. (**c**) The last row of plots is used to identify the newly described and rare subset termed stem memory T cell (TSCM). This T cell is the least differentiated and is the memory subset, which is the most close to a naïve T cell. It can be distinguished from the naïve-like T cell gate using the memory marker CD95 and then gating on this population ($CCR7^+CD45RO^-CD95^+$) to detect the CD122 positive cells. Percentages displayed are the percentage of cells from the parent population inside each gate

Table 4
Phenotypic characteristics of $CD8^+$ T cell subsets

	$CD8^+$ T cell subset					
Surface marker	Naive	TSCM	CM	TM	EM	EMRA
CD45RA	+	+	–	–	–	+
CD45RO	–	–	+	+	+	–
CCR7	+	+	+	–	–	–
CD62L	+	+	+	–	–	–
CD27	+	+	+	+	+/–	–
CD28	+	+	+	+	+/–	–
CD95	–	+	+	+	+	+
CD122	–	+	+	+	+	+

activated, with activation causing an upregulation of both HLA-DR and CD38 (Fig. 4a). Additionally, $CD8^+$ T cell populations can be characterized by immune checkpoint markers (regulators of the immune system) such as PD-1, CTLA-4, and LAG-3 (Table 1, Fig. 4b). These markers will be expressed differently across the different $CD8^+$ T cells populations [10] and will commonly be upregulated on activated $CD8^+$ T cells compared to unactivated $CD8^+$ T cells (Fig. 4b).

4 Notes

1. Instead of BSA and EDTA, you can use 2% FCS in FACS buffer.
2. A fixative of 1–4% paraformaldehyde will work on most cell types. If using infectious samples, please check your institution's requirements as often a 4% fixative is required.
3. Need to use flow cytometry tubes which make a good seal on your flow cytometer.
4. *A good cell viability* is vital for all flow cytometry experiments. Include a viability marker to remove dead cells from your analysis. Dead cells can give false positives from autofluorescence and cause increased nonspecific antibody binding.
5. *Good compensation controls* for each fluorochrome are vital for multicolor flow cytometry. The emission spectra of different fluorochromes can overlap which results in detection in a different channel (even several channels). This is solved by

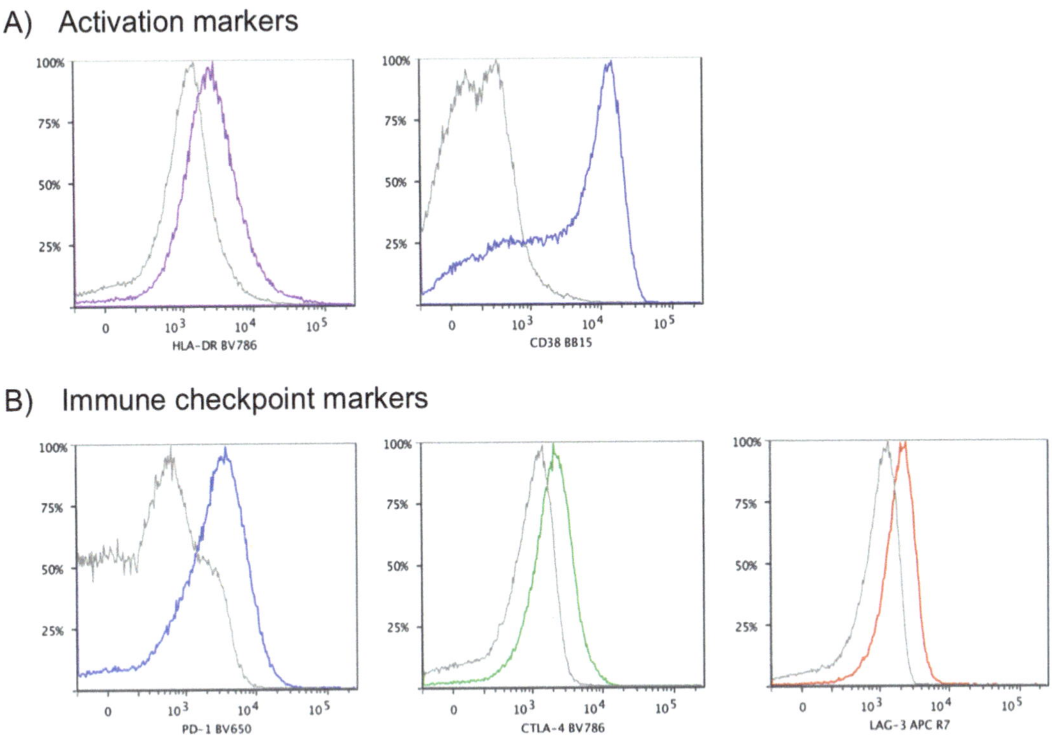

Fig. 4 Expression of activation and immune checkpoint markers on activated and unactivated CD8+ T cells. (**a**) The expression of activation markers including HLA-DR and CD38 is upregulated on CD8+ T cells shown in the top row. The unactivated CD8+ T cells are shown as a gray histogram with the activated cells shown in color. The increase in expression of CD38 is more pronounced than HLA-DR under these experiment conditions and for total CD8+ T cells. (**b**) The expression of immune checkpoint markers including PD-1, CTLA-4, and LAG-3 was upregulated upon activation shown in the bottom row. The unactivated CD8+ T cells are shown in gray with the activated cells shown in color. The increase in expression of PD-1 is more pronounced upon activation than CTLA-4 and LAG-3 under these experiment conditions and for total CD8+ T cells

compensation. If you don't have a good distinction of positive and negative populations perhaps due to rarity of a surface marker or dimness of a dye, then use compensation beads (described in this protocol). You will need to titrate each antibody on these beads to ensure your positive signal is not too bright. You will need to use the same flow antibodies in your single-color compensation controls that you use in your experiment.

6. *Other experimental controls.* If having difficulty with compensation and clear detection of positive populations for your gating strategy, fluorescence minus one (FMO) controls will provide a measure of spillover in a given channel. Each sample is stained with all flow cytometry antibodies but one. This will show the contribution of all the other flow cytometry antibodies into the unlabeled channel. This control also assists with correct gating

and selection of stained cells. Other controls you may need to consider are isotype controls which will account for nonspecific antibody binding and negative controls for your experiment to account for background staining.

7. *Titration of flow antibody* is a very important step as too little antibody can make it difficult to detect clear negative and positive populations. Too much antibody can cause both your negative and positive cell populations to move along the axis. This can cause difficulties for compensation, gating populations, and data analysis.
8. *Choice of fluorochrome* is important. Different fluorochromes have a different level of brightness which can be important for distinguishing cell populations. A dim fluorochrome may not provide the resolution you need to identify your cell populations, especially if a rare population. You also need to keep in mind the configuration of the flow cytometer to assist with fluorochrome choice (which lasers will detect which dyes) and detection of spillover into different channels.
9. *Rarity of surface marker* is also important for fluorochrome choice. If you have a surface marker which has a low antigen density, it may need a brighter fluorochrome for detection. Remember that antigen density can change on different cell types. A similar principle applies if the cell population of interest is a rare population and hard to detect.
10. *Spread of the surface marker* is an important consider too as some antigens have a clear separation of negative and positive populations with a high level of expression; other antigens have a spread of expression expressed at low, intermediate, and high levels (e.g., CD45RA or CD45RO); and some antigens have a lower level of expression (e.g., activation markers). Thus, those antigens with harder to distinguish populations may need a brighter fluorochrome with less spillover. You also need to consider the channels where spillover may occur, so you don't create false populations in your gating strategy.
11. Alternatively, and dependent upon your assay, you can coat the plates/flask with anti-CD3 (overnight at 4 °C or for 2 h at 37 °C) and add soluble CD28 to the cells in suspension prior to adding them to the well (this can give a better stimulation). You can also use IL-2 and PHA in the PBMC cell solution as an activation technique instead of anti-CD3 and CD28. I recommend, 10 U of IL-2 per mL of culture medium +1–10 mg/μL of PHA (depending upon the experimental design and cell number).
12. You only need one compensation tube for each fluorochrome. The panels in Tables 1 and 2 have quite a bit of overlap; you will

only need to ensure BB515 has compensation tubes as well as the antibodies in Table 1.

13. If having trouble getting good chemokine receptor staining, you can stain at room temperature. Can also try staining with the chemokine receptor antibodies first prior to the whole antibody panel.
14. Some fluorochromes, especially tandem dyes, need to be used with caution as they can be prone to uncoupling and sensitive to photobleaching and extended incubation in fixation buffers. In this panel, PE-Cy7 and APC-Cy7 can react with fixative and are more sensitive to light causing a loss of signal, thus washing off the fixative and keeping samples in the dark is required. Use of an H7 rather than Cy7 conjugate can improve stability.
15. Cell suspension volume will depend upon cell number and type of flow cytometer used for data analysis.
16. Ideally you wish to collect >500,000 events to be able to clearly detect rare populations of cells including the TSCM cells.
17. Please note $CD8^+$ T cell populations will move slightly between different donors and disease state (if examining clinical cohorts), so you will need to make sure your gates are correct and adjust them as required for each sample you examine.

References

1. Gulzar N, Copeland KF (2004) CD8+ T-cells: function and response to HIV infection. Curr HIV Res 2(1):23–37
2. Berg RE, Forman J (2006) The role of CD8 T cells in innate immunity and in antigen non-specific protection. Curr Opin Immunol 18(3):338–343
3. Barber DL, Wherry EJ, Ahmed R (2003) Cutting edge: rapid in vivo killing by memory CD8 T cells. J Immunol 171(1):27–31
4. Bohana-Kashtan O, Civin CI (2004) Fas ligand as a tool for immunosuppression and generation of immune tolerance. Stem Cells 22 (6):908–924
5. Flynn JK, Dore GJ, Hellard M, Yeung B, Rawlinson WD, White PA et al (2011) Early IL-10 predominant responses are associated with progression to chronic hepatitis C virus infection in injecting drug users. J Viral Hepat 18 (8):549–561
6. Cashin K, Paukovics G, Jakobsen MR, Ostergaard L, Churchill MJ, Gorry PR et al (2014) Differences in coreceptor specificity contribute to alternative tropism of HIV-1 subtype C for CD4 + T-cell subsets, including stem cell memory T-cells. Retrovirology 11 (1):97
7. Flynn JK, Gorry PR (2014) Stem memory T cells (TSCM)—their role in cancer and HIV immunotherapies. Clin Transl Immunology 3: e20
8. Flynn JK, Paukovics G, Cashin K, Borm K, Ellett A, Roche M et al (2014) Quantifying susceptibility of CD4+ stem memory T-cells to infection by laboratory adapted and clinical HIV-1 strains. Viruses 6(2):709–726
9. Flynn JK, Paukovics G, Moore MS, Ellett A, Gray LR, Duncan R et al (2013) The magnitude of HIV-1 resistance to the CCR5 antagonist maraviroc may impart a differential alteration in HIV-1 tropism for macrophages and T-cell subsets. Virology 442(1):51–58
10. Mahnke YD, Brodie TM, Sallusto F, Roederer M, Lugli E (2013) The who's who of T-cell differentiation: human memory T-cell subsets. Eur J Immunol 43(11):2797–2809

Chapter 2

Flow Cytometry Analysis to Identify Human CD4$^+$ T Cell Subsets

Jacqueline Flynn and Paul Gorry

Abstract

Flow cytometry is a powerful tool, which uses lasers to analyze a wide range of different characteristics of cells. It is commonly used to determine the expression of cell surface markers and intracellular molecules to define cells into different populations using cell size, granularity, and fluorescently labeled antibodies. Thus, flow cytometry enables simultaneous and mutliparameter analysis of single cells.

During the staining procedure, a single cell suspension is created for staining with flow cytometry antibodies for analysis on the flow cytometer. The staining methods from this technique can be used for different cell types by changing the surface marker targeted by the flow antibody, provided all antibodies are titrated prior to use, and are chosen with knowledge of the density of surface molecule for detection and brightness of fluorochrome to guide antibody selection.

This chapter's protocol has been designed specifically for detection of human CD4$^+$ T cell subsets defining naïve and memory subpopulations by surface marker phenotyping.

Key words CD4$^+$ T cells, Flow cytometry, Naïve T cells, Memory T cells, CD markers, T cell subsets

1 Introduction

Human CD4$^+$ T cells play a pivotal role in cell-mediated and humor immunity and have many subpopulations which can be defined by multiparameter flow cytometry. Flow cytometry techniques enable the identification of different CD4$^+$ T cell subsets through three main ways; these include the analysis of surface marker expression (by phenotyping surface markers), immune function (by intracellular staining of cytokines), and differentiation pathway (intracellular staining of transcription factors).

This chapter focuses on the first technique, surface marker staining, and uses this to identify populations of naïve and memory CD4$^+$ T cells. During an immune response to an invading pathogen or virus, for instance, a naïve T cell is stimulated by foreign antigen presented by antigen presenting cells. The naïve T cells will then undergo proliferative expansion and differentiation into different

Shin Kaneko (ed.), *In Vitro Differentiation of T-Cells: Methods and Protocols*, Methods in Molecular Biology, vol. 2048, https://doi.org/10.1007/978-1-4939-9728-2_2,

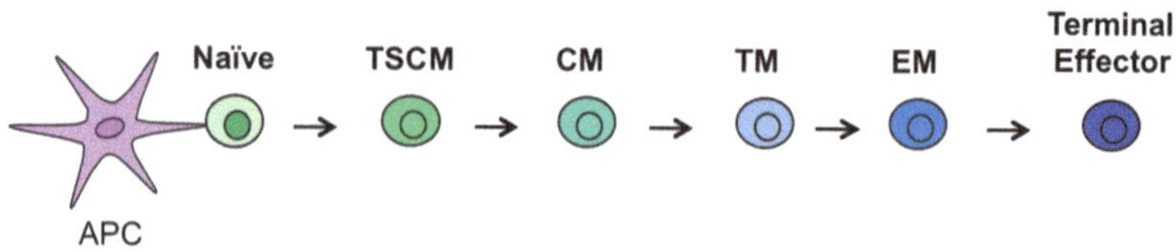

Fig. 1 Differentiation of CD4$^+$ T cell subsets. As a naïve T cell encounters antigen from an antigen presenting cell, the naïve T cells will undergo proliferative expansion and differentiation into a memory CD4$^+$ T cell. The differentiation process continues with the transition from naïve to stem memory T cells (TSCM), central memory (CM), transitional memory (TM) through to effector memory (EM) T cells. As differentiation occurs, the memory cells will progressively acquire the ability to respond to stimuli including homeostatic cytokines, tissue homing receptors, and antiapoptotic molecules and acquire different effector functions such as IFN-gamma production. This occurs while losing the expression of lymph node homing receptors (CCR7 and CD62L) and the capacity for self-renewal, survival, and IL-2 production

memory CD4$^+$ T cell subsets with different effector functions [1] (Fig. 1). Transition from naïve through to effector memory (EM) T cells allows the memory cells to progressively acquire the ability to respond to stimuli including homeostatic cytokines, tissue homing receptors, and antiapoptotic molecules and acquire different effector functions. This occurs while losing the expression of lymph node homing receptors (CCR7 and CD62L) and the capacity for proliferation and IL-2 production [1, 2]. These changes in the memory cell's differentiation state and effector function can be used as tools to identify different cell types, especially by surface marker expression.

Some of the CD4$^+$ T cell subsets require more surface markers for detection. For example, stem memory T cells (TSCM) share common phenotypic characteristics with naïve T cells as they are CD45RA$^+$, CD45RO$^-$, CCR7$^+$, and CD27$^+$; however, they can be distinguished from naïve T cells by a high expression of the memory marker CD95 and IL-2 receptor CD122 [3–6]. Also transitional memory (TM) T cells, which have functional and transcriptional characteristics between central memory (CM) and EM T cells, can be distinguished by the additional use of CD27 [7–9] (Table 1).

Importantly, the staining techniques used below can be used to define other cell populations by altering flow cytometry antibody selection with the use of three important steps: good titration of flow antibodies, knowledge of density of surface molecule for detection, and good fluorochrome choice.

2 Materials

1. FACS buffer: 500 mL sterile PBS (Ca^{2+}/Mg^{2+} free), 0.1% BSA, 1 mM EDTA (*see* **Note 1**). Filter-sterilize any unsterile components to keep buffer sterile. Keep at 4 °C.

Table 1
Phenotypic characteristics of CD4+ T cell subsets

	CD4+ T cell subset					
Surface marker	**Naive**	**TSCM**	**CM**	**TM**	**EM**	**EMRA**
CD45RA	+	+	–	–	–	+
CD45RO	–	–	+	+	+	–
CCR7	+	+	+	–	–	–
CD62L	+	+	+	–	–	–
CD27	+	+	+	+	+/–	–
CD28	+	+	+	+	+/–	–
CD95	–	+	+	+	+	+
CD122	–	+	+	+	+	+

2. FACS FIX: 500 mL sterile PBS (Ca^{2+}/Mg^{2+} free), 1% paraformaldehyde (*see* **Note 2**). Keep at 4 °C.
3. Flow cytometry antibodies: CCR7 AF-647, CD3 APC Cy7, CD4 FITC, CD122 PerCP EF710, CD45RO EF450, viability dye EF506, CCR5 PE, CD95 PE CF594, CXCR4 PE Cy5, CD27 PE Cy7.
4. Compensation beads: OneComp eBeads.
5. Flow tubes: 5 mL polypropylene tubes capped (*see* **Note 3**).
6. Range of pipettes: 0.3 μL to 1 mL.

3 Methods

Important notes to read prior to starting your surface marker staining which discuss the use of viability markers, compensation controls, experimental controls, and titration of flow antibodies have been added to the end of Subheading 4 (*see* **Notes 4–7**). Useful considerations for flow cytometry panel design have also been added to the end of Subheading 4 (*see* **Notes 8–10**).

3.1 Flow Cytometry Staining

Carry out all procedures in a class II biological hazard cabinet to maintain a sterile environment unless described otherwise.

Using thawed peripheral blood mononuclear cells (PBMCs) or freshly isolated PBMCs (from blood using Ficoll-Paque gradient), perform a cell count and resuspend cells in FACS buffer to 10×10^6/mL.

Table 2
Flow cytometry CD4⁺ T cell subset panel antibody amounts

Flow antibody	Fluorochrome	Amount per compensation tube	Amount per cell sample	Amount for a ten-sample tube master mix
CCR7	AF-647	2 μL	12 μL	120 μL
CD3	APC Cy7	3 μL	3 μL	30 μL
CD4	FITC	5 μL	10 μL	100 μL
CD122	PerCP EF710	0.3 μL	1 μL	10 μL
CD45RO	EF450	5 μL	10 μL	100 μL
Viability Dye	EF506	N/A	5 μL	50 μL
CCR5	PE	1 μL	10 μL	100 μL
CD95	PE CF594	0.3 μL	2 μL	20 μL
CXCR4	PE Cy5	3 μL	10 μL	100 μL
CD27	PE Cy7	0.5 μL	0.7 μL	7 μL

Please note, depending upon the supplier of flow antibody you use, you may need to titrate the amount of antibody for your experiments. Add the amount per master mix and divide by the number of tubes, for example, for ten tubes, as shown in the last column of the table, add 54.7 μL of the master mix per tube

3.1.1 Compensation Tubes

1. Label a flow tube with each antibody and fluorochrome for your compensation control tubes (one tube for each fluorochrome, Table 2).
2. Vortex compensation beads well.
3. Add one drop of the compensation bead solution to each labeled compensation tube.
4. Add the appropriate amount of antibody to each compensation tube (Table 2).
5. Stain the beads for 30 min at 4 °C in the dark (in the fridge, covered with foil) (*see* **Note 3**).
6. Add 1 mL of FACS buffer to each compensation tube and spin beads at 400 × *g* for 10 min.
7. Remove supernatant carefully and add 200 μL of FACS FIX.
8. Fix the beads for 30 min at 4 °C in the dark (in the fridge, covered in foil).
9. Add 1 mL of FACS buffer to each compensation tube and spin beads at 400 × *g* for 10 min.
10. Remove supernatant carefully and resuspend in 200 μL of FACS buffer.
11. Store flow tubes at 4 °C covered in foil (in the fridge). The compensation tubes can be stored once fixed for up to 1 week.

3.1.2 CD4+ T Cell Tubes

1. Label flow cytometry tubes with CD4+ T cell panel, keep one tube labeled as cells alone.
2. Aliquot 200 μL of cell suspension into labeled FACS tubes (approximately one million cells per tube).
3. Set aside the cells alone tube (unstained by antibodies) to allow for correct setup of voltages on the flow cytometer.
4. Make a master mix of titrated flow antibodies (Table 2), and to the tubes labeled CD4+ T cell panel, add antibody master mix to each tube. Resuspend well.
5. Stain cells for 30 min at 4 °C in the dark (in the fridge, covered in foil) (*see* **Notes 3** and **11**).
6. Add 1 mL of FACS buffer to each flow tube and spin cells at 400 × *g* for 10 min.
7. Remove supernatant carefully and add 200 μL of FACS FIX.
8. Fix cells for 30 min at 4 °C in the dark (in the fridge, covered in foil).
9. Add 1 mL of FACS buffer to each flow tube and spin cells at 400 × *g* for 10 min (*see* **Note 12**).
10. Remove supernatant carefully and resuspend in 200–400 μL of FACS buffer (*see* **Note 13**).
11. Store flow tubes at 4 °C covered in foil (in the fridge) for up to 1 week.

3.2 Flow Cytometer Data Acquisition

Please note, you will need training to use the flow cytometer to ensure setup and data collection is correct. Contact your flow cytometry department for assistance.

Brief notes for setting up an LSR II flow cytometer:

1. Start the machine and switch on the computer.
2. Ensure adequate sheath fluid levels and that the waste tank is empty (except for 5% bleach for human cells).
3. Prime instrument twice and remove any air bubbles in the system. Remove any air bubbles remaining by gently releasing the relief value at the bubble filter and bleeding the lines into a waste container.
4. Open the software program and set up your templates.
5. Configure the instrument by selecting the lasers and fluorochrome combinations used (an example for the antibodies used in this protocol and configurations for an LSR II flow cytometer is in Table 3).
6. Using your cells alone, tube set your voltages for all fluorochromes.
7. Run your compensation control tubes collecting at least 10,000 events.

Table 3
Flow cytometry CD4$^+$ T cell subset panel and LSR II instrument configuration

Flow antibody	Fluorochrome	LSR laser	Description
CCR7	AF-647	640 nm	R 670/14
CD3	APC Cy7	640 nm	R 780/60
CD4	FITC	488 nm	B 530/30
CD122	PerCP EF710	488 nm	B 710/50
CD45RO	EF450	405 nm	V 450/50
Viability dye	EF506	405 nm	V 525/50
CCR5	PE	561 nm	YG 582/15
CD95	PE CF594	561 nm	YG 610/20
CXCR4	PE Cy5	561 nm	YG 685/35
CD27	PE Cy7	561 nm	YG 780/60

8. Run your T cell panel tubes uncompensated collecting all events (*see* **Note 14**). Compensation will be done post-data acquisition.
9. Export your data and follow the flow cytometry department cleaning and shut down procedure. Turn off the computer and flow cytometer.

3.3 Data Analysis

Given the multiple fluorochromes are used in this experiment, it is easier to compensate post-acquisition. With a good panel design and good software, this is not too difficult to do post-acquisition compensation, and this also means if you need to change any compensation applied, you can create a new compensation matrix and apply it to your data set. You need to ensure however that your compensation is done correctly for each fluorochrome. An example of correct compensation for the CD4 FITC cell surface antibody is shown in Fig. 2. The compensated populations have the median of each cell population in line with each other for each fluorochrome used in the experiment. An example of incorrect compensated populations is shown in Fig. 3. Once compensation is complete and you have saved your compensation matrix, you are ready to been analysis.

Load your FSC files into your software program and apply the compensation matrix, so all files will now be compensated. Below is a strategy of how to set up your gates to define different CD4$^+$ T cell subset via the parameters listed in Table 1 using the panel of CD4$^+$ T cell subset flow cytometry antibodies shown in Table 2 and Fig. 4. Select the statistics you wish to display in your software

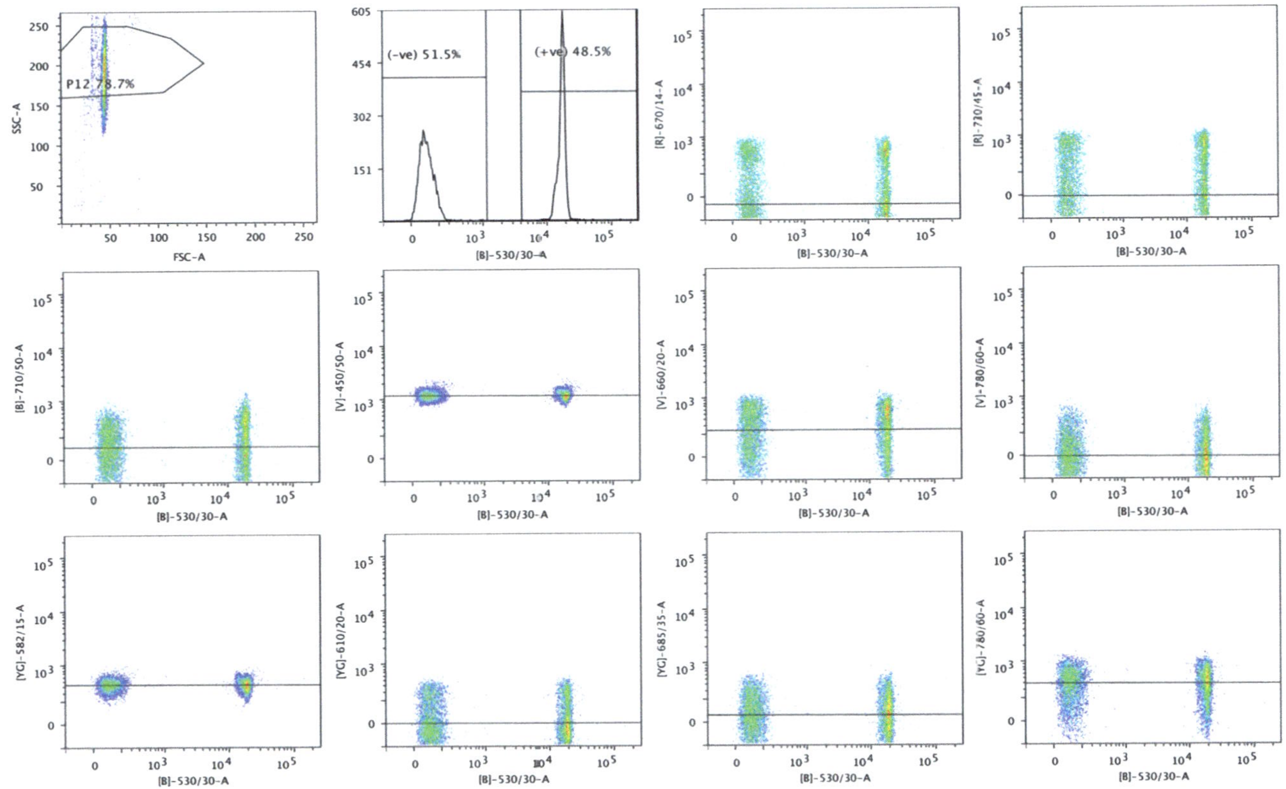

Fig. 2 Compensation matrix for CD4 FITC. Compensation can be performed post-acquisition using flow cytometry software. Here OneComp eBeads have been stained with CD4 FITC and analyzed on an LSR II flow cytometer. In the software program, the beads are selected on an FSC-A and SSC-A plot with a gate and then represented as histograms of the negative and positive populations, shown by the markers and percentage. This then creates a series of dot plots showing the median of the negative (or unlabeled beads) and positive populations for each fluorochrome used in the experiment against each other. This figure shows a representation of only the FITC plots (B 530/30-A) against each fluorochrome. For correct compensation, each plot needs to show the median fluorescence intensity of both the negative and positive populations (shown by the line) for each fluorescence control is aligned to ensure no spillover into other channels occurs. You can adjust the alignment of the median fluorescence intensity of the populations in the analysis software if required by fluorescence subtraction

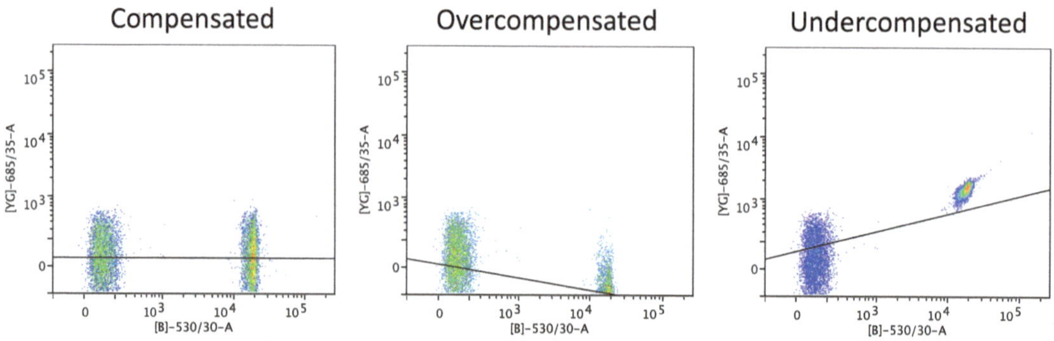

Fig. 3 Correct compensation example. The left plot shows correct compensation of FITC (B 530/30-A) versus Pe-Cy5 (YG 685/35-A) where the median fluorescence intensity (MFI) of the negative and positive populations are equal. The middle plot shows an example of overcompensation, where too much compensation also termed fluorescence subtraction has been applied to the FITC population. Here the Pe-Cy5 MFI of the negative FITC population is greater than the MFI of the positive FITC population. The right dot plot shows an example of under-compensation, where not enough compensation/fluorescence subtraction has been applied. Here the Pe-Cy5 MFI of the positive FITC population is greater than that of the negative FITC population

program, and calculate your parameters of interest, for example, percentage of each $CD4^+$ T cell subtype from your clinical samples. Save your experiment and gating template in your software program for future experiments and data analysis (*see* **Note 15**).

4 Notes

1. Instead of BSA and EDTA, you can use 2% FCS in FACS buffer.
2. A fixative of 1–4% paraformaldehyde will work on most cell types. If using infectious samples, please check your institution's requirements as often a 4% fixative is required.
3. Need to use flow cytometry tubes which make a good seal on your flow cytometer.
4. *A good cell viability* is vital for all flow cytometry experiments. Include a viability marker to remove dead cells from your analysis. Dead cells can give false positives from autofluorescence and cause increased nonspecific antibody binding.
5. *Good compensation controls* for each fluorochrome are vital for multicolor flow cytometry. The emission spectra of different fluorochromes can overlap which results in detection in a different channel (even several channels). This is solved by compensation. If you do not have a good distinction of positive and negative populations perhaps due to rarity of a surface marker or dimness of a dye, then use compensation beads (described in this protocol). You will need to titrate each antibody on these

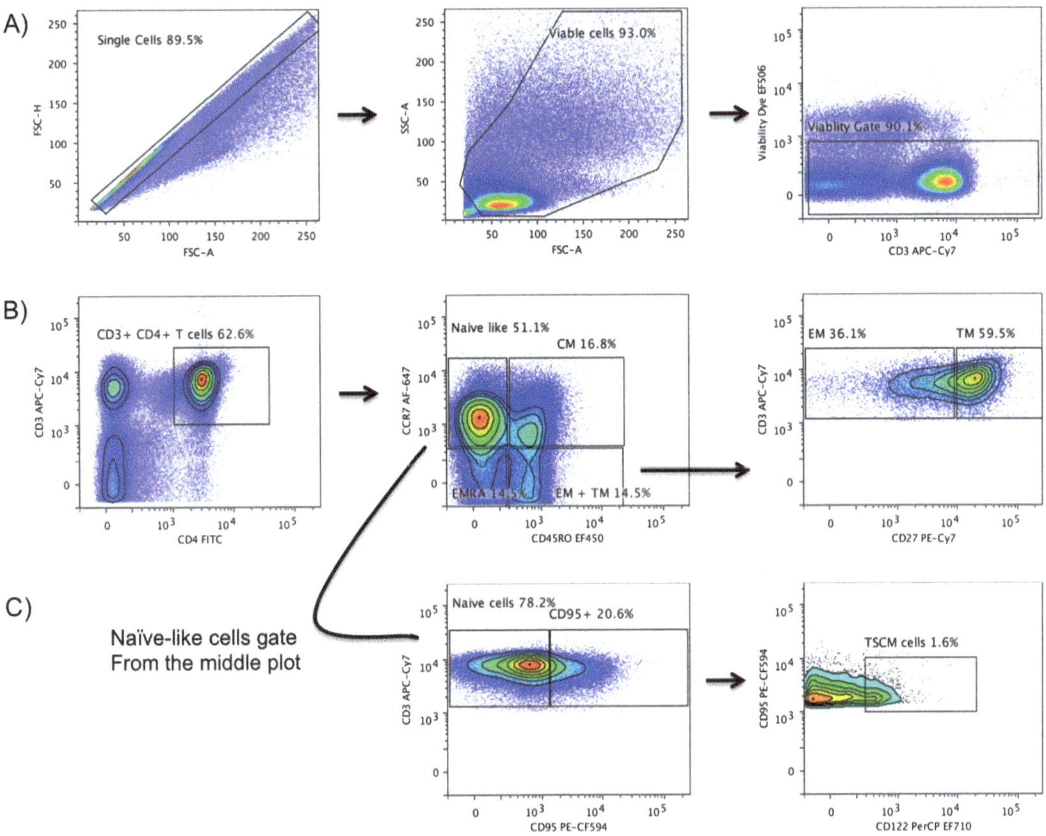

Fig. 4 $CD4^+$ T cell subset gating strategy. To identify $CD4^+$ T cells from peripheral blood mononuclear cells (PBMCs), the cells are first gated using a series of gates to ensure viable single cell populations. (**a**) Moving from left to right on the top row, these gates include single cells (FSC-A versus FSC-H) then a general viable cell gate to exclude apoptotic and dying cells (FSC-A versus SSC-A) and distinguish cells based on size and granularity. Next is the viable cell gate using the fixable viability dye EF506, which is shown against CD3 APC-Cy7. Viable cells are able to exclude the dye and are thus negative. (**b**) The second row shows a series of plots to identify the $CD4^+$ T cells. Firstly, the $CD4^+$ T cells are identified as the $CD3^+CD4^+$ double-positive population using CD4 FITC and CD3 APC-Cy7. The middle plot shows the $CD4^+$ T cells being distinguished using CD45RO EF450 to identify naïve-like T cells from memory subsets. This is shown against CCR7 AF-647 which is a chemokine receptor expressed on naïve and the less differentiated central memory cells (CM) and is used for homing to secondary lymphoid organs. Loss of this receptor is associated with a more differentiated memory T cell, and thus it is a useful marker to distinguish $CD4^+$ T cells subsets. The third plot in the second row uses CD27 to enable separation of effector memory (EM) from transition memory (TM) $CD4^+$ T cell subsets. (**c**) The last row of plots is used to identify the newly described and rare subset termed stem memory T cell (TSCM). This T cell is the least differentiated and is the memory subset, which is the closest to a naïve T cell. It can be distinguished from the naïve-like T cell gate using the memory marker CD95 then gating on this population ($CCR7^+CD45RO^-CD95^+$) to detect the CD122 positive cells. Percentages displayed are the percentage of cells from the parent population inside each gate

beads to ensure your positive signal is not too bright. You will need to use the same flow antibodies in your single color compensation controls that you use in your experiment.

6. *Other experimental controls.* If having difficulty with compensation and clear detection of positive populations for your gating strategy, fluorescence minus one controls (FMO) controls will provide a measure of spillover in a given channel. Each sample is stained with all flow cytometry antibodies but one. This will show the contribution of all the other flow cytometry antibodies into the unlabeled channel. This control also assists with correct gating and selection of stained cells. Other controls you may need to consider are isotype controls which will account for nonspecific antibody binding and negative controls for your experiment to account for background staining.
7. *Titration of flow antibody* is a very important step as too little antibody can make it difficult to detect clear negative and positive populations. Too much antibody can cause your both your negative and positive cell populations to move along the axis. This can cause difficulties for compensation, gating populations, and data analysis.
8. *Choice of fluorochrome* is important. Different fluorochromes have a different level of brightness which can be important for distinguishing cell populations. A dim fluorochrome may not provide the resolution you need to identify your cell populations, especially if a rare population. You also need to keep in mind the configuration of the flow cytometer to assist with fluorochrome choice (which lasers will detect which dyes) and detection of spillover into different channels.
9. *Rarity of surface marker* is also important for fluorochrome choice. If you have a surface marker which has a low antigen density, it may need a brighter fluorochrome for detection. Remember that antigen density can change on different cell types. A similar principle applies if the cell population of interest is a rare population and hard to detect.
10. *Spread of the surface marker* is an important consideration too as some antigens have a clear separation of negative and positive populations with a high level of expression; other antigens have a spread of expression expressed at low, intermediate, and high levels (e.g., CD45RA or CD45RO); and some antigens have a lower level of expression (e.g., activation markers). Thus, those antigens with harder to distinguish populations may need a brighter fluorochrome with less spillover. You also need to consider the channels where spillover may occur, so you do not create false populations in your gating strategy.
11. If having trouble getting good chemokine receptor staining, can stain at 4 °C. Can also try staining with the chemokine receptor antibodies first prior to the whole antibody panel.

12. Some fluorochromes, especially tandem dyes, need to be used with caution as they can be prone to uncoupling and sensitive to photobleaching and extended incubation in fixation buffers. In this panel, PE-Cy7 and APC-Cy7 can react with fixative and are more sensitive to light causing a loss of signal, thus washing off the fixative and keeping samples in the dark are required. Use of an H7 rather than Cy7 conjugate can improve stability.
13. Cell suspension volume will depend upon cell number and type of flow cytometer used for data analysis.
14. Ideally you wish to collect >500,000 events to be able to clearly detect rare populations of cells including the TSCM cells.
15. Please note $CD4^+$ T cell populations will move slightly between different donors and disease state (if examining clinical cohorts) so you will need to make sure your gates are correct and adjust them as required for each sample you examine.

References

1. Flynn JK, Gorry PR (2014) Stem memory T cells (TSCM)—their role in cancer and HIV immunotherapies. Clin Transl Immunology 3: e20
2. Lanzavecchia A, Sallusto F (2005) Understanding the generation and function of memory T cell subsets. Curr Opin Immmunol 17:326–332
3. Flynn JK, Paukovics G, Cashin K, Borm K, Ellett A et al (2014) Quantifying susceptibility of CD4 + stem memory T-cells to infection by laboratory adapted and clinical HIV-1 strains. Viruses 6(2):709 726
4. Gattinoni L, Lugli E, Ji Y, Pos Z, Paulos CM et al (2011) A human memory T cell subset with stem cell-like properties. Nat Med 17 (10):1290–1297
5. Mahnke YD, Brodie TM, Sallusto F, Roederer M, Lugli E (2013) The who's who of T-cell differentiation: human memory T-cell subsets. Eur J Immunol 43(11):2797–2809
6. Flynn JK, Gorry PR (2015) T cell therapies-are T memory stem cells the answer? Ann Transl Med 3(17):251
7. Cashin K, Paukovics G, Jakobsen MR, Østergaard L, Churchill MJ et al (2014) Differences in coreceptor specificity contribute to alternative tropism of HIV-1 subtype C for CD4+ T-cell subsets, including stem cell memory T-cells. Retrovirology 11(1):97
8. Flynn JK, Paukovics G, Moore MS, Ellett A, Gray LR et al (2013) The magnitude of HIV-1 resistance to the CCR5 antagonist maraviroc may impart a differential alteration in HIV-1 tropism for macrophages and T-cell subsets. Virology 442(1):51–58
9. Chomont N, El-Far M, Ancuta P, Trautmann L, Procopio FA et al (2009) HIV reservoir size and persistence are driven by T cell survival and homeostatic proliferation. Nat Med 15 (8):893–900

Chapter 3

Gene Modification and Immunological Analyses for the Development of Immunotherapy Utilizing T Cells Redirected with Antigen-Specific Receptors

Toshiki Ochi, Masaki Maruta, and Naoto Hirano

Abstract

Cancer immunotherapy has been developed and established as a new treatment modality. Recently, adoptive transfer therapy using T cells redirected with antigen-specific antitumor receptors, such as T-cell receptor (TCR) and chimeric antigen receptor (CAR), has demonstrated clinical benefits even in patients with refractory malignancies. To advance this treatment modality, both generation of gene-modified T cells and evaluation of their reactivity with high quality in vitro are required. To achieve this, it is important to establish the ways (1) to generate optimal viral particle for T-cell transduction, (2) to transduce antitumor receptors into T cells and expand redirected T cells efficiently, and (3) to assess the functionality of antigen-specific gene-modified T cells precisely. Here, we summarize established protocols to generate and analyze antitumor receptor-transduced T cells. These procedures help to further assess characteristics of gene-modified T cells, resulting in promotion of translational research for cancer immunotherapy.

Key words T-cell receptor, Chimeric antigen receptor, Gene transduction, T-cell expansion, T-cell functional assays

1 Introduction

Cancer immunotherapy has been developed and established as a new treatment modality [1]. Clinical trials of immunotherapy have demonstrated that antitumor T cells can suppress tumor growth even in patients with refractory tumors. To obtain better clinical responses while minimizing undesirable adverse events, it is crucial to employ optimal antitumor receptors which endow T cells with cancer specificity. A T-cell receptor (TCR) plays an important role in the recognition of tumor cells by T cells. TCRs specific for tumor antigens deliver activation signals into T cells to induce antitumor T-cell responses. Clinical trials of adoptive transfer therapy using T cells transduced with antitumor TCR (TCR-T-cell therapy) have demonstrated the effectiveness of this therapeutic modality

Shin Kaneko (ed.), *In Vitro Differentiation of T-Cells: Methods and Protocols*, Methods in Molecular Biology, vol. 2048, https://doi.org/10.1007/978-1-4939-9728-2_3,

[2–7]. Moreover, recent technical advances allow us to generate antibody-based antitumor receptors, such as chimeric antigen receptor (CAR) which can activate T cells with cancer specificity. Clinical trials of adoptive therapy using T cells redirected by CAR (CAR-T-cell therapy), particularly for the treatment of refractory CD19-positive hematological malignancies, have shown great promise [8–14].

So far, we have been focusing on T cells redirected by antitumor receptors to further develop T-cell-based immunotherapy for malignancies [15–19]. To promote both TCR-T-cell therapy and CAR-T-cell therapy and translate them into the clinic, it requires optimal generation of gene-modified T cells and evaluation of their target-specific reactivity in vitro. To achieve this, it is important to establish ways (1) to generate optimal viral particle for T-cell transduction, (2) to transduce antitumor receptors into T cells and expand redirected T cells efficiently, and (3) to assess the functionality of gene-modified T cells precisely. Here, we summarize our established protocols to generate and analyze antitumor receptor-transduced T cells. These procedures allow accurate characterization of gene-modified T cells, which can significantly advance translational research for cancer immunotherapy.

2 Materials

2.1 Cells

1. K562 cells: Purchased from the American Type Culture Collection (ATCC) and cultured with RPMI1640 medium supplemented with 10% fetal calf serum (10%FCS/RPMI) (*see* **Note 1**).
2. T2 cells: Purchased from ATCC and cultured with 10%FCS/RPMI (*see* **Note 2**).
3. Jurkat 76 cells: Generous gift from Dr. Heemskerk, Leiden University Medical Center. The cells are cultured with 10% FCS/RPMI (*see* **Note 3**).
4. Plat-A cells: Kindly provided by Dr. Kitamura, the Institute of Medical Science, the University of Tokyo. The cells are maintained with DMEM medium supplemented with 10%FCS (10% FCS/DMEM) containing 1 μg/mL puromycin and 10 μg/mL blasticidin (*see* **Note 4**).
5. PG13 cells: Purchased from ATCC and cultured with 10% FCS/DMEM (*see* **Note 5**).
6. Human peripheral blood mononuclear cells (PBMCs): Isolated individually. Store them at −80 °C or a liquid nitrogen tank until use.

2.2 Genes

All genes with or without a tag gene are individually integrated into a pMX retroviral vector [20].

1. T-cell receptor (TCR): Genetically synthesized, or isolated from an antigen-specific T-cell clone.
2. Chimeric antigen receptor (CAR): Genetically synthesized (*see* **Note 6**).
3. Nerve growth factor receptor (NGFR): Used as a tag gene to detect and isolate gene-modified T cells (*see* **Note 7**).

2.3 Gene Transduction and T-Cell Culture

1. Transfection/transduction reagents: TransIT293, polybrene, polystyrene tubes, 0.45 μm SFCA filter.
2. Antibody and cytokines for stimulation: Antihuman CD3 monoclonal antibody (mAb) (clone OKT3), human IL-2, and human IL-15.
3. T-cell culture media: RPMI1640 medium, 10% human AB serum (10%HS/RPMI1640).
4. Peptides: Synthesized with the following condition. Lyophilized, free ends, and more than 80% purity. Dissolved with DMSO basically (*see* **Note 8**).
5. MACS beads system: Utilized fluorochrome-conjugated antihuman NGFR mAb (clone ME20.4), anti-fluorochrome microbeads, LS or MS column, MACS buffer (PBS, 1%FCS, 2 mM EDTA), according to the brochure's recommendation (Miltenyi Biotec).

2.4 Flow Cytometry

1. Antibodies: All the mAbs are labeled with fluorochrome. Antihuman CD8 mAb (clone B9.11), antihuman CD4 mAb (clone OKT4), antihuman NGFR mAb (clone C40-1457, ME20.4), antihuman TNF-α mAb (clone MAb11), antihuman IL-2 mAb (clone MQ1-17H12), antihuman IFN-γ (clone B27), and antihuman CD69 mAb (clone FN50).
2. HLA/peptide tetramers: Purchased biotinylated monomers to generate fresh tetramers by multimerization with fluorochrome-conjugated streptavidin (*see* **Note 9**).
3. Dasatinib: Dissolved at the concentration of 50 μM in DMSO. Store at −20 °C.
4. Blocking buffer: PBS, 20% murine serum. Pass through 0.45 μm filter for use.
5. FACS buffer: PBS, 2% FCS.
6. Buffers for intracellular staining: Utilized a BD Cytofix/Cytoperm Fixation/Permeabilization Kit which includes both Fix/-Perm buffer and Perm/Wash buffer.
7. Dead cell staining buffer: Utilized a LIVE/DEAD Fixable Dead Cell Stain Kit.

2.5 ELISA

1. High protein-binding ELISA plates: Nunc MaxiSorp 96-well flat bottom plates.
2. Antibodies and reagents: Recombinant human IFN-γ, recombinant human IL-2, antihuman IFN-γ capture mAb (clone 1D1K), biotin-conjugated antihuman IFN-γ detection mAb (clone 7-B6-1), antihuman IL-2 capture mAb (clone MT2A91/2C95), biotin-conjugated antihuman IL-2 detection mAb (clone MT8G10), streptavidin-conjugated alkaline phosphatase (SA-ALP), 1-step *p*-nitrophenyl phosphate disodium salt (pNPP) solution, and stop solution (2 N NaOH).
3. Coating buffer: PBS, 0.1% bovine serum albumin (BSA).

2.6 Chromium Release Assay

1. Chromium-51 (^{51}Cr): 37 MBq/mL sodium chromate in normal saline ($Na_2{}^{51}CrO_4$), pH 8.0 to 10.0.
2. Permeabilization buffer: PBS, 0.2% Triton X-100.

3 Methods

Culture or incubate the cells at 37 °C, 5% CO_2 incubator unless otherwise specified. Subheadings 3.1 and 3.2 are also summarized in the article [15].

3.1 Generation of Retrovirus for Transduction of Antigen-Specific Receptors into T Cells

1. Seed 3.0 × 10^6 Plat-A cells to a T75 flask 3 days before transfection. It will reach 70–80% confluency on the day of transfection (Day 0). One flask is used for one transfection of a pMX retroviral plasmid which encodes an antigen-specific receptor.
2. On Day 0, mix 50 μL transIT293 and 1.5 mL serum-free DMEM in a polystyrene tube. Vortex the tube and incubate for 10 min. Add 20 μg pMX retroviral plasmid into the mixture generated as above, mix by gentle pipetting, and incubate for 15 min. Change the supernatant of Plat-A cells with 8.5 mL fresh 10%FCS/DMEM. Add 1.5 mL plasmid/transIT293/DMEM into those cells. Shake the flasks back and forth, and left and right, and keep culturing.
3. On Day 1, change the supernatant of transfected Plat-A cells with 10 mL fresh 10%FCS/DMEM. Seed 3.0 × 10^5 PG13 cells to a T75 flask for transduction.
4. On Day 2, collect the supernatant of transfected Plat-A cells and add 10 mL fresh 10%FCS/DMEM to keep culturing the cells in the morning. Pass through a 0.45 μm SFCA filter. Mix 5 mL filtered retrovirus with fresh 5 mL 10%FCS/DMEM containing 2 μg/mL polybrene (*see* **Note 10**). Exchange PG13 culture supernatant with this viral mixture for transduction. In the evening, add remaining 5 mL Plat-A-derived supernatants into PG13 cells.

5. On Day 3, repeat **step 4**.
6. On Day 4, change the supernatant of PG13 transfectants with 10 mL fresh 10%FCS/DMEM.
7. On Day 5, collect PG13 supernatant which contains pseudo-typed retroviruses for transduction into T cells, and add 10 mL fresh 10%FCS/DMEM to keep culturing PG13 transfectants. Pass through a 0.45 μm SFCA filter. Directly use freshly collected and filtered supernatant for T-cell transduction. If not used immediately, the supernatant should be stored at −80 °C. If necessary, repeat this step for 3–4 continuous days.

3.2 Establishment of Antigen-Specific Receptor-Transduced T Cells

1. Isolate 2.0×10^7 PBMCs from healthy donors or patients. Suspend PBMCs with 4 mL T-cell culture media containing 50 ng/mL antihuman CD3 mAb, 100 IU/mL human IL-2, and culture for 2 days before transduction (Day −2) (*see* **Note 11**).
2. On Day 0, mix 5 mL retroviruses obtained as above with 5 mL fresh T-cell culture media to make 10 mL virus-containing media. Culture 5.0×10^5 stimulated T cells with 1.5 mL virus-containing media together with 100 IU/mL human IL-2 in a 24-well plate. Prepare six wells per plate for transduction. Centrifuge the plates for 1 h at 1000 × *g* at 32 °C (*see* **Note 12**).
3. From Day 1 to Day 3, aspirate 1 mL supernatant from each well and add 1 mL virus-containing media with 100 IU/mL human IL-2 into each well. Centrifuge the plates for 1 h at 1000 × *g* at 32 °C.
4. On Day 4, collect T cells, centrifuge, and aspirate the supernatants. Resuspend 1.0×10^6 T cells with 1.5 mL virus containing media supplemented with 100 IU/mL human IL-2 in a 24-well plate. Prepare six to ten wells per plate for transduction. Centrifuge the plates for 1 h at 1000 × *g* at 32 °C.
5. On Day 5, repeat **step 3**. Transduce an antigen-specific receptor into T cells 6 times in total.
6. On Day 9 or 10, measure the expression of antigen-specific receptor in gene-modified T cells using flow cytometer.

3.3 Expansion of Antigen-Specific Gene-Modified T Cells

1. Prepare appropriate stimulator cells, for example, K562-based artificial antigen-presenting cells (aAPCs) (*see* **Note 13**).
2. Isolate gene-modified T cells using MACS beads system (if needed), and resuspend T cells at 2.0×10^6/mL in T-cell culture media.
3. Suspend HLA-expressing aAPCs with serum-free RPMI1640. Pulse 10 μg/mL peptides of interest onto aAPCs for 6 h at

room temperature, followed by irradiation at 200 Gy. Vortex occasionally during incubation (*see* **Note 14**).

4. Wash them twice with RPMI1640 supplemented with 1% human serum, resuspend them with T-cell culture media to achieve 1.0×10^5 cells/mL. Put them on ice (*see* **Note 15**).
5. Mix 2.0×10^6 T cells and 1.0×10^5 aAPCs, and seed 2 mL per well in a 24-well plate.
6. Next day, add 40 μL T-cell culture media supplemented with 20 IU/mL human IL-2 and 20 ng/mL human IL-15 into each well (2 mL/well) (*see* **Note 16**). Replenish cytokines every 3 days. Restimulate T cells every 7–10 days, depending on T-cell condition.

3.4 Functional Assays

3.4.1 Staining of T-Cell Transfectants with HLA/Peptide Tetramers

1. Prepare and transfer 2.0×10^5 T cells to a 96-well v-bottom plate for staining. Centrifuge for 1 min at $1000 \times g$ at 4 °C. Discard the supernatant.
2. Add 200 μL 10%FCS/RPMI1640 supplemented with 50 nM dasatinib into each well. Incubate the cells for 30 min at 37 °C (*see* **Note 17**).
3. Wash each well with 200 μL FACS buffer and centrifuge for 1 min at $1000 \times g$ at 4 °C. Discard the supernatant.
4. Add 10 μL blocking buffer, vortex the plate, and incubate for 10 min at room temperature. Wash each well with 200 μL FACS buffer and centrifuge for 1 min at $1000 \times g$ at 4 °C. Discard the supernatant.
5. Add 10 μL fluorochrome-conjugated fresh tetramers into each well (*see* **Note 9**). Vortex the plate and incubate for 20 min at room temperature by shading the plate with foil.
6. Add 10 μL fluorochrome-conjugated antihuman antibody cocktails into each well. Vortex the plate and incubate for 10 min at 4 °C.
7. Wash each well with 200 μL FACS buffer and centrifuge for 1 min at $1000 \times g$ at 4 °C. Discard the supernatant. Stain dead cells using dead cell staining buffer. Centrifuge the plate for 1 min at $1000 \times g$ at 4 °C. Discard the supernatant.
8. Wash each well with 200 μL FACS buffer and centrifuge for 1 min at $1000 \times g$ at 4 °C. Analyze the cells with flow cytometer (Fig. 1).

3.4.2 Intracellular Cytokine Assays to Assess Antigen-Specific T-Cell Reactivity

1. Incubate 3.0×10^5 T cells with 5.0×10^4 specific target cells in a 96-well flat bottom plate for 6 h at 37 °C (*see* **Notes 18** and **19**). Transfer the cells into a 96-well v-bottom plate.
2. Wash each well with 200 μL FACS buffer and centrifuge for 1 min at $1000 \times g$ at 4 °C. Discard the supernatant.

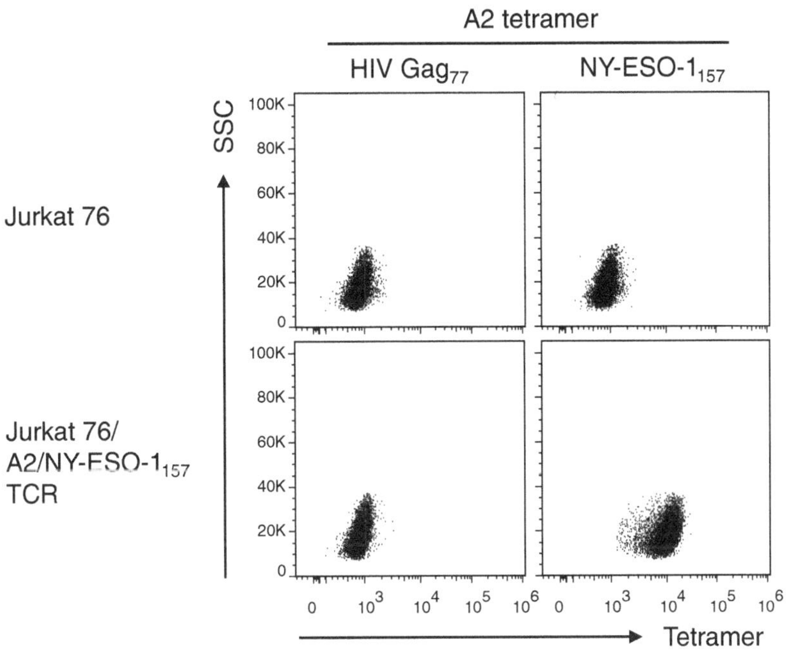

Fig. 1 Tetramer staining of gene-modified T cells. Jurkat 76 cells were transduced with HLA-A∗02:01 (A2)-restricted NY-ESO-$1_{157\text{-}165}$ (NY-ESO-1_{157})-specific TCR (clone 1G4) [28] following Subheadings 3.1 and 3.2. The Jurkat 76 transfectant was isolated with CD3 microbeads and stained with indicated fresh A2 tetramers (5 μg/mL) following Subheading 3.4.1. The transfectant was successfully stained with A2/NY-ESO-1_{157} tetramer but not control A2/HIV Gag_{77} tetramer (bottom). On the other hand, control cells were stained with neither A2/NY-ESO-1_{157} tetramer nor A2/HIV Gag_{77} tetramer (top)

3. Add 10 μL blocking buffer, vortex the plate, and incubate for 10 min at room temperature. Wash each well with 200 μL FACS buffer and centrifuge for 1 min at 1000 × *g* at 4 °C. Discard the supernatant.
4. Add 10 μL fluorochrome-conjugated antihuman antibody cocktails into each well (*see* **Note 20**). Vortex the plate and incubate for 15 min at 4 °C.
5. Wash each well with 200 μL FACS buffer and centrifuge for 1 min at 1000 × *g* at 4 °C. Discard the supernatant.
6. Vortex the plate, add 100 μL Fix/Perm buffer into each well. Incubate for 20 min at 4 °C.
7. Centrifuge the plate for 1 min at 1000 × *g* at 4 °C. Discard the supernatant.
8. Vortex the plate, wash each well with 200 μL Perm/Wash buffer and centrifuge for 1 min at 1000 × *g* at 4 °C. Discard the supernatant. Repeat twice.
9. Add 10 μL fluorochrome-conjugated antihuman cytokine antibody cocktails (*see* **Notes 20** and **21**). Vortex the plate and

incubate for 30 min at room temperature by shading the plate with foil.

10. Wash each well with 200 μL Perm/Wash buffer and centrifuge for 1 min at 1000 × *g* at 4 °C. Repeat twice. Resuspend the cells with FACS buffer and analyze them using flow cytometer.

3.4.3 CD69 Upregulation Assays to Assess Antigen-Specific T-Cell Reactivity

1. Incubate 3.0×10^5 T cells with 5.0×10^4 specific target cells in a 96-well flat bottom plate for 5–6 h at 37 °C (*see* **Note 18**). Transfer the cells into a 96-well v-bottom plate.
2. Wash each well with 200 μL FACS buffer and centrifuge for 1 min at 1000 × *g* at 4 °C. Discard the supernatant.
3. Add 10 μL blocking buffer, vortex the plate, and incubate for 10 min at room temperature. Wash each well with 200 μL FACS buffer and centrifuge for 1 min at 1000 × *g* at 4 °C. Discard the supernatant.
4. Add 10 μL fluorochrome-conjugated antihuman antibody cocktails containing the antihuman CD69 mAb into each well (*see* **Note 20**). Vortex the plate and incubate for 15 min at 4 °C.
5. Wash each well with 200 μL FACS buffer and centrifuge for 1 min at 1000 × *g* at 4 °C. Discard the supernatant. Stain dead cells using dead cell staining buffer. Centrifuge the plate for 1 min at 1000 × *g* at 4 °C. Discard the supernatant.
6. Wash each well with 200 μL FACS buffer and centrifuge for 1 min at 1000 × *g* at 4 °C. Analyze the cells with flow cytometer (Fig. 2).

3.4.4 ELISA Assays to Assess Antigen-Specific T-Cell Reactivity

1. Incubate 1.0×10^5 T cells with 2.0×10^4 specific target cells in a 96-well flat bottom plate for 20 h at 37 °C (*see* **Note 18**). Collect the supernatants and stored at −80 °C until use.
2. One day before the experiments, coat a high protein-binding ELISA plate with 35 μL antihuman cytokine capture antibody with appropriate concentration. Incubate the plate for overnight at 4 °C.
3. Wash each well twice with 200 μL PBS.
4. Add 200 μL coating buffer into each well. Incubate for 1 h at room temperature.
5. Wash each well five times with 200 μL PBS. Prepare cytokine standard diluents and samples diluted with appropriate volume of PBS.
6. Add 35 μL standard diluents and diluted samples into each well and incubate for 2 h at room temperature.
7. Wash each well five times with 200 μL PBS, and add 35 μL biotinylated antihuman cytokine detection antibody diluted

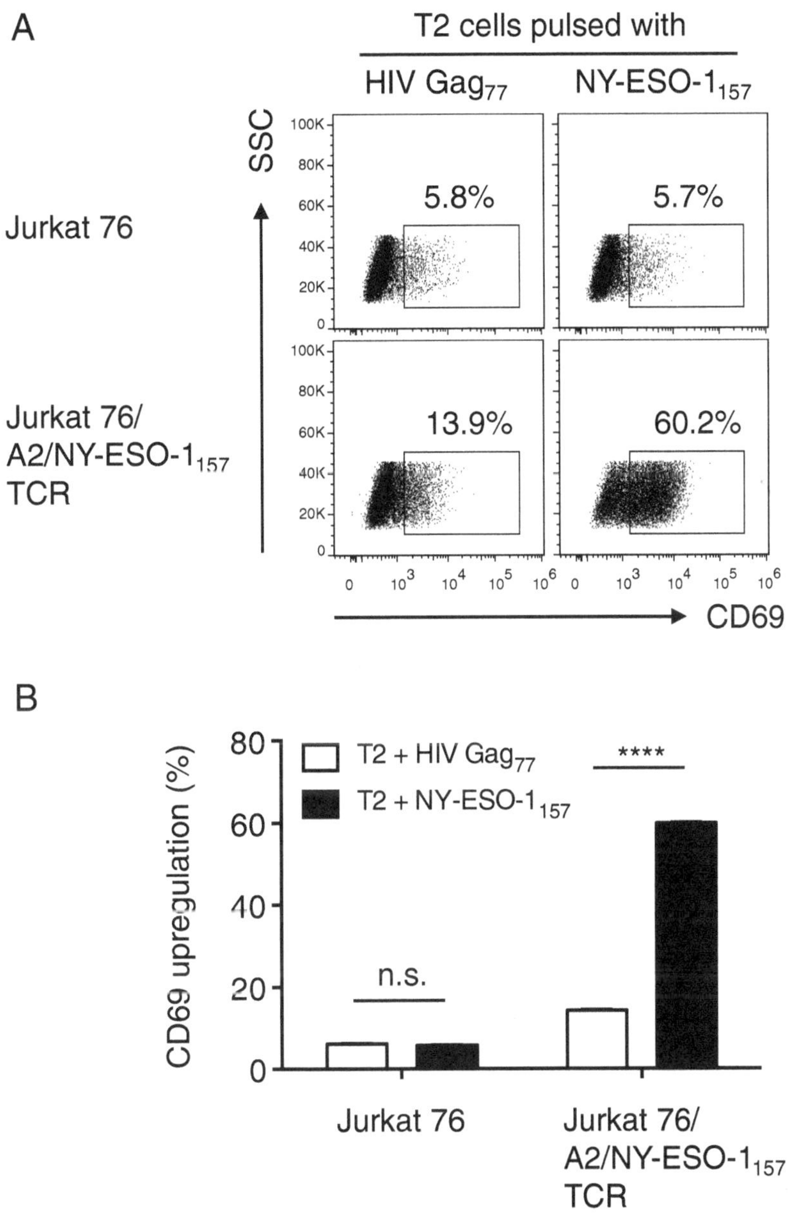

Fig. 2 Target-specific reactivity of gene-modified T cells. A2/NY-ESO-1_{157}-specific TCR-transduced Jurkat 76 cells established above were used for further experiments to assess their functionality and target-specificity utilizing CD69 upregulation assay following Subheading 3.4.3. Three hundred thousand Jurkat 76/A2/NY-ESO-1_{157} TCR cells or control Jurkat 76 cells were cocultured with 5.0×10^4 HLA-A∗02:01-positive T2 cells pulsed with 10 μg/mL NY-ESO-1_{157} peptide or 10 μg/mL irrelevant HIV Gag_{77} peptide. (**a**) Jurkat 76/A2/NY-ESO-1_{157} TCR cells upregulated CD69 molecule by recognition of A2/NY-ESO-1_{157} but not A2/HIV Gag_{77} (bottom). Control cells did not upregulate CD69 molecule (top). Representative dot plots are shown. (**b**) Results are displayed as mean and SEM values obtained from experiments with triplicate. ∗∗∗∗, $P < 0.0001$, *n.s.* not significant

with appropriate volume of PBS into each well. Incubate the plate for 1 h at room temperature.

8. Wash each well five times with 200 μL PBS, and add 35 μL of 1000 times-diluted SA-ALP with PBS into each well. Incubate the plate for 1 h at room temperature.
9. Wash each well five times with 200 μL PBS, and add 50 μL 1-step pNPP solution into each well. Incubate the plate for 20 min at room temperature.
10. Add 25 μL stop solution into each well. If needed, mix thoroughly by gently agitating the plate. Measure absorbance of each well at 405 nm.

3.4.5 Chromium Release Assays to Assess Antigen-Specific T-Cell Reactivity

1. Suspend 5.0×10^5 specific target cells with 100 μL 10%FCS/RPMI1640 in a 15 mL tube. Add 25–50 μL ^{51}Cr into each tube and incubate for 1.5 h at 37 °C. Tap tubes gently every 30 min to loosen the cell pellet and mix the cells with ^{51}Cr well.
2. Wash them 3 times with 1 mL 10%FCS/RPMI1640. Suspend the cells with 10 mL 10%FCS/RPMI to make 5.0×10^4 cells/mL, and seed 5.0×10^3 cells/100 μL into a 96-well round-bottom plate (*see* **Note 18**).
3. Add 100 μL responder cells suspended with 10%FCS/RPMI1640 with various E:T ratio onto target cells (*see* **Note 22**). Use same volume of 10%FCS/RPMI1640 or permeabilization buffer to obtain spontaneous release and maximal release, respectively. Incubate the plate for 5–6 h at 37 °C.
4. Collect 100 μL supernatants from each well and measure cpm counts. The percentage of specific lysis is calculated as follows: (experimental release cpm − spontaneous release cpm)/(maximal release cpm − spontaneous release cpm) × 100 (%).

4 Notes

1. K562 cells lack the expression of endogenous HLA molecules. K562-based aAPCs are established by transduction of an HLA allele of interest or a target surface antigen along with CD80 and CD83 as costimulatory molecules into K562 cells [21].
2. T2 cells do not express transporter associated with antigen processing (TAP), thereby have a potential to present exogenously pulsed peptides together with HLA molecules efficiently.
3. Jurkat 76 cells lack the expression of endogenous TCR and CD3 [22].
4. Plat-A cell line is a retrovirus packaging cell line based on 293 T cells to generate amphotropic retrovirus [23].

5. PG13 cell line is a retrovirus packaging cell line derived from NIH/3T3 cells and is based on the gibbon ape leukemia virus (GaLV).
6. Variable regions derived from immunoglobulin light and heavy chains of a mAb are linked with a linker sequence to generate a single chain fragment variable (scFv). Generally, a CAR encodes an scFv followed by transmembrane region of CD28 and/or 4-1BB and intracellular domain of CD3ζ chain.
7. An antigen-specific receptor gene can be fused with a truncated form of NGFR gene via a furin cleavage site, an SGSG spacer sequence, and a codon-optimized P2A sequence [24, 25].
8. Synthesized peptides can bind to HLA class I or HLA class II molecules recognized by gene-modified T cells.
9. Biotinylated monomers should be prepared and stored at −80 °C until use. One day before the experiments, fresh biotinylated monomers are mixed with fluorochrome-conjugated streptavidin following the procedure as indicated in http://tetramer.yerkes.emory.edu/support/protocols#10.
10. Final concentration of polybrene is 1 μg/mL.
11. Ten million T cells can be obtained 2 days after stimulation of 2.0×10^7 PBMCs with antihuman CD3 mAb in the presence of human IL-2.
12. When transduce antigen-specific receptors into a T-cell line, for example, Jurkat 76 cell line, maintain cells in appropriate culture medium such as 10%FCS/RPMI1640, without cytokines.
13. Following protocol is optimized for stimulation of T cells with aAPCs [16–18, 26].
14. When stimulate and expand CAR-transduced T cells which recognize a surface antigen but not an HLA/peptide complex, use aAPCs expressing a target surface antigen. There is no need to pulse exogenous peptides onto aAPCs expressing HLA molecules.
15. After removal of floating peptides, aAPCs should be put on ice to avoid unloading of bound peptides from HLA molecules.
16. Final concentrations of human IL-2 and human IL-15 are 10 IU/mL and 10 ng/mL, respectively.
17. Dasatinib can block downregulation of receptors induced by binding between antigen-specific receptors and HLA/peptide tetramers, resulting in enhancement of tetramer staining intensity of gene-modified T cells [27].
18. To determine peptide specificity of T cells, pulse appropriate amounts of peptides onto target cells and incubate for 1–1.5 h at 37 °C before adding T cells, which allows target cells to

present exogenously pulsed peptides together with endogenously expressed HLA molecules.

19. Brefeldin A should be added into each well after 2 h to inhibit extracellular release of cytokines, and the cells are further cultured for 4 h.

20. After discarding the supernatants by flipping the 96-well v-bottom plate, 10 μL will remain in each well. Appropriate amounts of antihuman mAbs are mixed together in 10 μL FACS buffer and used as antibody cocktails. Consequently, 1.0×10^6 cells are stained with 1 μg fluorochrome-conjugated antihuman mAb in 20 μL.

21. Fluorochrome-conjugated antihuman cytokine antibody cocktails should be prepared with Perm/Wash buffer but not FACS buffer.

22. If needed, ^{51}Cr-unlabeled K562 cells should be added to each well at the target cells: K562 cells ratio of 1:20 to suppress nonspecific NK-like reactivity of induced gene-modified T cells.

Acknowledgments

This work was supported by Grant-in-Aid for Research Activity Start-up of JSPS (to T.O.) and Research Scholarship of The Uehara Memorial Foundation (to T.O.). Plat-A cell line is kindly provided by Dr. Toshio Kitamura, The Institute of Medical Science, The University of Tokyo. Jurkat 76 cell line is a generous gift from Dr. Mirjam Heemskerk, Leiden University Medical Center. We thank Drs. Hiroshi Fujiwara and Masaki Yasukawa for helpful discussions.

References

1. Couzin-Frankel J (2013) Breakthrough of the year 2013. Cancer immunotherapy. Science 342:1432–1433
2. Morgan RA, Dudley ME, Wunderlich JR et al (2006) Cancer regression in patients after transfer of genetically engineered lymphocytes. Science 314:126–129
3. Johnson LA, Morgan RA, Dudley ME et al (2009) Gene therapy with human and mouse T-cell receptors mediates cancer regression and targets normal tissues expressing cognate antigen. Blood 114:535–546
4. Parkhurst MR, Yang JC, Langan RC et al (2011) T cells targeting carcinoembryonic antigen can mediate regression of metastatic colorectal cancer but induce severe transient colitis. Mol Ther 19:620–626
5. Robbins PF, Morgan RA, Feldman SA et al (2011) Tumor regression in patients with metastatic synovial cell sarcoma and melanoma using genetically engineered lymphocytes reactive with NY-ESO-1. J Clin Oncol 29:917–924
6. Rapoport AP, Stadtmauer EA, Binder-Scholl GK et al (2015) NY-ESO-1-specific TCR-engineered T cells mediate sustained antigen-specific antitumor effects in myeloma. Nat Med 21:914–921
7. Tawara I, Kageyama S, Miyahara Y et al (2017) Safety and persistence of WT1-specific T-cell receptor gene-transduced lymphocytes in

patients with AML and MDS. Blood 130:1985–1994

8. Maude SL, Frey N, Shaw PA et al (2014) Chimeric antigen receptor T cells for sustained remissions in leukemia. N Engl J Med 371:1507–1517
9. Kochenderfer JN, Dudley ME, Kassim SH et al (2015) Chemotherapy-refractory diffuse large B-cell lymphoma and indolent B-cell malignancies can be effectively treated with autologous T cells expressing an anti-CD19 chimeric antigen receptor. J Clin Oncol 33:540–549
10. Schuster SJ, Svoboda J, Chong EA et al (2017) Chimeric antigen receptor T cells in refractory B-cell lymphomas. N Engl J Med 377:2545–2554
11. Park JH, Riviere I, Gonen M et al (2018) Long-term follow-up of CD19 CAR therapy in acute lymphoblastic leukemia. N Engl J Med 378:449–459
12. Khalil DN, Smith EL, Brentjens RJ et al (2016) The future of cancer treatment: immunomodulation, CARs and combination immunotherapy. Nat Rev Clin Oncol 13:273–290
13. Brudno JN, Kochenderfer JN (2018) Chimeric antigen receptor T-cell therapies for lymphoma. Nat Rev Clin Oncol 15:31–46
14. June CH, O'Connor RS, Kawalekar OU et al (2018) CAR T cell immunotherapy for human cancer. Science 359:1361–1365
15. Guo T, Ochi T, Nakatsugawa M et al (2016) Generating de novo antigen-specific human T cell receptors by retroviral transduction of centric hemichain. J Vis Exp (116). https://doi.org/10.3791/54697
16. Nakatsugawa M, Yamashita Y, Ochi T et al (2015) Specific roles of each TCR hemichain in generating functional chain-centric TCR. J Immunol 194:3487–3500
17. Ochi T, Nakatsugawa M, Chamoto K et al (2015) Optimization of T-cell reactivity by exploiting TCR chain centricity for the purpose of safe and effective antitumor TCR gene therapy. Cancer Immunol Res 3:1070–1081
18. Chamoto K, Guo T, Imataki O et al (2016) CDR3beta sequence motifs regulate autoreactivity of human invariant NKT cell receptors. J Autoimmun 68:39–51
19. Kagoya Y, Tanaka S, Guo T et al (2018) A novel chimeric antigen receptor containing a JAK-STAT signaling domain mediates superior antitumor effects. Nat Med 24:352–359
20. Kitamura T (1998) New experimental approaches in retrovirus-mediated expression screening. Int J Hematol 67:351–359
21. Butler MO, Hirano N (2014) Human cell-based artificial antigen-presenting cells for cancer immunotherapy. Immunol Rev 257:191–209
22. Heemskerk MH, Hoogeboom M, de Paus RA et al (2003) Redirection of antileukemic reactivity of peripheral T lymphocytes using gene transfer of minor histocompatibility antigen HA-2-specific T-cell receptor complexes expressing a conserved alpha joining region. Blood 102:3530–3540
23. Morita S, Kojima T, Kitamura T (2000) Plat-E: an efficient and stable system for transient packaging of retroviruses. Gene Ther 7:1063–1066
24. Yang S, Cohen CJ, Peng PD et al (2008) Development of optimal bicistronic lentiviral vectors facilitates high-level TCR gene expression and robust tumor cell recognition. Gene Ther 15:1411–1423
25. Kim JH, Lee SR, Li LH et al (2011) High cleavage efficiency of a 2A peptide derived from porcine teschovirus-1 in human cell lines, zebrafish and mice. PLoS One 6:e18556
26. Hirano N, Butler MO, Xia Z et al (2006) Engagement of CD83 ligand induces prolonged expansion of CD8+ T cells and preferential enrichment for antigen specificity. Blood 107:1528–1536
27. Lissina A, Ladell K, Skowera A et al (2009) Protein kinase inhibitors substantially improve the physical detection of T-cells with peptide-MHC tetramers. J Immunol Methods 340:11–24
28. Robbins PF, Li YF, El-Gamil M et al (2008) Single and dual amino acid substitutions in TCR CDRs can enhance antigen-specific T cell functions. J Immunol 180:6116–6131

Chapter 4

In Vitro Conversion of Activated T Cells into Stem Cell Memory-Like T Cells

Taisuke Kondo, Yuki Imura, Makoto Ando, Shunsuke Chikuma, and Akihiko Yoshimura

Abstract

Adoptive T cell therapy is an attractive strategy in tumor immunotherapy. The transfer of in vitro expanded tumor-associated antigen (TAA)-specific T cells from patients may effectively destroy the original tumor cells. One of the limitations is a rapid acquisition of tolerant (anergy, deletion, dysfunctional, and/or exhausted) phenotypes. We and others found that stem cell memory T (T_{SCM}) cells are strongly resistant to tolerance, showing strong expansion and persistence in vivo and providing long-lasting antitumor effects. We previously established that phenotypically T_{SCM} cells (iT_{SCM}) can be induced using a simple coculture of activated T cells with OP9 stroma cells expressing a Notch ligand. Here, we describe a defined protocol for generating human iT_{SCM} cells, including reagents, culture setting, and procedure.

Key words Adoptive T cell therapy, Stem cell memory T cells, Notch signaling, Coculture with feeder cells, Cancer immunotherapy

1 Introduction

Adoptive T cell-based therapy is a well-known strategy that is a recent technological breakthrough for cancer therapy [1, 2]. The concept includes (1) collection of potential tumor-associated antigen (TAA)-specific T cells from tumor tissues and peripheral blood of patients, (2) expansion of T cells in vitro, and (3) infusion of the expanded T cells into the patient. Several technical limitations remain unsolved. In vitro expansion of T cells provides strong effector functions, but it often contributes to terminally differentiated phenotypes and easily induces dysfunctionality upon transfer because of repeated antigen stimulation [3]. On the other hand, less differentiated T cells showing naïve T cell or central memory phenotypes are known to govern to the outcomes of adoptive T cell therapy [4, 5]. Naïve T cell phenotypes, which are shared by conventional naïve T cells and stem cell memory T (T_{SCM}) cells, can provide superior antitumor ability [4, 6]. Recent studies focus on

Shin Kaneko (ed.), *In Vitro Differentiation of T-Cells: Methods and Protocols*, Methods in Molecular Biology, vol. 2048, https://doi.org/10.1007/978-1-4939-9728-2_4,

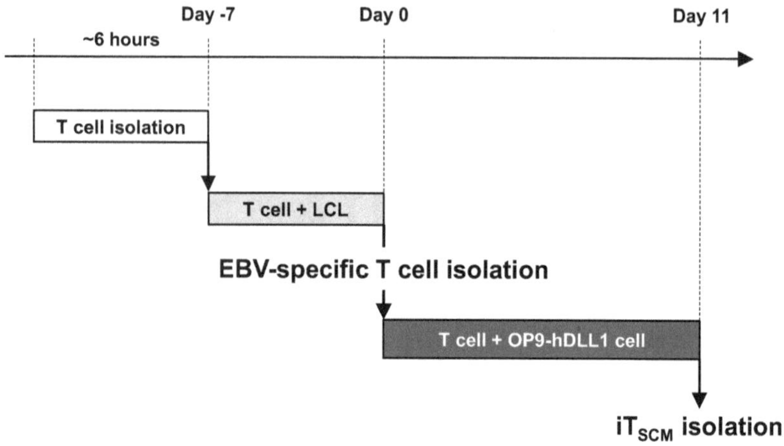

Fig. 1 Schematic for scheduling iT$_{SCM}$ cell generation. Peripheral CD8α^+ T cells are isolated at first and continuously cocultured with autologous lymphoblastoid cell line (LCL) for 7 days. Next, Epstein-Barr virus (EBV)-specific T cells are purified by cell sorting and transferred onto human delta-like 1-expressing feeder cells, OP9-hDLL1 cells, for 11 days. Evaluation of iT$_{SCM}$ is performed 11 days after OP9-hDLL1 coculture

methods to generate TAA-specific T$_{SCM}$ cells from naïve T cells in combination with an artificial antigen receptor, i.e., the chimeric antigen receptor [7–11]. However, age-related changes in the immune system result in the reduction of the naïve T cell compartment, and it increases the memory T cell pool [12]. The methods for T$_{SCM}$ generation from naïve T cells would be limited in aged patients. Therefore, direct conversion of activated T cells, but not naïve T cells, into T$_{SCM}$-like (iT$_{SCM}$) cells would be beneficial. We previously reported the methods for generating iT$_{SCM}$ cells from endogenous memory-phenotype T cells [13]. Briefly, activated T cells were converted into iT$_{SCM}$ cells by coculturing with Notch ligand-expressing feeder cells.

In this report, we describe a step-by-step protocol for generating human iT$_{SCM}$ cells, including reagents, experimental setting, and detailed procedures. First, whole CD8 T cells are isolated from peripheral blood mononuclear cells (PBMCs) and stimulated by the coculture with autologous lymphoblastoid cell line (LCL) cells. After stimulation, proliferated LCL-specific T cells are transferred onto a Notch ligand-expressing feeder (OP9-hDLL1) layer and cocultured for 11 days (Fig. 1). The coculture with OP9-hDLL1 cells induces iT$_{SCM}$ cells.

2 Materials

2.1 Cells

1. OP9 cells were purchased from the American Type Culture Collection (ATCC; Manassas, VA, USA). We generated an OP9 cell line expressing the human delta-like 1 (DLL1) molecule, a Notch ligand. We first cloned human DLL1 into a

lentiviral vector and transduced the human DLL1 coding insert into OP9 cells.

2. Autologous LCLs were prepared, as described in previous reports (*see* **Note 1**) [14, 15].
3. PBMCs were used (*see* **Note 2**); *see* Subheading 3.

2.2 Cell Culture

1. Density gradient media, i.e., Ficoll-Paque and Lymphoprep.
2. Phosphate buffer of saline (PBS).
3. Pooled normal human male AB serum, which is obtained from healthy donors with serotype AB (Innovative Research, Inc.) (*see* **Note 3**).
4. Fetal bovine serum for LCL culture (Thermo Fisher, Waltham, MA, USA).
5. Fetal bovine serum for OP9-hDLL1 cell culture (Sigma-Aldrich, St. Louis, MO, USA) (*see* **Note 4**).
6. Human interleukin-7 (IL-7) (Peprotech, Rocky Hill, NJ, USA).
7. $CD8^+$ T cell isolation kit, human (Miltenyi Biotec, Bergisch Gladbach, Germany).
8. LS column (Miltenyi Biotec).
9. MACS buffer: 0.5% [w/v] bovine serum albumin (BSA), 2 mM ethylenediaminetetraacetic acid (EDTA), PBS.
10. OP9-hDLL1 cell medium: 20% fetal bovine serum (FBS), antibiotics, minimum essential medium alpha (αMEM).
11. T cell prime medium: 10% AB serum, antibiotics, 1 mM sodium pyruvate, 1 mM HEPES buffer solution, MEM nonessential amino acid solution, 55 μM 2-mercaptoethanol, αMEM.
12. iT_{SCM} induction medium: 20% FBS, antibiotics, 55 μM 2-mercaptoethanol, and 10 ng/mL human IL-7 in αMEM.

2.3 Antibodies (Ab) and Fluorophores

1. Antihuman CD8α PerCP-Cy5.5 (BioLegend, San Diego, CA, USA).
2. Antihuman CD45RA PE-Cy7 (BioLegend).
3. Antihuman CCR7 Alexa 647 (BioLegend).
4. CellTrace CFSE cell proliferation kit (Thermo Fisher Scientific).

2.4 Equipment

1. QuadroMACS Separator (Miltenyi Biotec).
2. Cell incubator (temperature 37 °C, CO_2 level 5%).
3. Water bath (temperature 37 °C).
4. Centrifuge (TOMY, Tokyo, Japan).

5. X-ray irradiator: CellRad (Faxitron, Tucson, US-AZ, USA).
6. FACS Canto II cytometer (BD Biosciences, Franklin Lakes, US-NJ, USA).
7. FACS Aria II cell sorter (BD Biosciences).
8. FlowJo software v10 (Tree Star, Ashland, OR, USA).

3 Methods

Appropriate institutional regulatory board approval must be obtained before starting experiments. Written informed consent must be obtained from all individuals.

3.1 Human CD8$^+$ T Cell Isolation

1. Collect 40 mL of whole blood from an EBV-seropositive healthy donor.
2. Separate PBMC from 40-mL samples of whole blood by centrifugation (900 × *g*, 35 min, 20 °C, slow acceleration, and slow deceleration) using Lymphoprep (Alere Technologies AS, Oslo, Norway) or other density centrifugation media (e.g., Ficoll-Paque [GE Healthcare Life Science]), according to the manufacturer's instructions (Fig. 2) (*see* **Note 5**).
3. Collect the PBMC band between the serum and Lymphoprep layers.
4. Wash the banded PBMC twice in MACS buffer and resuspend in 10% AB serum containing RPMI 1640 media in 15-mL centrifuge tubes until the next step is started (*see* **Note 6**).
5. Obtain a cell count for PBMC.
6. Enrich human CD8$^+$ T cells by negative selection (Fig. 2). We use a CD8$^+$ T cell isolation kit, human from Miltenyi Biotec, and below is described the protocol for its use.
7. Remove the supernatant by centrifugation (400 × *g*, 5 min, 4 °C).
8. Resuspend the PBMC pellet in 40 μL of MACS buffer per 10^7 total PBMC.
9. Add 10 μL of biotin-antibody cocktail per 10^7 total PBMC.
10. Mix the contents well and incubate for 5 min in the refrigerator (4 °C).
11. Add an additional 30 μL of MACS buffer per 10^7 total PBMC.
12. Add an additional 20 μL of CD8$^+$ T cell microbead cocktail per 10^7 total PBMC.
13. Mix the cell suspension well and incubate for 10 min in the refrigerator (4 °C).

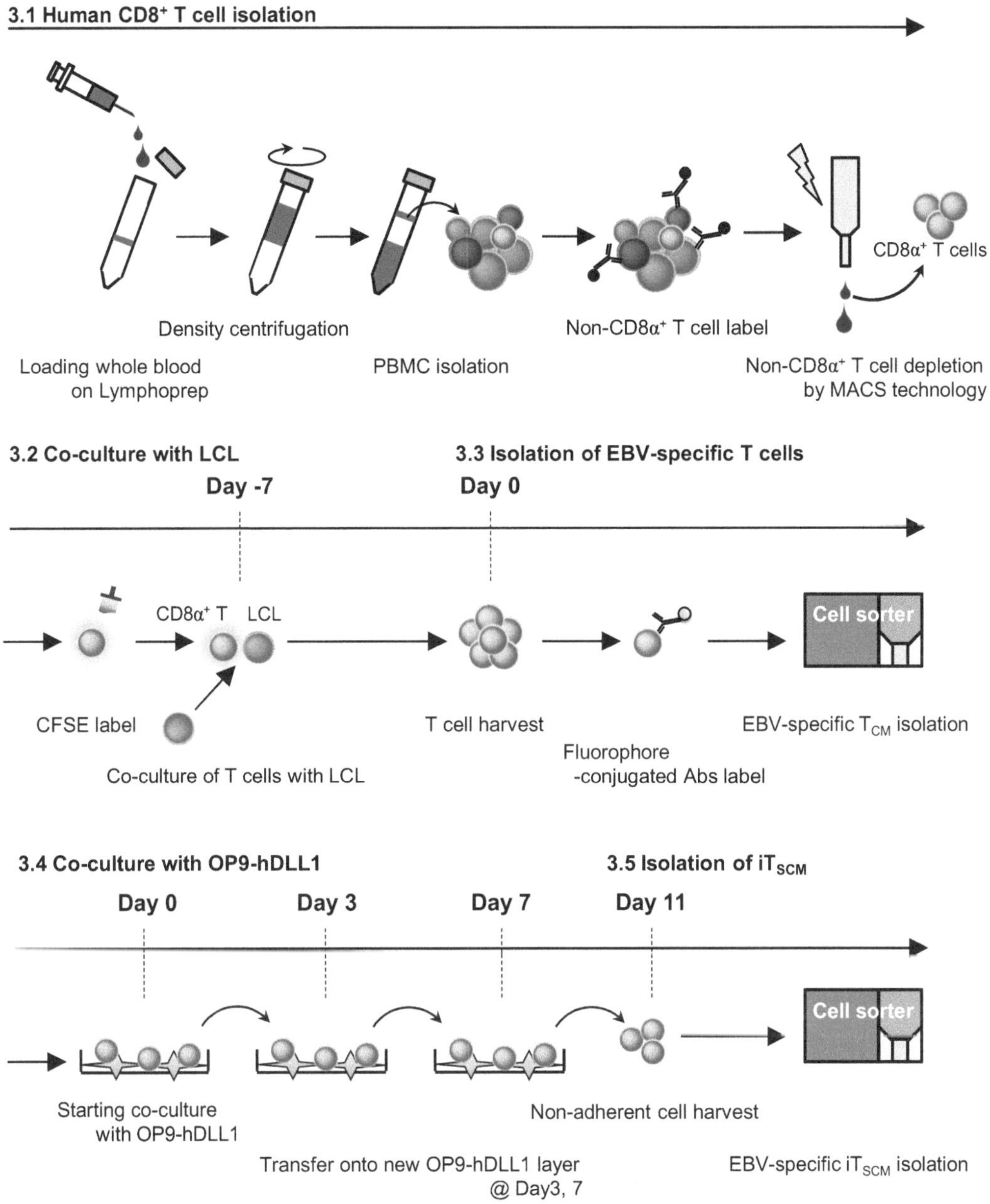

Fig. 2 Detailed protocols for iT_{SCM} generation. (Top) Start with $CD8\alpha^+$ T cell isolation on Day −7 of the prime step. Whole blood cells from healthy donors are loaded on Lymphoprep reagent and density-based centrifuged. Next, peripheral blood mononuclear cell (PBMC) layer is carefully isolated from the centrifuge tube. Peripheral $CD8\alpha^+$ T cells are then negatively isolated from the PBMCs by $CD8\alpha^+$ T cell isolation kit. (Middle) Isolated $CD8\alpha^+$ cells are labeled using cell trace dye (e.g., carboxyfluorescein diacetate succinimidyl ester [CFSE]), followed by the addition of labeled $CD8\alpha^+$ T cells to irradiated autologous LCL (T-to-LCL ratio of 4:1) and the start of coculture in a 96-well round-bottom plate (100 μL per well). EBV-specific T cells with memory phenotypes, which are defined as cell trace dye-diluted $CD45RA^-CCR7^+CD8\alpha^+$ cells, are purified by cell sorting (also *see* Fig. 3a). (Bottom) Purified EBV-specific T cells (1×10^5 cells/mL) are cocultured with OP9-hDLL1 cells (6-, 12-, 24-, 48-well flat bottom plate or 10-cm dish) in the presence of human IL-7 (10 ng/mL). Harvesting $CD8\alpha^+$ T cells, adjusting the cell density (1×10^5 cells/mL), and transferring them onto a new OP9-hDLL1 layer are performed on Days 3 and 7. Coculture with OP9-hDLL1 cells for 11 days induced iT_{SCM} cells, defined as $CD8\alpha^+CD45RA^+CCR7^+$ cells (also *see* Fig. 3b)

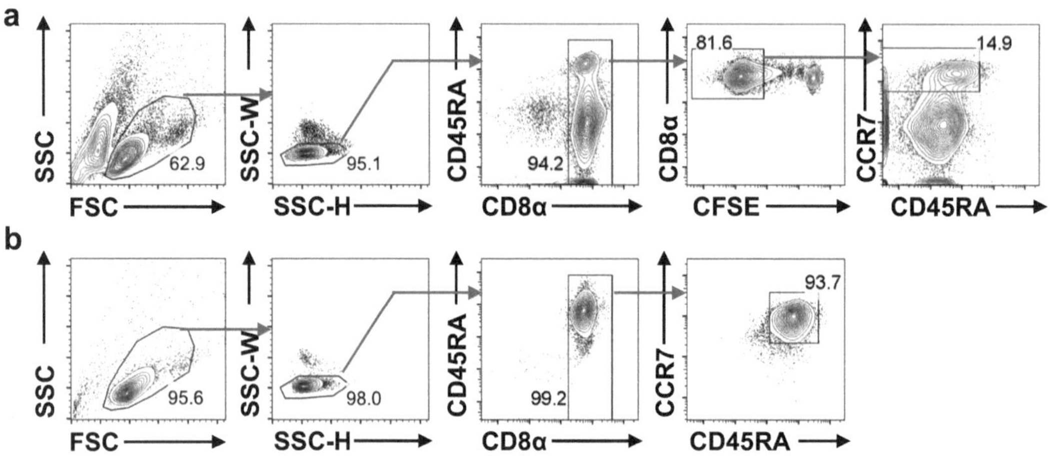

Fig. 3 Gating strategy in EBV-specific iT_{SCM} generation. (**a**) CFSE-labeled $CD8^+$ T cells were cocultured with 40 Gy irradiated EBV-transformed autologous LCL for 7 days. EBV-specific activated T cells mainly showed T_{EM} ($CD8\alpha^+CFSE^{lo/-}CCR7^-$) cell phenotypes and T_{CM} ($CD8\alpha^+CFSE^{lo/-}CCR7^-$) cell phenotypes (Day 0). (**b**) T_{CM} cells were sorted and then cocultured with OP9-hDLL1 cells for 11 days. Flow cytometry analysis of $CD8\alpha^+$ cells 11 days after OP9-hDLL1 cell coculture

14. Add an additional 10 mL of MACS buffer and centrifuge (400 × *g*, 5 min, 4 °C).
15. Place a QuadroMACS Separator on a clean bench.
16. Place an LS column in the magnetic field of the QuadroMACS Separator.
17. Equilibrate the LS column by rinsing with 3 mL of MACS buffer.
18. Resuspend the pellet in 1 mL of MACS buffer (*see* **Note 7**).
19. Apply cell the suspension onto the LS column.
20. Collect negatively isolated flow-through that contains enriched $CD8^+$ T cells (Fig. 2).
21. Rinse a 15-mL centrifuge tube twice with 1 mL of MACS buffer, and also apply the buffer onto the column.
22. Wash the column twice with 3 mL of MACS buffer and collect all the unlabeled cells in the flow-through.
23. Collect a total of 9 mL of cell suspension that contains enriched $CD8^+$ T cells.
24. Remove the supernatant by centrifugation (400 × *g*, 5 min, 4 °C).
25. Resuspend the pellet in RPMI 1640 media containing AB serum in a 15-mL centrifuge tube until the next step is started.

3.2 Coculturing of Human $CD8^+$ T Cells with Autologous LCL

1. Obtain a count for human $CD8^+$ T cells.
2. Prepare T cell prime medium, RPMI 1640 medium, using 10% AB serum.
3. Stain T cells in a 37 °C water bath using cell trace dye (e.g., CFSE), according to the manufacturer's instructions (Fig. 2) (*see* **Note 8**).
4. Harvest autologous LCL.
5. Irradiate autologous LCL at a dose of 40 Gy.
6. Remove the LCL supernatant by centrifugation (400 × *g*, 5 min, 4 °C).
7. Resuspend the T cell prime medium and obtain a cell count for LCL.
8. Transfer the T cells and LCL (T cell-to-LCL ratio of 4:1) into fresh T cell prime medium at 5×10^5 mL^{-1} of T cells and 1.25×10^5 mL^{-1} of LCL in 100 μL medium in a 96-well round-bottom plate (Fig. 2).
9. Spin down the T cell culture plate that contains T cells and LCL (*see* **Note 9**).
10. Place the tissue culture plate in a humidified CO_2 incubator at 37 °C and start the incubation.
11. Change the medium and resuspend the T cells in T cell prime medium every 2–3 days and incubate at 37 °C.
12. Scale up the culture medium volume and culture wells. T cells should be cultured at 5×10^5 mL^{-1}.
13. Analyze and isolate EBV-specific T cells using flow cytometry analysis. EBV-specific T cells can be detected by cell trace dye-diluted T cells (Fig. 2).

3.3 Isolation of EBV-Specific T Cells with Central Memory Phenotypes

1. Harvest activated T cells from the culture plate and transfer the cell suspension into 1.5-mL tubes at around 7 days after activation (Fig. 2) (*see* **Note 10**).
2. Harvest the supernatant-containing T cells.
3. Transfer cell suspension to 15-mL or 50-mL centrifuge tubes.
4. Remove the supernatant by centrifugation (400 × *g*, 5 min, 4 °C).
5. Prepare fluorophore-conjugated Ab staining solution.

   Briefly, Ab staining solution contains fluorophore-conjugated anti-CD8α, anti-CD45RA, and anti-CCR7 antibodies (dilution factor; 1:100) in MACS buffer.
6. Resuspend the pellet in 100 μL Ab staining solution and stain cells using fluorophore-conjugated Ab for 15 min at 4 °C (Fig. 2).

7. Add 10 mL of MACS buffer and centrifuge (400 × *g*, 5 min, 4 °C).
8. Remove supernatant and resuspend cells in 1 mL of MACS buffer.
9. Isolate phenotypically T_{CM} cells using a cell sorter (we normally use a FACS Aria sorter) gated using $CD8\alpha^{+}CD45RA^{-}CCR7^{+}$ cells (Figs. 2 and 3a) (*see* **Note 11**).

3.4 Coculturing of EBV-Specific T Cells with OP9-hDLL1 Cells

1. Prepare iT_{SCM} induction medium containing 10 ng/mL recombinant human IL-7 during cell sorting.
2. Centrifuge the collected cell suspension immediately (400 × *g*, 5 min, 4 °C).
3. Resuspend pellet in iT_{SCM} induction medium.
4. Obtain the cell count for isolated T_{CM} cells.
5. Resuspend T cells into fresh iT_{SCM} induction medium at 1×10^5 mL^{-1} (*see* **Note 12**).
6. Remove the supernatant of OP9-hDLL1 culture plate or dish and immediately transfer T cell suspension onto OP9-hDLL1 cells (Fig. 2) (*see* **Note 13**).
7. Place the culture plate or dish into a humidified CO_2 incubator at 37 °C and start the incubation.
8. Change medium and resuspend T cells in iT_{SCM} induction medium, replate T cells onto new OP9-hDLL1 layer, and incubate at 37 °C to passage T cells on new feeder cells every 3–4 days (Fig. 2) (*see* **Notes 14** and **15**).

3.5 Isolation and Analysis of iT_{SCM} Cells

Anticipated results by this method are described in Subheading 4 (*see* **Note 16**)

1. Gently harvest the supernatant of the coculture with OP9-hDLL1 cells, and place it into a 50-mL centrifuge tube. It will contain enriched iT_{SCM} cells (*see* **Note 17**).
2. Remove the supernatant by centrifugation (400 × *g*, 5 min, 4 °C).
3. Prepare fluorophore-conjugated Ab staining solution.
4. Resuspend the pellet in 100 μL Ab staining solution and stain cells using fluorophore-conjugated Ab for 15 min at 4 °C.
5. Add 10 mL of MACS buffer and centrifuge (400 × *g*, 5 min, 4 °C).
6. Remove supernatant and resuspend cells in 1 mL of MACS buffer.
7. Analyze using flow cytometry (e.g., FACS Canto II) or isolate phenotypically iT_{SCM} cells using a cell sorter (we normally use a

FACS Aria sorter) that is gated using $CD8\alpha^{+}CD45RA^{+}CCR7^{+}$ cells (Figs. 2 and 3b).

8. Centrifuge the collected cell suspension immediately ($400 \times g$, 5 min, 4 °C).
9. Obtain the cell count for isolated iT_{SCM} cells and continuously use iT_{SCM} cells for appropriate experiments.
10. All data obtained by the flow cytometer are analyzed using FlowJo software.

4 Notes

1. LCL is a B cell line that was transformed by Epstein-Barr (EB) virus. Established cell lines generally express the CD19 molecule and present EB virus-specific antigens. Coculturing of memory T cells with autologous LCLs induces endogenous EB virus antigen-specific T cells. Additionally, the maintenance of a biosafety level 2 (BSL-2) facility is required to start experiments involving the EB virus. More information and guidelines on biosafety and practices are provided by the CDC and NIH [16]. This study was performed using BCL-2 equipment in a BCL-2 facility.
2. Checking for Epstein-Barr virus seropositivity is required before starting experiments. Endogenous EBV-specific T cells in the memory T cell pool could react with EBV-specific antigens that are present on the LCL cell surface.
3. Because serum quality is critical for primary T cell culture, testing several lots of serum is necessary for culture optimization and experimental quality.
4. Because serum quality is critical for the OP9 feeder cell culture, testing several lots of serum is necessary for culture optimization and experimental quality.
5. Centrifugation at the appropriate temperature is critical because low or high temperatures can change the density of the centrifugation media. Slow acceleration and deceleration are also critical for creating a clear PBMC band. We use 15-mL centrifuge tubes made of polyethylene terephthalate (PET). The PBMC band is clearly visible in PET tubes.
6. Centrifugation speed is higher than $500 \times g$ during the first wash step to avoid contaminating the density gradient medium. Centrifugation using low gravity is not sufficient to generate a pellet.
7. It is necessary to use more than 1 mL of MACS buffer when resuspending a pellet if there are more than 5×10^{7} PBMC to avoid clogging the column.

8. A small number of EBV-specific T cells are contained in the overall amount of $CD8^+$ T cells. To distinguish between EBV-specific T cells and non-responding T cells, a cell proliferation detection system is required.
9. Accumulation of both T cells and LCL at the bottom of the 96-well round-bottom plate is necessary for strongly responding EBV-specific T cells.
10. It is necessary to prepare OP9-hDLL1 cells. Harvest and replate 2–4 × 10^4 mL^{-1} OP9-hDLL1 cells into a 6-well tissue culture plate or a 10-cm culture dish the day before harvesting activating T cells.
11. Maintain the collection tube at 4 °C. Chilling collected cells is critical for cell survival.
12. Concentration is critical for iT_{SCM} induction. If the OP9-hDLL1 layer is overloaded, iT_{SCM} induction may not succeed.
13. Immediate removal of the OP9-hDLL1 cell supernatant and transfer of the T cell suspension onto the OP9-hDLL1 layer are critical to avoid desiccation of the OP9-hDLL1 plate.
14. During the induction step, T cells are robustly expanded threefold compared to the initial cell count. Count the number of T cells at every passage and maintain the T cell density at 1 × 10^5 mL^{-1}. Maintaining the T cell culture at a low density is essential for iT_{SCM} induction because contact between single T cells and OP9-hDLL1 cells is critical.
15. Do not allow OP9-hDLL1 cells to become over confluent. OP9-hDLL1 cells easily die after reaching over confluence, and this causes T cell damage. Before OP9-hDLL1 reaches over confluence, transfer T cells onto a new OP9-hDLL1 layer.
16. If isolation of all $CD8^+$ T cells has been performed without any difficulties, the recovered cell number and purity should be greater than 0.5–3 × 10^6 cells (dependent on donors) and 90%, respectively, after the $CD8^+$ enrichment procedure using 10 mL of whole blood from a single healthy donor. If $CD8^+$ T cell activation was also performed without any difficulties, more than 60% of T cells can be activated by TCR activation on Day 0 (Fig. 3a), and 15% of activated T cells will show T_{CM} phenotypes. Given that complete $CD8^+$ T cell recovery from the $CD8^+$ T cell enrichment procedure is 3 × 10^6 cells (Day −7) and the phenotypically T_{CM} cell recovery from cell sorting is 15% (Day 7), the number of T_{CM} cells recovered from 10 mL of whole blood from a healthy donor should be 1–2 × 10^6 cells. After OP9-hDLL1 coculture for 11 days without any difficulties, the number of iT_{SCM} cells recovered

would be 0.5–2 × 10^7 cells, and the purity of iT_{SCM} cells would be greater than 80% (Fig. 3b).

17. Do not aggressively pipette the supernatant in the coculture dish to prevent OP9-hDLL1 contamination into the T cell suspension.

References

1. June CH (2007) Adoptive T cell therapy for cancer in the clinic. J Clin Invest 117 (6):1466–1476
2. Rosenberg SA, Dudley ME (2009) Adoptive cell therapy for the treatment of patients with metastatic melanoma. Curr Opin Immunol 21 (2):233–240
3. Gattinoni L et al (2005) Acquisition of full effector function in vitro paradoxically impairs the in vivo antitumor efficacy of adoptively transferred CD8+ T cells. J Clin Invest 115 (6):1616–1626
4. Hinrichs CS et al (2009) Adoptively transferred effector cells derived from naive rather than central memory CD8+ T cells mediate superior antitumor immunity. Proc Natl Acad Sci U S A 106(41):17469–17474
5. Berger C et al (2008) Adoptive transfer of effector CD8+ T cells derived from central memory cells establishes persistent T cell memory in primates. J Clin Invest 118(1):294–305
6. Gattinoni L et al (2011) A human memory T cell subset with stem cell-like properties. Nat Med 17(10):1290–U325
7. Alvarez-Fernandez C et al (2016) A short CD3/CD28 costimulation combined with IL-21 enhance the generation of human memory stem T cells for adoptive immunotherapy. J Transl Med 14(1):214
8. Cieri N et al (2013) IL-7 and IL-15 instruct the generation of human memory stem T cells from naive precursors. Blood 121(4):573–584
9. Hurton LV et al (2016) Tethered IL-15 augments antitumor activity and promotes a stem-cell memory subset in tumor-specific T cells. Proc Natl Acad Sci U S A 113(48): E7788–E7797
10. Scholz G et al (2016) Modulation of mTOR signalling triggers the formation of stem cell-like memory T cells. EBioMedicine 4:50–61
11. Sabatino M et al (2016) Generation of clinical-grade CD19-specific CAR-modified CD8+ memory stem cells for the treatment of human B-cell malignancies. Blood 128 (4):519–528
12. Nikolich-Zugich J (2014) Aging of the T cell compartment in mice and humans: from no naive expectations to foggy memories. J Immunol 193(6):2622–2629
13. Kondo T et al (2017) Notch-mediated conversion of activated T cells into stem cell memory-like T cells for adoptive immunotherapy. Nat Commun 8:15338
14. Kubuschok B et al (2002) Use of spontaneous Epstein-Barr virus-lymphoblastoid cell lines genetically modified to express tumor antigen as cancer vaccines: mutated p21 ras oncogene in pancreatic carcinoma as a model. Hum Gene Ther 13(7):815–827
15. Hui-Yuen J et al (2011) Establishment of Epstein-Barr virus growth-transformed lymphoblastoid cell lines. J Vis Exp (57)
16. Chosewood LC et al (2009) Biosafety in microbiological and biomedical laboratories, 5th edn. U.S. Dept. of Health and Human Services, Public Health Service, Centers for Disease Control and Prevention, National Institutes of Health, Washington, DC

Chapter 5

Human iPSC Generation from Antigen-Specific T Cells

Toshinobu Nishimura, Yukiko Murmann, and Hiromitsu Nakauchi

Abstract

The discovery and development of induced pluripotent stem cells (iPSCs) opened a novel venue for disease modeling, drug discovery, and personalized medicine. Additionally, iPSCs have been utilized for a wide variety of research and clinical applications without immunological and ethical concerns that encounter embryonic stem cells. Adoptive T cell immunotherapy is a form of cellular immunotherapy that involves transfusion of functional T cells. However, this approach requires T cell expansion and the process causes T cell exhaustion. As a result, highly expanded T cells have not proven to be particularly effective for treatments. This exhaustion issue could be overcome due to rejuvenation of T cells by reprogramming to pluripotency and redifferentiation to T cells. This is a potential therapeutic strategy for combating various types of cancer.

Key words Cancer immunotherapy, Pluripotent stem cell, T cell

1 Introduction

Cancer Immunotherapy is an entirely new class of cancer treatment, with unique characteristics that distinguish it from other kinds of cancer therapies. It exploits the fact that cancer cells often have subtly different antigens/molecules that the immune system can detect. Immunotherapy is used to provoke the immune system into attacking the tumor cells, with these antigens/molecules as targets. Recently, immune checkpoint inhibitor for programmed cell death protein 1/programmed death ligand 1 (PD-1/PD-L1) or cytotoxic T lymphocyte antigen (CTLA)-4 have shown dramatic clinical efficacies against some solid tumors such as lung cancer treatment [1, 2]. The other eye-catching immunotherapy is antigen-specific targeting of cancer by chimeric antigen receptor (CAR) transferred T cells against leukemia [3, 4].

To date, adoptive T cell immunotherapy has shown promise in treating melanoma and other cancers [5, 6]; however, these cytotoxic T cells can become exhausted, with loss of efficacy during

Shin Kaneko (ed.), *In Vitro Differentiation of T-Cells: Methods and Protocols*, Methods in Molecular Biology, vol. 2048, https://doi.org/10.1007/978-1-4939-9728-2_5,

ex vivo expansion before T cell infusion [7]. To overcome this obstacle, a novel system has developed recently, in which antigen-specific T cells are reprogrammed to pluripotent stem cells (T-iPSCs) using Yamanaka factors. These T-iPSCs are redifferentiated to functional T cells in vitro while retaining their original antigen specificity. These redifferentiated T cells display a naïve/central memory T cell phenotype, with longer telomeres and higher proliferative activity. They thus are "rejuvenated" through these reprogramming and redifferentiation processes [8, 9]. In this chapter, we describe methods to yield T-iPSCs from antigen-specific T cells, especially exhausted cytotoxic T lymphocyte (CTL) clones.

2 Materials

2.1 Cells

1. Antigen-specific T cells.

2.2 Equipment

1. Biosafety cabinet.
2. Hemocytometer.
3. Tube revolver/rotator.
4. Tissue culture incubator.

2.3 Plasticware and Other Disposables

1. 6-well culture plate (Thermo Fisher scientific).
2. 24-well culture plate (Thermo Fisher scientific).
3. Cryotube (Thermo Fisher Scientific).

2.4 Reagents and Media

1. Phosphate-buffered saline (PBS) (Gibco): PBS used in the chapter is free from both Ca^{2+} and Mg^{2+}.
2. RPMI-1640 medium + GlutaMaxI (Gibco).
3. Heat-inactivated human AB serum (Innovative Research).
4. Dynabeads Human T-Activator CD3/CD28 (Invitrogen).
5. RetroNectin (Takara).
6. Recombinant human IL-2 (Peprotech).
7. Recombinant human IL-7 (Peprotech).
8. Recombinant human IL-15 (Peprotech).
9. StemFit AK02N medium (Ajinomoto, Inc.).
10. iMatrix-511 Stem Cell Culture Substrate (Stemgent, Inc.).
11. CytoTune-iPS Sendai Reprogramming Kit (Life Technologies).
12. Sendai virus encoding SV40 large T antigen (SV40T) (ID Pharma).
13. TrypLE Select Enzyme (1×) (Gibco).

14. ROCK inhibitor (Y-27632) (Tocris).
15. Recombinant human FGF-basic (bFGF, 146aa) (Peprotech).
16. Dimethyl sulfoxide (DMSO) (Sigma-Aldrich).

3 Methods

3.1 Activate Antigen-Specific T Cells

1. T cell medium: RPMI-1640 medium + GlutaMaxI supplemented with 5% human AB serum, 50 U/ml IL-2, 10 ng/ml IL-7, and 10 ng/ml IL-15.
2. Treat a well of 24-well plate by RetroNectin coating solution (50 μg/ml; diluted with PBS) at RT for 2 h.
3. Count cell number of antigen-specific T cells and separate 1.0×10^6 cells into a conical tube. Wash cells in PBS and centrifuge them at $300 \times g$ for 3 min.
4. Resuspend the cell pellet in 1 ml of T cell medium in a 1.5 ml micro tube, then add 1.0×10^6 CD3/CD28 dynabeads. Rotate the tube to move the media inside up and down at RT for 45 min.
5. Pipette the cell suspension slowly and gently transfer whole 1 ml into a RetroNectin-coated well of 24-well plate.
6. Incubate at 37 °C, 5% CO_2, 20% O_2, for 72 h (Fig. 1a, *see* **Notes 1** and **2**).

3.2 Infection of Sendai Virus

1. Refresh half volume of culture medium (~0.5 ml) by T cell medium.
2. Transduce the cell with Sendai virus encoding Yamanaka reprogramming factors and SV40T at MOI = 2. Follow the manufacturer's instruction step by step. Then, spin the plate at $1000 \times g$ for 1 h.
3. Incubate at 37 °C, 5% CO_2, 20% O_2, for 72 h.

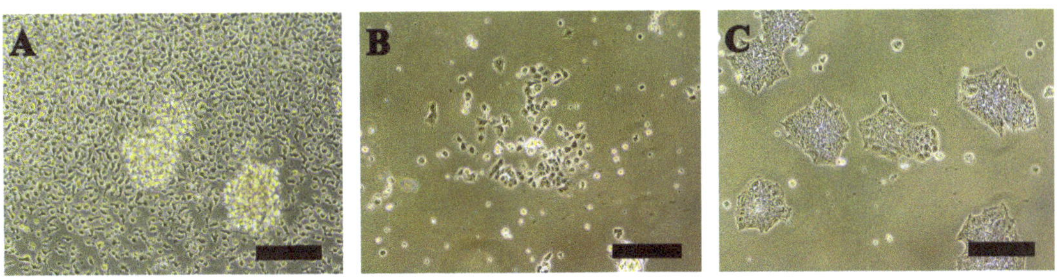

Fig. 1 Reprogramming of antigen-specific T cells to T-iPSCs. (**a**) Activated T cells cultured on RetroNectin-coated plate. Three days after activation. Scale bar = 100 μm. (**b**) Reprogramming T cells on iMatrix-511-coated plate. Ten days after activation. Scale bar = 200 μm. (**c**) Fully iPS-reprogrammed cells forming iPS colonies. Twenty-four days after activation. Scale bar = 200 μm

3.3 Reprogramming

1. StemFit medium: StemFit AK02N supplemented with 20 ng/ml bFGF.
2. Treat several wells of 6-well plate by iMatrix-511 coating solution (5 μg/ml; diluted with PBS) at 37 °C for 1 h.
3. Take half volume of culture medium (~0.5 ml), then add T cell medium to dilute into 0.5×10^6 cell/ml. Gently pipette the infected T cells up and down to resuspend in culture medium. Transfer T cells into iMatrix-511-coated wells at 1.0×10^6 cell/well (2 ml) density.
4. Incubate at 37 °C, 5% CO_2, 20% O_2, for 24 h.
5. Take half volume of culture medium (~1.0 ml), then add 1.0 ml StemFit medium. Repeat this step every 24 h to gradually replace culture medium with StemFit medium (*see* **Note 3**).
6. Replace whole 2 ml with StemFit medium. Refresh culture medium every other day (Fig. 1b, *see* **Note 4**).
7. Primary T-iPS colonies are manually picked and transferred onto an iMatrix-511-coated well of 6-well plate in 1.5 ml StemFit medium with 10 μM Y-27632 (Fig. 1c).

3.4 Maintenance of T-iPSCs

1. Replace culture medium every other day with 1.5 ml of StemFit medium (*see* **Note 5**).
2. When the T-iPSCs are 80–90% confluent, aspirate medium and wash well with 2 ml of PBS. Aspirate PBS and add 1 ml of TrypLE Select. Incubate cells at 37 °C for 4 min. Remove the TrypLE Select and create a single cell suspension.
3. Count cells and plate 2×10^4 cells/well in an iMatrix-511-coated 6-well plate.

3.5 Freezing T-iPSCs

1. Freezing solution: 90% StemFit medium, 10% DMSO. Prepare 5 ml freezing solution for T-iPSCs per well. Pre-cool the solution on ice.
2. Passage the cells as described in **step 2** of the previous subsection, using 5 ml of freezing medium in place of StemFit medium. Disaggregate the target cells into a single cell suspension. Aliquot 1 ml cell suspension per cryovial.
3. Freeze the cells at −80 °C for 24 h, then transfer them to liquid nitrogen for longer storage.

4 Notes

1. Antigen-specific T cells or T cell clones usually have less growth potential than freshly isolated T cells from peripheral or umbilical cord blood.

2. Sendai virus infection efficiency varies clone to clone. Optimization of infection schedule may be necessary for some T cell clones. Generally, the highest infection efficiency is achieved between day 2 and day 6 after activation.
3. Most cells are floating in medium at this time point. Pay extra attention not to aspirate floating cells as possible.
4. Reprogramming efficiency depends on proliferative potential and/or viability of antigen-specific T cells. Less proliferative cells tend to take longer time to be reprogrammed into pluripotency (e.g., ~40 days after activation).
5. StemFit PSC medium is designed to enable medium changes every other day. Other media such as mTeSR (STEMCELL Technologies) must be replaced daily. Be sure to check the appropriate feeding routine for your medium of choice.

Acknowledgments

The study was supported by the fellowship to T.N. from Japanese Society for the Promotion of Science (JSPS) and The Quest Award to H.N. from California Institute for Regenerative Medicine (CIRM).

References

1. Zou W, Wolchok JD, Chen L (2016) PD-L1 (B7-H1) and PD-1 pathway blockade for cancer therapy: mechanisms, response biomarkers, and combinations. Sci Transl Med 8:328rv324. https://doi.org/10.1126/scitranslmed.aad7118
2. Naidoo J et al (2016) Autoimmune bullous skin disorders with immune checkpoint inhibitors targeting PD-1 and PD-L1. Cancer Immunol Res 4:383–389. https://doi.org/10.1158/2326-6066.CIR-15-0123
3. Morello A, Sadelain M, Adusumilli PS (2016) Mesothelin-targeted CARs: driving T cells to solid tumors. Cancer Discov 6:133–146. https://doi.org/10.1158/2159-8290.CD-15-0583
4. Adusumilli PS et al (2014) Regional delivery of mesothelin-targeted CAR T cell therapy generates potent and long-lasting CD4-dependent tumor immunity. Sci Transl Med 6:261ra151. https://doi.org/10.1126/scitranslmed.3010162
5. Wu R et al (2012) Adoptive T-cell therapy using autologous tumor-infiltrating lymphocytes for metastatic melanoma: current status and future outlook. Cancer J 18:160–175. https://doi.org/10.1097/PPO.0b013e31824d4465
6. Tran E et al (2016) T-cell transfer therapy targeting mutant KRAS in cancer. N Engl J Med 375:2255–2262. https://doi.org/10.1056/NEJMoa1609279
7. Ghoneim HE, Zamora AE, Thomas PG, Youngblood BA (2016) Cell-intrinsic barriers of T cell-based immunotherapy. Trends Mol Med 22:1000–1011. https://doi.org/10.1016/j.molmed.2016.10.002
8. Nishimura T et al (2013) Generation of rejuvenated antigen-specific T cells by reprogramming to pluripotency and redifferentiation. Cell Stem Cell 12:114–126. https://doi.org/10.1016/j.stem.2012.11.002
9. Vizcardo R et al (2013) Regeneration of human tumor antigen-specific T cells from iPSCs derived from mature CD8(+) T cells. Cell Stem Cell 12:31–36. https://doi.org/10.1016/j.stem.2012.12.006

Chapter 6

In Vitro Differentiation of T Cells: From Human Embryonic Stem Cells and Induced Pluripotent Stem Cells

Shoichi Iriguchi and Shin Kaneko

Abstract

In this chapter, we describe a protocol for hematopoietic differentiation of human pluripotent stem cells (PSCs), including embryonic stem cells (ESCs) and induced pluripotent stem cells (iPSCs) derived from non-T cells, followed by the differentiation of the T-cell lineage. Derivation of T cells from PSCs involves three steps: induction of PSCs to hematopoietic progenitor cells (HPCs), differentiation of HPCs into progenitor T cells, and maturation of progenitor T cells into mature T cells (CD8 single-positive (SP) or CD4 SP).

Key words T cells, Human pluripotent stem cells, In vitro differentiation, Hematopoietic progenitor cells

1 Introduction

Protocols to generate of T cells from pluripotent stem cells (PSCs) could be exploited to study human lymphopoiesis as well as serve as an alternative cell source for T-cell immunotherapy [1–3]. Derivation of T cells from PSCs involves three steps: induction of PSCs to hematopoietic progenitor cells (HPCs), differentiation of HPCs into progenitor T cells, and maturation of progenitor T cells into mature T cells (CD8 single-positive (SP) or CD4 SP). Protocols for hematopoietic lineage cell induction from ESCs and iPSCs are well established. In general, induction can be achieved by coculturing ESCs and iPSCs with stromal feeder cells supporting hematoendothelial differentiation or by means of embryoid body formation accompanied by a stepwise addition of key cytokines for hematoendothelial differentiation. Either way, a $CD34^{+}CD43^{low}$ population, which represents multipotent hematopoietic progenitors, can be detected over a period of 10–14 days or more after initiation via mesoderm intermediates followed by hematoendothelial cells [4–6]. The resulting $CD34^{+}$ cell population was further differentiated into T cells by coculture on a DLL-1- or DLL-4-expressing

Shin Kaneko (ed.), *In Vitro Differentiation of T-Cells: Methods and Protocols*, Methods in Molecular Biology, vol. 2048, https://doi.org/10.1007/978-1-4939-9728-2_6,

OP9 cell line in the presence of the lymphopoietic cytokines IL-7 and Flt3L, a culture system widely used to differentiate $CD34^+$ hematopoietic stem/progenitor cells [7]. After differentiating for 1 week in the presence of Notch ligands, the $CD34^+$ cells acquired CD7 expression, and a fraction of these $CD7^+$ cells acquired CD5 expression and eventually CD4 and CD8 expressions, which gave rise to CD4 and CD8 double-positive (DP) T cells. As in normal development, the ESC-derived DP T cells expressed CD3-TCR complexes on their surfaces, a feature that defines T lymphocyte lineage determination. Importantly, those DP cells displayed a variety of TCR repertoires, suggesting that they underwent genomic TCR rearrangement during the differentiation culture [8, 9]. The differentiation culture generates both $\alpha\beta TCR^+$ and $\gamma\delta TCR^+$ T cells. The final maturation step, where CD4 or CD8 SP T cells emerge via TCR stimulation, appears possible in vitro, but controlling for the desired subset (either CD4 SP only or CD8 SP only) requires further investigation. Generation of SP T cells from DP T cells requires the positive selection, a process by which DP T cells expressing TCRs with intermediate affinity and/or avidity for self-peptide-MHC complexes presented by cortical thymic epithelial cells (cTECs) are induced to differentiate into SP T cells. When iPSCs from T-cell clones were induced to redifferentiate, few SP T cells, mostly CD8 SP T cells, can be obtained, but majority of the cells remain as DP T cells during the OP9-DLL-1 cultures. These cells can be induced to differentiate into CD8 SP T cells by TCR stimulation methods with an anti-CD3 monoclonal antibody or PHA-P in the presence of inactivated allogenic peripheral blood mononuclear cell feeders [10, 11]. To date, there has been no report that has demonstrated regeneration of CD4 SP T cells from human ESCs/iPSCs. In this chapter of the collection, we will cover a protocol to differentiate T cells from human ESCs and iPSCs derived from non-T cells.

2 Materials

2.1 Reagents

1. hESC/iPSCs.
2. Mouse embryonic fibroblasts (MEFs) (*see* **Note 1**).
3. OP9-DLL1 mouse bone marrow stromal cell line (RIKEN BRC) (*see* **Note 2**).
4. DMEM High glucose (Nacalai).
5. DMEM/nutrient mixture F-12 (DMEM/F-12, SIGMA).
6. α-MEM basal medium powder (Thermo Fischer) (*see* **Note 2**).
7. Iscove's modified Dulbecco's medium powder (IMDM, Thermo Fischer).
8. StemPro-34 SFM (Thermo Fischer).

9. KnockOut Serum Replacement (KSR) for ES cells (Thermo Fischer).
10. Insulin-Transferrin-Selenium Solution (ITS), 100× (Thermo Fischer).
11. Ascorbic acid 2-phosphate (SIGMA).
12. L-Glutamine, penicillin, streptomycin (SIGMA).
13. MEM nonessential amino acid (NEAA, 100× solution, Gibco-Invitrogen).
14. 2-Mercaptoethanol (Thermo Fischer).
15. Mitomycin C (Wako).
16. $CaCl_2$.
17. Sodium bicarbonate (Nacalai).
18. NaOH (Nacalai).
19. D-PBS (Ca^{2+} and Mg^{2+} free, Nacalai).
20. Gelatin from porcine skin, type A (SIGMA).
21. 2.5% trypsin (Gibco).
22. EDTA (0.5 M, pH 8.0, Nacalai).
23. Monothioglycerol (MTG, Nacalai).
24. FBS, defined (HyClone, GE Healthcare or Corning) (*see* **Note 2**).
25. TrypLE Select (Thermo Fischer).
26. Trypsin/EDTA (0.05% (wt/vol)/0.5 mM, SIGMA).
27. Human basic FGF (Wako).
28. Human BMP4 (R&D Systems).
29. Human Flt3 ligand (FLT3L) (Peprotech).
30. Human SCF (R&D Systems).
31. Human VEGF (R&D Systems).
32. Human IL-7 (Peprotech).
33. Y-27632 (Tocris) (*see* **Note 3**).
34. Trypan blue solution (0.4%, Nacalai).
35. BSA fraction V (SIGMA).
36. Antibodies and sorting reagents (*see* Subheading 2.4).
37. Distilled water.

2.2 Equipment

1. Cell strainer (70 μm, Thermo Fischer Scientific).
2. Nalgene disposable bottle top filter (polyethersulfone membrane with 0.2-μm pore size, Thermo Fischer).
3. T75 tissue culture flask (250 ml, nonpyrogenic polystyrene).
4. Tissue culture dishes (polystyrene, TPP).
5. Tissue culture 6-well plate (polystyrene flat bottom, TPP).

6. Polystyrene round-bottom tube (5 ml, with 35-μm cell strainer cap, BD Biosciences).
7. Polypropylene conical tubes (15, 50 ml).
8. Serological pipette (nonpyrogenic).
9. P1000G pipette tip (MBP).
10. 1-ml syringe (Terumo).
11. 21-G needle (Terumo).
12. Hemocytometer.
13. Water bath.
14. Microcentrifuge.
15. Sterile biosafety cabinet.
16. CO_2 incubator (37 °C/5%, Thermo Scientific).
17. Inverted microscope (Olympus).
18. Milli-Q water purification system (Millipore).
19. Pipet-Aid (Drummond).
20. Flow cytometer (Becton Dickinson).
21. Microcentrifuge tube (1.5 ml).
22. Bright-field microscope.

2.3 Reagent Setup

1. *Gelatin solution (0.1% (wt/vol))*: Add 500 mg of gelatin to 500 ml of Milli-Q. Solubilize and sterilize by autoclaving for 20 min at 121 °C. Store the solution at 4 °C for up to 6 months.
2. *Mitomycin C solution*: Dissolve 10 mg of lyophilized mitomycin C to 25-ml D-PBS. Mix well and divide into 1.5-ml aliquots. Store at −20 °C.
3. *Human ES/iPS dissociation solution (TK)*: Add 20-ml 2.5% trypsin, 40-ml KSR, and 2-ml 100-mM $CaCl_2$ to 138-ml D-PBS (−). Mix well and divide into 10-ml aliquots and store at −20 °C. After thawing, store at 4 °C for 1 week.
4. *α-MEM basal medium*: Prepare according to the manufacturer's instruction. Divide into 210-ml aliquots and store at 4 °C for up to 2 weeks.
5. *FACS buffer:* FACS buffer contains 2% (vol/vol) FBS in D-PBS. Store at 4 °C for up to 6 months.
6. *MTG solution*: Add 87 μl of MTG to 10 ml of endotoxin-free reagent-grade distilled water. Mix well and divide into 500-μl aliquots. Store for up to 6 months at −20 °C.
7. *Ascorbic acid solution*: Add 50 mg of ascorbic acid to 1 ml of endotoxin-free reagent-grade distilled water. Dissolve completely, divide into 100-μl aliquots, and store for up to 6 months at −20 °C.

8. *BSA/PBS solution (0.1% (wt/vol))*: Dissolve 25 mg of bovine serum albumin fraction V in 25 ml of PBS. Sterilize the solution by filtration using a 0.22-μm membrane filter and store for up to 6 months at 2–8 °C.
9. *Reconstitution of cytokines*: Reconstitute cytokines according to the product information provided by manufacturer (*see* Subheading 2.5).

2.4 Antibodies Used to Analyze Differentiation of HPCs and Pro-T Cells

1. *Antihuman CD34 (Abcam)*: PE/Cy7, 4H10.
2. *Antihuman CD43 (BD Biosciences)*: APC, 1G10.
3. *Antihuman CD45 (BioLegend)*: Brilliant Violet 510, HI30.
4. *Antihuman CD7 (BioLegend)*: APC, CD7-6B7.
5. *Antihuman CD5 (eBioscience)*: PE/Cy7, UCHT2.
6. *Antihuman CD3 (BioLegend)*: APC/Cy7, UCHT1.
7. *Antihuman αβTCR (eBioscience)*: FITC, IP26, eBioscience.
8. *Antihuman CD4 (BioLegend)*: Brilliant Violet 421, OKT4.
9. *Antihuman CD8αβ (Beckman Coulter)*: PE, 2ST8.5H7.
10. *Antihuman CD8α (BioLegend)*: PerCP/Cy5.5, SK1.

2.5 Reconstitution of Cytokines

1. *Human bFGF*: 100 μg/ml, 0.1% (wt/vol) BSA/D-PBS solution, aliquot 50 μl into 1.5-ml tubes, and store at −80 °C.
2. *Human VEGF*: 100 μg/ml, 0.1% (wt/vol) BSA/D-PBS solution, aliquot 50 μl into 1.5-ml tubes, and store at −80 °C.
3. *Human BMP4*: 50 μg/ml, 0.1% (wt/vol) BSA/D-PBS 4-mM HCl, aliquot 50 μl into 1.5-ml tubes, and store at −80 °C.
4. *Human SCF*: 50 μg/ml, 0.1% (wt/vol) BSA/D-PBS solution, aliquot 50 μl into 1.5-ml tubes, and store at −80 °C.
5. *Human TPO*: 30 μg/ml, 1% (wt/vol) BSA/D-PBS solution, aliquot 50 μl into 1.5-ml tubes, and store at −80 °C.
6. *Human FLT3L*: 200 μg/ml, 0.1% (wt/vol) BSA/D-PBS solution, aliquot 50 μl into 1.5-ml tubes, and store at −80 °C.
7. *Human IL-7*: 100 μg/ml, 0.1% (wt/vol) BSA/D-PBS solution, aliquot 50 μl into 1.5-ml tubes, and store at −80 °C.

2.6 Media Composition

1. *MEF growth medium*: DMEM 500 ml, FBS 55.5 ml 10%, PSG 5.5 ml 1%.
2. *10 T1/2 medium*: BME 500 ml, FBS 55.5 ml 10%, PSG 5.5 ml 1%.
3. *OP9 medium*: αMEM 210 ml, FBS 37.5 ml 15%, PSG 2.5 ml 1%.
4. *hES/iPSC culture medium*: DMEM/F-12500 ml, KSR 125 ml 20%, PSG 6.25 ml 1%, NEAA (100×) 6.25 ml 1×, 2ME (55 mM) 1.2 ml, bFGF 50 μl 5 ng/ml.

5. *Hematopoietic differentiation medium*: StemPro34 basal medium 50 ml, StemPro34 supplement (40×) 1.25 ml 1×, PSG 0.5 ml 1%, ITS (100×) 0.5 ml 1×, MTG 44.4 μl 400 μM, ascorbic acid 2-phosphate 50 μl 50 μg/ml.
6. *T-cell differentiation medium*: OP9 medium 20 ml, IL-7 1 μl 5 ng/ml, FLT3L 1 μl 10 ng/ml.

3 Methods

3.1 MEF Preparation for Human ES/iPSC Culture

1. Thaw cryopreserved MEFs and seed into a 100-mm dish at a density of 3×10^6 cells/dish in MEF growth medium. Incubate at least overnight at 37 °C/5% CO_2.
2. Add 300 μl of mitomycin C solution to the overnight culture dish and inactivate MEFs by incubating 90 min in a 37 °C incubator.
3. While inactivating MEFs, prepare gelatin-coated 60-mm dishes by adding 2 ml 0.1% gelatin solution per dish. Incubate the dishes at a 37 °C incubator for at least 30 min.
4. After the incubation, aspirate medium and wash MEFs twice with 10-ml D-PBS.
5. Add 1 ml of trypsin/EDTA solution and incubate for 5 min at a 37 °C incubator.
6. Add 9 ml of MEF growth medium and collect MEFs by pipetting.
7. Transfer cell suspension into a 15-ml centrifuge tube and centrifuge for 5 min at $300 \times g$ at room temperature.
8. Aspirate supernatant and resuspend the pellet in 1 ml of MEF growth medium.
9. Determine a viable cell count by trypan blue.
10. Adjust the volume of cell suspension to the density of 1.5×10^5 cells/ml.
11. Aspirate the gelatin solution from the precoated dishes, and add 2-ml cell suspension per dish (3×10^5 cells/dish).
12. Distribute MEFs evenly and incubate MEF dishes in a 37 °C CO_2 incubator. It is important to make dishes with evenly distributed MEFs for efficient differentiation (*see* **Note 4**).

3.2 hES/iPSC Culture

1. Aspirate spent hPSC medium from a dish.
2. Wash cells with 3 ml of D-PBS.
3. Add 1 ml per 60-mm dish of TK solution and incubate at 37 °C in a CO_2 incubator for 5 min.
4. After the incubation, gently tap the edges of the dish to remove MEFs and aspirate TK solution.

5. Add 3 ml of hES/iPSC culture medium and break up the colonies into small cell clumps by pipetting. Use P1000G pipette tip.
6. Transfer cell clumps to a 15-ml centrifuge tube.
7. Centrifuge at 170 × *g* for 5 min at room temperature.
8. Aspirate the supernatant, gently tap the bottom of the tube to loosen the pellet, and resuspend the pellet in 3 ml of hES/iPSC culture medium.
9. Aspirate MEF growth medium from Subheading 3.1 and add 3 ml of hES/iPSC medium.
10. Add 0.5 ml of cell suspension to the MEF dishes (a 1:5 split). Note, the split ratio may require optimization depending on PSC lines.
11. Feed hPSCs daily by 3 ml of fresh hES/iPSC medium.
12. Passage hPSCs weekly (*see* **Note 5**).

3.3 Culture of Mouse OP9-DLL1 Cells

1. Aspirate spent medium and wash cells twice with 10-ml D-PBS.
2. Add 1 ml of trypsin/EDTA solution and incubate for 5 min in a 37 °C incubator.
3. Add 9 ml of OP9 medium and collect OP9-DLL1 by pipetting.
4. Transfer cell suspension into a 15-ml centrifuge tube and centrifuge for 5 min at 300 × *g* at room temperature.
5. Aspirate supernatant and resuspend the pellet in 1 ml of OP9 medium.
6. Determine a viable cell count by trypan blue.
7. Add 10 ml of OP9 medium to a 100-mm dish for maintenance.
8. Add 2.0×10^5 cells to the 100-mm dish and place it in a CO_2 incubator at 37 °C. Passages should occur every 2–3 days. It is crucial to avoid overconfluent during maintenance culture as it greatly compromises T-cell differentiation efficiency.
9. For T-cell differentiation, add 1.5×10^5 cells to 60-mm dishes supplemented with 3 ml of OP9 medium. Dishes are ready for use typically in 2 days (*see* **Note 6**).

3.4 Hematopoietic Differentiation by EB Formation

3.4.1 Day 1: Generation of EB from hPSCs

1. Aspirate spent hPSC medium from a dish.
2. Wash cells with 3 ml of D-PBS.
3. Add 1 ml per 60-mm dish of TK solution and incubate at 37 °C in a CO_2 incubator for 5 min.
4. After the incubation, gently tap the edges of the dish to remove MEFs and aspirate TK solution.
5. Add 3 ml of hES/iPSC culture medium and break up the colonies into small cell clumps by pipetting. Use P1000G pipette tip.

6. Transfer cell clumps to a 15-ml centrifuge tube.
7. Centrifuge at 170 × *g* for 5 min at room temperature.
8. Aspirate the supernatant, gently tap the bottom of the tube to loosen the pellet, and resuspend the pellet in 3 ml of hES/iPSC culture medium.
9. Add 3 ml of hES/iPSC culture medium to a well of ultralow attachment 6-well plate. Add 3 μl of 10-mM Y27632 to the well.
10. Seed 300 μl of cell suspension to the well prepared in **step 9** (*see* **Note 7**).
11. Incubate the cells in a 37 °C 5% CO_2/5% O_2 incubator (hypoxia incubator) for 24 h.

3.4.2 Day 2: Induction of Mesoderm

1. Harvest EBs to a 15-ml tube.
2. Let it stand for 10 min and allow EBs to settle to the bottom of the tube (*see* **Note 8**).
3. Carefully aspirate the medium.
4. Resuspend the pellet in 1 ml of hematopoietic differentiation medium and add to the same well.
5. Wash the tube with 1 ml of hematopoietic differentiation medium and add to the same well.
6. Add 2 μl of BMP4, 1 μl of bFGF, and 2 μl of ascorbic acid 2-phosphate (PAA) to the well. If you work with multiple wells, prepare cytokine premixture, and add 5 μl per well.
7. Incubate the EBs in a hypoxia incubator.

3.4.3 Day 5: Induction of Hematopoietic Progenitor Cells

1. Harvest EBs to a 15-ml tube.
2. Let it stand for 10 min to allow EBs to settle to the bottom of the tube, or centrifuge 200 × *g* for 5 min at room temperature.
3. Carefully aspirate the medium.
4. Resuspend the pellet in 1 ml of hematopoietic differentiation medium and add to the same well.
5. Wash the tube with 1 ml of hematopoietic differentiation medium and add to the same well.
6. Add 1 μl of VEGF, bFGF, 2 μl of SCF, and PAA to the well. If you work with multiple wells, prepare cytokine premixture, and add 6 μl per well.
7. Incubate the EBs in a hypoxia incubator.

3.4.4 Day 8: Expansion of Hematopoietic Progenitor Cells

1. Harvest EBs to a 15-ml tube.
2. Centrifuge 300 × *g* for 5 min at room temperature.
3. Carefully aspirate the medium.

4. Resuspend the pellet in 1 ml of hematopoietic differentiation medium and add to the same well.
5. Wash the tube with 1 ml of hematopoietic differentiation medium and add to the same well.
6. Add 1 μl of VEGF, bFGF, and 2 μl of SCF, TPO, FLT3L, and PAA to the well. If you work with multiple wells, prepare cytokine premixture, and add 10 μl per well.
7. Incubate the EBs in a 37 °C 5% CO_2/20% O_2 incubator (*see* **Note 9**).

*3.4.5 Day 11–14: Harvesting and Dissociation of EBs (See **Note 10**)*

1. Harvest EBs to a 15-ml tube.
2. Centrifuge 300 × *g* for 5 min at room temperature.
3. Carefully aspirate the medium.
4. Add 300 μl of TrypLE Select and incubate the tube in a 37 °C water bath for 8 min.
5. Add 1 ml of hematopoietic differentiation medium.
6. Dissociate the EBs by passing them six times through a 1-ml syringe with a 21-gauge needle.
7. Centrifuge 300 × *g* for 5 min at room temperature.
8. Aspirate the supernatant and resuspend in 1-ml FACS buffer.
9. Determine a viable cell count.
10. Cells are ready to be used in downstream applications (flow cytometry and 3.6 T-cell differentiation). For flow cytometry, use 3.5×10^5 cells for flow cytometer setup (5×10^4 cells/fluorochrome) and evaluation of hematopoietic differentiation efficacy (1×10^5 cells/sample). Hematopoietic differentiation efficiency can be evaluated by analysis of CD34, CD43, and CD45 expressions.

3.5 T-Cell Differentiation on OP9-DLL1

1. After the EB dissociation, prepare T-cell differentiation medium as in Subheading 2.6.
2. Aliquot 3 ml of T-cell differentiation medium to a 15-ml tube and add 5×10^5 cells/dish.
3. Remove spent medium from a dish of OP9-DLL1 cells at subconfluent culture (refer to Subheading 3.3).
4. Add the cell suspension to the OP9-DLL1 dish.
5. Incubate in a 37 °C incubator.
6. On day 3, add 3 ml of T-cell differentiation medium. Prepare a new OP9-DLL1 dish for cell transfer.
7. On day 5, harvest the cells by pipetting and centrifuge at 300 × *g* for 5 min at room temperature. Aspirate the supernatant and resuspend the cells in 3 ml of T-cell differentiation medium.

8. Aspirate the spent medium from the new OP9-DLL1 dish and add the differentiating cells.
9. On day 8, add 3 ml of T-cell differentiation medium. Prepare a new OP9-DLL1 dish for cell transfer.
10. On day 10, transfer differentiating cells as did on day 5.
11. On day 13, add 3 ml of T-cell differentiation medium. Prepare a new OP9-DLL1 dish for cell transfer.
12. On day 15, transfer differentiating cells as did on day 5 (*see* **Note 11**).
13. On day 18, add 3 ml of T-cell differentiation medium.
14. On day 20, harvest the cells by pipetting and centrifuge at 300 × *g* for 5 min at room temperature. Aspirate the supernatant and resuspend the cells in 1 ml of FACS buffer.
15. Determine cell number and evaluate T-cell differentiation by flow cytometric analysis of CD3, CD4, CD5, CD7, 8α, 8αβ, CD45, and αβTCR expressions. Evaluation of T-cell differentiation can be assessed by genomic TCR rearrangements.

4 Notes

1. For efficient PSC differentiation, it is important to maintain PSC in an appropriate condition. This largely depends on the condition of feeder cells used in maintenance culture. Our lab uses mitomycin C-treated MEFs, freshly isolated from ICR embryos and cryopreserved at passage 2.
2. Condition of OP9-DLL1 feeder is the most important parameter of successful T-cell differentiation. This could be achieved by selection of an appropriate serum lot, careful culture of OP9-DLL1 cells during maintenance, and use of OP9 medium prepared from powder. The most critical composition of the differentiation culture is serum. Our lab spends considerable amount of time to select an appropriate lot of serum. To select an appropriate serum, HPCs are induced to differentiate into T cells in the various test lots and are evaluated for the efficiency of T-cell differentiation by analysis of CD7, CD5, CD4, and CD8 expressions after 20 days differentiation. Test lots are also evaluated for its ability to support OP9-DLL1 cells by monitoring cell viability, growth, and morphology. Although OP9-DLL1 feeders get overconfluent during 5 days differentiation, OP9-DLL1 cells for maintenance culture should not get overconfluent as it reduces the T-cell differentiation ability. Thus, it is important to perform passaging before it gets confluent, typically 2–3 days when cultured in an appropriate serum lot.

3. For EB formation, it is important to add Y-27632, a ROCK inhibitor, to the hES/iPS culture medium when hPSCs are dissociated into small aggregates to improve their survival and efficiency of EB formation.
4. The inactivated MEF feeders should be used by 4 days after preparation.
5. Some hES/iPSC lines may require passaging shorter than day 7 if some colonies undergo spontaneous differentiation. If earlier passages cannot reduce the numbers of differentiated cells, the hES/iPSC line should be subcloned by manual colony picking.
6. If OP9-DLL1 feeder dishes are required the next day, passage cells at 2×10^5 cells/dish instead of 1.5×10^5 cells/dish.
7. The amount of cell suspension may require optimization by hES/iPSC lines you use.
8. Centrifugation results in extensive EB aggregation and compromised differentiation and, therefore, should be avoided at this stage.
9. The cultures are now placed in a normal incubator.
10. EBs can be harvested and dissociated for further application from day 11 to 14. Some hES/iPSC lines differentiate slower than the others and require longer period of time for sufficient hematopoietic differentiation.
11. Differentiating cells get confluent around day 15 after T-cell differentiation. If this happens, split the culture at the ratio of 1:5–10. Overconfluent culture may result in reduced T-cell differentiation efficiency.

Acknowledgments

We thank Mr. Shuichi Kitayama and Ms. Ayako Kumagai for help in the preparation. This work is supported in part by KAKENHI to S.I. and S.K.

References

1. Themeli M, Rivière I, Sadelain M (2015) New cell sources for T cell engineering and adoptive immunotherapy. Cell Stem Cell 16:357–366. https://doi.org/10.1016/j.stem.2015.03.011
2. Karagiannis P, Iriguchi S, Kaneko S (2016) Reprogramming away from the exhausted T cell state. Semin Immunol 28:35–44. https://doi.org/10.1016/j.smim.2015.10.007
3. Brauer PM, Singh J, Xhiku S, Zúñiga-Pflücker JC (2016) T cell genesis: in vitro veritas est? Trends Immunol 37:1–13. https://doi.org/10.1016/j.it.2016.09.008
4. Kardel MD, Eaves CJ (2012) Modeling human hematopoietic cell development from pluripotent stem cells. Exp Hematol 40:601–611. https://doi.org/10.1016/j.exphem.2012.04.001
5. Kennedy M, D'Souza SL, Lynch-Kattman M et al (2007) Development of the hemangioblast defines the onset of hematopoiesis in

human ES cell differentiation cultures. Blood 109:2679–2687. https://doi.org/10.1182/blood-2006-09

6. Vodyanik MA, Thomson JA, Slukvin II (2006) Leukosialin (CD43) defines hematopoietic progenitors in human embryonic stem cell differentiation cultures. Blood 108:2095–2105. https://doi.org/10.1182/blood-2006-02-003327
7. La Motte-Mohs RN, Herer E, Zúñiga-Pflücker JC (2005) Induction of T-cell development from human cord blood hematopoietic stem cells by Delta-like 1 in vitro. Blood 105:1431–1439. https://doi.org/10.1182/blood-2004-04-1293
8. Timmermans F, Velghe I, Vanwalleghem L et al (2009) Generation of T cells from human embryonic stem cell-derived hematopoietic zones. J Immunol 182(11):6879–6888
9. Kennedy M, Awong G, Sturgeon CM et al (2012) T lymphocyte potential marks the emergence of definitive hematopoietic progenitors in human pluripotent stem cell differentiation cultures. Cell Rep 2:1722–1735. https://doi.org/10.1016/j.celrep.2012.11.003
10. Vizcardo R, Masuda K, Yamada D et al (2013) Regeneration of human tumor antigen-specific T cells from iPSCs derived from mature CD8 (+) T cells. Cell Stem Cell 12:31–36. https://doi.org/10.1016/j.stem.2012.12.006
11. Nishimura T, Kaneko S, Kawana-Tachikawa A et al (2013) Generation of rejuvenated antigen-specific T cells by reprogramming to pluripotency and redifferentiation. Cell Stem Cell 12:114–126. https://doi.org/10.1016/j.stem.2012.11.002

Chapter 7

Redifferentiation of Adaptive Naïve-Like CTL from T-Cell-Derived iPSC

Yohei Kawai and Shin Kaneko

Abstract

In this chapter, we describe redifferentiation procedures from iPSCs to CD8αβ$^+$ cytotoxic T cells in 10 T1/2 and OP9/DL1 feeder condition. iPSC used here is derived from T-cell clone (T-iPSC), which has lost naïve phenotype and acquired exhaustion/senescence phenotype during cloning process (**Note 1**). On the other hand, redifferentiated T cells (T-iPSC-Ts) reacquire naïve phenotype (CD45RA$^+$C-D45RO$^-$CCR7$^+$CD62L$^+$), which are reportedly critical for in vivo persistence of infused T cells and greatly affect therapeutic efficacy of adoptive immunotherapy. Indeed, T-iPSC-Ts exhibit much superior proliferative capacity while retaining equivalent effector function compared to parental T-cell clones. Here, we demonstrate the methodology to produce naïve-like T-iPSC-Ts, which could be potent cell source for adoptive immunotherapy.

Key words iPSC, CTL, Adaptive, Naïve, Nkp44, CCR7

1 Introduction

Adoptive immunotherapy is attracting increasing attention, but there is major challenge in mass preparation of target-reactive T cells because exhaustion and senescence inevitably arise during cloning/expansion process and profoundly impair replication capacity [1, 2]. Exhaustion/senescence hamper infused T cells not only quantitatively but also qualitatively by impairing effector function and in vivo persistence. Therefore, it is being established that less-differentiated unexhausted naïve T cell is preferable cell source compared to more-differentiated effector cells [3–5]. To meet demand for massive number of unexhausted target-specific T cells, several reports have proposed novel strategy to rejuvenate exhausted/senescent T-cell clone by iPSC technology [6, 7]. They have provided the proof of concept by showing the same TCR sequence, and antigen specificity was inherited into T-iPSC and redifferentiated T-iPSC-T, while exhausted/senescent phenotype such as shortened telomere length was reversed. At present,

Shin Kaneko (ed.), *In Vitro Differentiation of T-Cells: Methods and Protocols*, Methods in Molecular Biology, vol. 2048, https://doi.org/10.1007/978-1-4939-9728-2_7, © Springer Science+Business Media, LLC, part of Springer Nature 2019

however, the character of T-iPSC-T is considerably varied among research teams [8, 9] especially in (1) naïve phenotype (expression of naïve-associated markers and replication potential), (2) CD8β expression, and (3) ectopic NK activity. Some conventional T-iPSC-T express only CD8αα homodimer as a co-receptor, but not CD8αβ heterodimer, which was indispensable for optimal binding to HLA-peptide complex [10]. That is why 10- to 100-fold more peptide pulse was needed to stimulate conventional T-iPSC-T than $CD8\alpha\beta^{+}$ parental T-cell clone. Ectopic NK activity had a risk of inducing graft-versus-host disease (GVHD) in recipients. Therefore, the redifferentiation protocol described here was optimized based on following markers: naïve-associated markers (CD27, CD28, CCR7, CD45RA), NK-associated markers (CD56, NKp44, NKp30), and CD8αβ. After a stepwise redifferentiation via hematopoietic progenitor cell (HPC) and CD4/CD8 double-positive (DP) stages, naïve-like ($CD27^{+}CD28^{+}CCR7^{+}CD45RA^{+}CD45RO^{-}$) T-iPSC-T was induced.

2 Materials

2.1 Generation of Hematopoietic Progenitor Cells (HPCs) from T-iPSCs

1. T-iPSCs, cultured on irradiated MEF in iPSC medium.
2. iPSC medium: DMEM nutrient mixture F-12 HAM (Sigma-Aldrich) supplemented with 20% knockout serum replacement (KSR, Invitrogen), 1% L-glutamine-penicillin-streptomycin solution (PSG, Sigma-Aldrich), 0.1 mM nonessential amino acid (Invitrogen), 0.1 mM 2-mercaptoethanol (Invitrogen), 5 ng/ml basic fibroblast growth factor (b-FGF, Wako, Osaka, Japan).
3. C3H10T1/2 murine stromal cell line (RIKEN BioResource Center, Tsukuba, Japan).
4. C3H10T1/2 medium: Basal Eagle's medium (BME, Invitrogen) supplemented with 10% FBS (Biological Industries, Kibbutz Beit Haemek, Israel), 1% PSG.
5. 0.05% trypsin-EDTA (Sigma-Aldrich).
6. PBS (Nacalai Tesque, Kyoto, Japan).
7. Dissociation solution: PBS supplemented with 0.25% human trypsin (Invitrogen), 1 mM $CaCl_2$ (Nacalai Tesque, Kyoto, Japan), 20% KSR.
8. HPC development medium: IMDM (Sigma-Aldrich) with 15% FBS (AusGeneX, Oxenford, Australia), 10 μg/ml human insulin, 5.5 μg/ml human transferrin, 6.7 ng/ml sodium selenite (1× ITS, Gibco), 1% PSG, 50 μg/ml ascorbic acid (Nacalai Tesque, Kyoto, Japan), 0.45 μM α-monothioglycerol (Nacalai

Tesque, Kyoto, Japan), 20 ng/ml human recombinant vascular endothelial growth factor (R&D Systems).

9. Human recombinant SCF (Peprotech).
10. Human recombinant Flt3L (Peprotech).
11. Cell strainer size 70 μm (Corning).

2.2 Generation of CD4/CD8 DP Cells from HPCs

1. OP9/DL1 murine stromal cell line (RIKEN BioResource Center, Tsukuba, Japan).
2. OP9/DL1 medium: α-MEM (Invitrogen) supplemented with 20% FBS (BioWest, Nuaillé, France) and 1% PSG.
3. T-cell development medium: α-MEM (Invitrogen) supplemented with 20% FBS (BioWest, Nuaillé, France), 1% PSG, 1× ITS, 50 μg/ml ascorbic acid 2-phosphate, 1 ng/ml human recombinant interleukin-7 (Peprotech), 10 ng/ml human recombinant Flt3L (Peprotech).
4. 0.05% trypsin-EDTA (Sigma-Aldrich).
5. PBS (Nacalai Tesque, Kyoto, Japan).

2.3 Generation of CD8 Single-Positive (SP) T-iPSC-T from DP Cells

1. Maturation medium: α-MEM (Invitrogen) supplemented with 20% FBS (BioWest, Nuaillé, France), 1% PSG, 1× ITS, 50 μg/ml ascorbic acid 2-phosphate, 10 ng/ml human recombinant interleukin-7 (Peprotech).
2. Mouse antihuman CD3 antibody (OKT3, eBioscience).
3. RetroNectin™ (TAKARA, Otsu, Japan).
4. FACS Aria.

3 Methods

3.1 Generation of HPCs from T-iPSCs

1. Prepare three dishes (10 cm) of confluent 10 T1/2, which is seeded 1–2 days before (untreated with growth inhibitors such as mitomycin C).
2. Harvest T-iPSCs from a confluent dish (day 7 after seeded). Remove iPSC medium, add iPSC dissociation solution, and incubate at 37 °C for 5 min for trypsinization. Exchange dissociation medium with HPC development medium, detach and dissociate iPSC colonies appropriately by mild pipetting with a p1000 cut tip (clump size should be nearly equal to that of a cut tip).
3. Seed 30–40 clumps of harvested iPSC onto one dish (10 cm) of 10 T1/2 (**Note 2**).
4. Culture with HPC development medium for 7 days in low oxygen condition (5% O_2) and 6 days in normal oxygen

condition. Change medium at day 4, 7, and 10, and supplement 20 ng/ml SCF and 20 ng/ml Flt3L only at day 7 and 10.

5. HPCs surrounded by sac-like structure appear on hillsides of colonies at around 10 days.
6. Harvest HPCs at day 13 of 10 T1/2 coculture. Nick cell layer of 10 T1/2 in several directions and scratch to peel up colonies together with 10 T1/2 stromal layer using tip of 10 ml disposable pipet. Dissociate HPCs from sac-like structure by vigorous 10-time pipetting with tip of pipet pushed onto the dish bottom.
7. Naturally drop cell aggregates for 5 min at room temperature, and pass the supernatant through 70 μm-pore-size cell strainer to harvest nonadherent HPCs.

3.2 Generation of CD4/CD8 DP Cells from HPCs

1. Prepare one dish (6 cm) of confluent OP9/DL1, which is seeded 1–2 days before, against three dishes (10 cm) of 10 T1/2 coculture (untreated with growth inhibitors such as mitomycin C).
2. Resuspend unfractionated harvested HPCs with T-cell development medium and seed onto OP9/DL1.
3. Change medium every 3 days and change OP9/DL1 at day 3 and every 6 days afterward. Both adherent and nonadherent cells should be passed for optimal cell recovery by trypsinization of cell layer containing strongly adherent T-iPSC-Ts (**Note 3**). Adhesion selection should be performed afterward to harvest only dissociated T-iPSC-Ts, but not old OP9/DL1.
4. CD4/CD8 DP cells appear at around 3 weeks in OP9/DL1 coculture and increase steadily until around 4–5 weeks (**Note 4**).

3.3 Generation of CD8 SP T-iPSC-T from DP Cells

1. Add equal amount of maturation medium containing 2× concentration of IL-7, Flt3L, and OKT3 to T-iPSC-T/OP9/DL1 coculture (final concentrations of IL-7, Flt3L, and OKT3 are 10 ng/ml, 10 ng/ml, and 1 ug/ml, respectively) and culture for 2 days.
2. Prepare RetroNectin-coated 24-well plate (1 well for 1 dish of OP/DL1 coculture) by incubating 5 μg/ml RetroNectin/PBS in a well at 4 °C overnight.
3. Harvest nonadherent T-iPSC-T and perform adhesion selection to remove residual OP9/DL1 completely (**Note 5**). After 3-time wash with medium, resuspend with maturation medium, transfer to RetroNectin-coated plates, and culture for 5 days.
4. Change medium at day 3 of maturation.
5. Check the surface marker.

4 Notes

1. The differentiation protocol described here is not optimized for normal iPSCs, but for T-iPSCs, which retain rearranged TCR of parental T-cell clones and can more efficiently progress through developmental checkpoints such as β-selection compared to normal iPSCs.
2. Size of iPSC clump seeded on 10 T1/2 profoundly affects yield of HPCs.
3. Not only floating but also highly adherent cells to OP9/DL1 should be passed because some highly competent progenitor cells preferentially adhere to OP9/DL1.
4. DP frequency is highly variable between different iPSCs and experimental lots unless an experimenter is skilled.
5. Complete removal of OP9/DL1 at maturation stage is important because the residual stromal cell hinders contact of developing iPSC-T to RetroNectin and impairs maturation efficacy.

References

1. Wherry EJ, Kurachi M (2015) Molecular and cellular insights into T cell exhaustion. Nat Rev Immunol 15:486–499
2. Akbar AN, Henson SM (2011) Are senescence and exhaustion intertwined or unrelated processes that compromise immunity? Nat Rev Immunol 11:289–295
3. Hinrichs CS et al (2009) Adoptively transferred effector cells derived from naive rather than central memory CD8+ T cells mediate superior antitumor immunity. Proc Natl Acad Sci U S A 106:17469–17474
4. Gattinoni L et al (2011) A human memory T cell subset with stem cell-like properties. Nat Med 17:1290–1297
5. Hinrichs CS et al (2011) Human effector CD8 + T cells derived from naive rather than memory subsets possess superior traits for adoptive immunotherapy Human effector CD8+ T cells derived from naive rather than memory subsets possess superior traits for adoptive immunotherapy. Blood 117:808–814
6. Nishimura T et al (2013) Generation of rejuvenated antigen-specific T cells by reprogramming to pluripotency and redifferentiation. Cell Stem Cell 12:114–126
7. Vizcardo R et al (2013) Regeneration of human tumor antigen-specific T cells from iPSCs derived from mature CD8+ T cells. Cell Stem Cell 12:31–36
8. Themeli M et al (2013) Generation of tumor-targeted human T lymphocytes from induced pluripotent stem cells for cancer therapy. Nat Biotechnol 31:928–933
9. Maeda T et al (2016) Regeneration of CD8ab T cells from T-cell-derived iPSC imparts potent tumor antigen-specific cytotoxicity. Cancer Res 76:6839–6850. https://doi.org/10.1158/0008-5472.CAN-16-1149
10. Cheroutre H, Lambolez F (2008) Doubting the TCR coreceptor function of CD8aa. Immunity 28:149–159

Chapter 8

In Vitro Differentiation of T Cell: From Human iPS Cells in Feeder-Free Condition

Yutaka Yasui, Yasumichi Hitoshi, and Shin Kaneko

Abstract

In order to differentiate T cells in vitro, co-culture systems with Notch ligand-expressing feeder cells have been in use for a long time. Here we describe a feeder-free culture condition for differentiating T cells from hematopoietic cells that are cultured on Fc-DLL4-coated plate with T-lineage cytokines. This condition is capable of efficiently differentiating hematopoietic progenitor cells (HPCs) to immature T cells expressing both CD4 and CD8. To mature those cells into functional T cells, further stimulation and culture is necessary.

Key words iPS cell, T cell, Feeder-free differentiation, Regenerative medicine, Immunotherapy

1 Introduction

In our body, T cell develops at thymus by receiving a signal via Notch and T-lineage cytokines IL-7, Flt3L, and SCF [1–4]. Notch ligand is expressing at thymic epithelial cell and triggering for T-lineage fate decision [5]. OP9-DLL1, mice stromal cells expressing Notch ligand delta-like 1, have been used for T cell differentiation in vitro from mice and human hematopoietic progenitor cells [6–8].

Induced pluripotent stem cell (iPS cell) is generated by transducing four transcriptional gene (Yamanaka factor) into the mature somatic cell [9, 10]. T-iPS cell, reprogrammed from T cell, has different characteristics from other iPS cell that T cell receptor gene is conserved from originating T cell. Therefore, redifferentiated T cell from T-iPS cell has the same antigen specificity with originated T cell and could be used for regenerated immunotherapy against cancer and infectious diseases [11–13]. On the other hand, conventional iPS cell, reprogrammed from non-T cell, can also be used for the same application by introducing TCR or chimeric antigen receptor (CAR).

Shin Kaneko (ed.), *In Vitro Differentiation of T-Cells: Methods and Protocols*, Methods in Molecular Biology, vol. 2048, https://doi.org/10.1007/978-1-4939-9728-2_8, © Springer Science+Business Media, LLC, part of Springer Nature 2019

For clinical application of differentiated T cells, OP9-DLL1 co-culture system should be avoided; because of that, the stromal cells were derived from mice. There is a possibility that it is adversely affected by unknown viruses and antigens contained in Xeno cells. Also, T cell differentiation efficiency depends on the condition of OP9-DLL1 cells and is very unstable. Therefore, feeder-free T cell differentiation methods were required for efficient and stable clinical manufacturing. For the purpose, we have developed the culture condition that differentiates T cell without feeder cell by culturing the hematopoietic progenitor cells on Notch ligand-coated plate.

2 Materials

2.1 Notch Ligand-Coating Solution

1. Fc-DLL4: The recombinant human DLL4 and Fc fusion protein can be purchased from Sino Biological Inc. (Cat# 10171-H02H). Reconstitute at 250 μg/mL in sterile water and store at −20 °C as stock solution.
2. RetroNectin®: RetroNectin® can be purchased from TAKARA BIO Inc. (Cat# T100A/T100B). Dilute the RetroNectin® with sterile water at 10 μg/mL, and store at −20 °C as stock/working solution.

2.2 Differentiation Culture Media

1. Basal media: MEM-alpha medium (e.g., Mem-alpha, nucleosides, powder, Cat# 11900024, Thermo Fisher Scientific Inc.) is supplemented with 10% FBS, 1× antibiotics penicillin and streptomycin (e.g., Cat# 15140122, Thermo Fisher Scientific Inc.), 55 μM 2-mercaptoethanol, 1× insulin-transferrin-selenium (e.g., Cat# 41400045, Thermo Fisher Scientific Inc.), and 50 ng/mL L-ascorbic acid 2-phosphate.
2. Cytokines: The cytokines are supplemented into the basal media along the following table. All the cytokines are reconstituted by appropriate solvent and stored along each product instructions.

	Final conc. in the culture media
SCF	50 ng/mL
TPO	100 ng/mL
Flt3L	50 ng/mL
IL-7	50 ng/mL

3 Methods

3.1 Preparation of Fc-DLL4-Coated Plates

1. One day before initiation of T cell differentiation, coat a plate with Fc-DLL4. Dissolve the Fc-DLL4 and RetroNectin stock solution at room temperature.
2. Add sterile water into the Fc-DLL4 stock solution as working solution at 10 μg/mL.
3. Mix the same volume of Fc-DLL4 and RetroNectin stock solution (final conc.: 5 μg/mL each), and dispense an appropriate volume of the mixture into each plate (e.g., 100 μL into each well of a 48-well plate).
4. Put the plate into a sealed container, or seal with Parafilm in not to dry up the coating solution.
5. Coat the plate at 4 °C overnight.

3.2 Culture for T Cell Differentiation

1. Remove the coating solution and dispense an appropriate volume of the differentiation culture media containing cytokines (e.g., 1 mL into each well of a 48-well plate).
2. Purify the hematopoietic cells expressing hematopoietic markers (e.g., CD34) by flow cytometry sorting or magnetic beads selection (*see* **Note 1**).
3. Put the cells into prepared wells with an appropriate density of the cells (2000 cells into each well of a 48-well plate).
4. Culture the cells at 37 °C, 5% CO_2 humid incubator.
5. Change the medium every 2 days, and transfer the differentiating cells onto newly prepared Fc-DLL4 coat well every week (*see* **Note 2** and **3**).
6. After 3 weeks, harvest the cells expressing both CD4 and CD8 by pipetting, and stimulate for maturation into matured T cells.

4 Notes

1. This feeder-free culture condition was developed for hematopoietic cells derived from iPS cells, but it also could be applicable for another hematopoietic cells, such as from bone marrow or cord blood.
2. During the culture, differentiating T cells grow 10–1000 times that depends on the condition of the cells. It is necessary to adjust the optimum culture volume and cell density.
3. For efficient T cell differentiation, the signal via Notch is exceedingly important. The differentiating T cells sink to the

bottom of the well, bind weakly to RetroNectin, and receive a Notch signal. If some adherent cells, such as iPS cell or undifferentiated cells, were contaminated into a purified hematopoietic cell, Notch signal is inhibited and cell differentiation efficiency decreases.

References

1. Radtke F, Wilson A, Mancini SJ, MacDonald HR (2004) Notch regulation of lymphocyte development and function. Nat Immunol 5 (3):247–253
2. Moore TA, Zlotnik A (1997) Differential effects of Flk-2/Flt-3 ligand and stem cell factor on murine thymic progenitor cells. J Immunol 158:4187–4192
3. Morrissey PJ, McKenna H, Widmer MB et al (1994) Steel factor (c-kit ligand) stimulates the in vitro growth of immature CD3-/CD4-/CD8- thymocytes: synergy with IL-7. Cell Immunol 157:118–131
4. Godfrey DI, Kennedy J, Mombaerts P, Tonegawa S, Zlotnik A (1994) Onset of TCR-b gene rearrangement and role of TCR-b expression during CD3-CD4-CD8-thymocyte differentiation. J Immunol 152:4783–4792
5. Hasserjian R, Aster J, Davi D, Weinberg D, Sklar J (1996) Modulated expression of Notch1 during thymocyte development. Blood 88:970–976
6. Jaleco AC, Neves H, Hooijberg E et al (2001) Differential effects of Notch ligands Delta-1 and Jagged-1 in human lymphoid differentiation. J Exp Med 194:991–1002
7. Schmitt TM, Zuniga-Pflucker JC (2002) Induction of T cell development from hematopoietic progenitor cells by delta-like-1 in vitro. Immunity 17:749–756
8. Schmitt TM, de Pooter RF, Gronski MA, Cho SK, Ohashi PS, Zuniga- Pflucker JC (2004) Induction of T cell development and establishment of T cell competence from embryonic stem cells differentiated in vitro. Nat Immunol 5:410–417
9. Takahashi K, Yamanaka S (2006) Induction of pluripotent stem cells from mouse embryonic and adult fibroblast cultures by defined factors. Cell 126(4):663–676
10. Takahashi K, Tanabe K, Ohnuki M, Narita M, Ichisaka T, Tomoda K, Yamanaka S (2007) Induction of pluripotent stem cells from adult human fibroblasts by defined factors. Cell 131 (5):861–872
11. Nishimura T, Kaneko S, Kawana-Tachikawa A, Tajima Y, Goto H, Zhu D, Nakayama-Hosoya-K, Iriguchi S, Uemura Y, Shimizu T, Takayama N, Yamada D, Nishimura K, Ohtaka M, Watanabe N, Takahashi S, Iwamoto A, Koseki H, Nakanishi M, Eto K, Nakauchi H (2013) Generation of rejuvenated antigen-specific T cells by reprogramming to pluripotency and redifferentiation. Cell Stem Cell 12(1):114–126
12. Kitayama S, Zhang R, Liu TY, Ueda N, Iriguchi S, Yasui Y, Kawai Y, Tatsumi M, Hirai N, Mizoro Y, Iwama T, Watanabe A, Nakanishi M, Kuzushima K, Uemura Y, Kaneko S (2016) Cellular adjuvant properties, direct cytotoxicity of re-differentiated Vα24 invariant NKT-like cells from human induced pluripotent stem cells. Stem Cell Reports 6 (2):213–227
13. Ueda N, Uemura Y, Zhang R, Kitayama S, Iriguchi S, Kawai Y, Yasui Y, Tatsumi M, Ueda T, Liu T-Y, Mizoro Y, Okada C, Watanabe A, Nakanishi M, Senju S, Nishimura Y, Kuzushima K, Kiyoi H, Naoe T, Kaneko S (2018) Generation of TCR-expressing innate lymphoid-like helper cells that induce cytotoxic T cell-mediated anti-leukemic cell response. Stem Cell Reports 10(6):1935–1946

Chapter 9

Differentiating CD8αβ T Cells from TCR-Transduced iPSCs for Cancer Immunotherapy

Atsutaka Minagawa and Shin Kaneko

Abstract

The use of induced pluripotent stem cells (iPSCs) as a cell source for producing cytotoxic T lymphocytes (CTLs) is expected to have advantages in the antigen specificity, rejuvenation profile, and reproducible number of CTLs. We have developed the way to differentiate CD8αβ T cells from TCR-transduced iPSCs (TCR-iPSCs). These T cells express monoclonal expression of the transduced TCR. Generating CD8αβ CTLs from TCR-iPSC could contribute to safe and effective allogeneic regenerative T cell immunotherapies.

Key words TCR transduction, iPSCs, T cell differentiation

1 Introduction

Antigen-receptor edited adoptive T cell immunotherapy modifies the antigen specificity of peripheral T cells to target cancer cells, making it the standard therapy for several types of cancer [1]. The use of induced pluripotent stem cells (iPSCs) as a cell source for producing cytotoxic T lymphocytes (CTLs) is expected to have advantages in the antigen specificity, rejuvenation profile, and reproducible number of CTLs [2–4]. In combination with antigen-specific receptor transduction, iPSCs are thought to be a good source of cancer antigen-specific T cells for allogeneic cancer immunotherapy [5]. Here we have established the ways to produce target antigen-specific CD8αβ T cells from TCR-transduced iPSCs (TCR-iPSCs).

2 Materials

2.1 Lentiviral Transduction to iPSCs

1. iMatrix-511 (laminin-511 E8).
2. 10 mM Y-27632.
3. iPSC culture medium (StemFit medium) (Ajinomoto).

Shin Kaneko (ed.), *In Vitro Differentiation of T-Cells: Methods and Protocols*, Methods in Molecular Biology, vol. 2048, https://doi.org/10.1007/978-1-4939-9728-2_9,

4. Lentiviral vectors.
5. 24-well plate.

2.2 T Cell Differentiation from TCR-iPSCs

1. α-MEM supplemented with 15% fetal bovine serum.
2. IMDM medium supplemented with 10% fetal bovine serum.
3. C3H10T1/2 feeder cells.
4. OP9-DLL1 stromal cells.
5. Cytokines: SCF, VEGF, FLT3-L, IL-7, IL-2, IL-15.
6. Anti-CD3 antibody (OKT-3).
7. Dexamethasone.
8. Ascorbic acid 2-phosphate.

3 Methods

3.1 Lentiviral Transduction to iPSCs

1. Coat 24-well plate with laminin-511 E8 in 300 μl PBS and mix well immediately (coating: 0.5 μg/cm^2).
2. Incubate at 37 °C, CO_2 5%, for at least 60 min.
3. Plate 5×10^5 iPSCs in one well of 24-well plate and culture the plate at 37 °C, CO_2 5% for 24 h in 300 μl of StemFit medium containing 10 mM Y-27632 (*see* **Note 1**).
4. The next day, change to regular iPSC culture medium (not containing Y-27632), add 1–15 μl lentiviral TCR expressing vector, and centrifuge culture plate at $1000 \times g$ for 30 min (*see* **Note 2**).
5. Perform medium change according to the iPSC density, and in about 3 days passage half of the iPSCs in 24-well plate to 6-well laminin-511 coated plate (*see* **Note 3**).
6. Sort the lentivirus-transduced cells by flow cytometry and expand iPSCs (*see* **Note 4**).

3.2 T Cell Differentiation from TCR-iPSCs

1. Clumps of TCR-iPSCs are transferred on to 10 cm dish confluent C3H10T1/2 feeder cells and cultured in IMDM medium containing rhVEGF and 50 ng/ml ascorbic acid. Cells are cultured under hypoxic condition (37 °C 5% O_2) (*see* **Note 5**).
2. On day 4, media are changed.
3. On day 7, media are changed adding 30 ng/ml rhSCF and 10 ng/ml FLT-3L. Cells are moved to normoxic condition.
4. On day 9 and day 11, media are changed.
5. On day 14, cells are scraped off from the 10 cm dish and transferred onto semi-confluent OP9-DLL1 cells and

co-cultured in α-MEM medium in the presence of 5 ng/ml rhIL-7 and 10 ng/ml rhFLT-3L. Feeder OP9-DLL1 cells are changed every 4 days (*see* **Note 6**).

6. On day 42, cells are stimulated by anti-CD3 antibody. 5×10^5 cells are collected and seeded to 48-well plate in the α-MEM medium in the presence of 500 ng/ml OKT3, 10 nM dexamethasone, 10 ng/ml IL-7, 10 ng/ml IL-2, and 50 μg/ml ascorbic acid 2-phosphate (*see* **Note 7**).
7. On day 45, media are changed to α-MEM medium with 10 ng/ml IL-7, 10 ng/ml IL-2, and 50 μg/ml ascorbic acid 2-phosphate. OKT3 and dexamethasone are removed.
8. On day 49, CD8αβ-positive T cells are sorted and expanded by mixing with irradiated PBMC in α-MEM medium with 10 ng/ml IL-7 and 5 ng/ml IL-15.

4 Notes

1. We seed a generous amount of iPSCs for lentiviral infection because viral infection shows severe cytotoxicity to iPSCs.
2. Lentiviral infection causes severe cytotoxicity to iPSCs. After the infection, according to the MOI of lentiviral vector, live iPSCs decrease. Severity of the cytotoxicity is different according to the vectors and lentiviral condition, so we recommend to infect lentivirus in several MOI conditions. About the lentiviral vector production, *see* ref. 6.
3. Before iPSC expansion, better to check the infection efficiency by fluorescent microscope if your expression vectors contain fluorescent protein. Preferably about half of the iPSCs are positive for fluorescent protein expression.
4. We recommend iPSCs to be treated with 10 mM Y-27632 for 1 h before sorting. If expression vector contains anti-antibiotic gene, selection can be performed by adding antibiotic (e.g., puromycin 0.2 μg/ml for 1 week).
5. C3H10T1/2 feeder cells are better to be very confluent at the time of starting co-culture.
6. OP9-DLL1 cells should be changed for fewer than 5 days. OP9-DLL1 cells cultured for a long time fail to differentiate. Adding some fresh culture media every other day is favorable.
7. Dexamethasone upregulates IL-7α receptor expression of premature T cells [7]. By adding dexamethasone, OKT3-stimulated cells show marked elevation of IL-7α receptor and increased survival of matured CD8αβ T cells.

References

1. Restifo NP, Dudley ME, Rosenberg SA (2012) Adoptive immunotherapy for cancer: harnessing the T cell response. Nat Rev Immunol 12 (4):269–281
2. Minagawa A, Kaneko S (2014) Rise of iPSCs as a cell source for adoptive immunotherapy. Hum Cell 27(2):47–50
3. Kitayama S et al (2016) Cellular adjuvant properties, direct cytotoxicity of re-differentiated Valpha24 invariant NKT-like cells from human induced pluripotent stem cells. Stem Cell Reports 6(2):213–227
4. Nishimura T et al (2013) Generation of rejuvenated antigen-specific T cells by reprogramming to pluripotency and redifferentiation. Cell Stem Cell 12(1):114–126
5. Sugita S et al (2016) Lack of T cell response to iPSC-derived retinal pigment epithelial cells from HLA homozygous donors. Stem Cell Reports 7(4):619–634
6. Hashizume O et al (2015) Epigenetic regulation of the nuclear-coded GCAT and SHMT2 genes confers human age-associated mitochondrial respiration defects. Sci Rep 5:10434
7. Abe A et al (2015) An enhancer of the IL-7 receptor alpha-chain locus controls IL-7 receptor expression and maintenance of peripheral T cells. J Immunol 195(7):3129–3138

Chapter 10

In Vitro Differentiation of T Cell: From CAR-Modified T-iPSC

Tatsuki Ueda and Shin Kaneko

Abstract

T cells engineered to express chimeric antigen receptor (CAR) against the B cell antigen CD19 are achieving remarkable clinical effects on hematological malignancies. Allogeneic transplantation approach is promising for broaden application of CART therapy. iPSCs are one of the ideal cell sources for this approach. CAR-engineered iPSCs are demonstrated to give rise to CAR-engineered T cell and exert their effector function. In this section, we describe the method to generate CAR-engineered iPSCs and differentiate them into T cells.

Key words Chimeric antigen receptor, Induced pluripotent stem cells, Immunotherapy, T cell differentiation, Lentiviral transduction

1 Introduction

Chimeric antigen receptor (CAR) is a synthetic receptor which provides T cells with antigen specificity. Chimeric antigen receptor consists of an extracellular single chain variable fragment (scFv) derived from an antibody and CD3z chain. CAR-engineered T cells (CART) recognize the target cells which express the target antigen of the original monoclonal antibody and exert cytotoxicity to the target cells. CART therapies proved their remarkable efficacy especially hematological malignancies [1–3], and the application for many patients is desired.

CART therapies require cell preparation for each patient. But for some cases such as patients who received chemotherapy repeatedly, HIV-induced immune deficiency, or small infants, it may be difficult to prepare enough number of CART cells [4]. Moreover, to broaden the applicability of CART therapy, this laborious manufacturing cell preparations for each patient would be troublesome. In this prospect of view, off-the-shelf therapies using allogeneic CART will take on increasing importance [5]. iPSCs are one of the promising cell sources for these approaches. First, iPSCs can produce unlimited number of T cells which enable us to prepare

Shin Kaneko (ed.), *In Vitro Differentiation of T-Cells: Methods and Protocols*, Methods in Molecular Biology, vol. 2048,
https://doi.org/10.1007/978-1-4939-9728-2_10,

enough number of CART cells to meet demand. Second, preparation of iPSC lines derived from HLA homologous donors or HLA deleted iPSC lines can overcome critical HLA barriers. There are several reports to generate T cells from iPSCs [6–8], and CAR-engineered iPSCs are reported to give rise to CAR-engineered T cell and exert their effector function [9].

In this section, we describe the method to generate CAR-engineered iPSCs and differentiate them into T cells.

2 Materials

2.1 Reagent

1. Dulbecco's Modified Eagle's Medium (Nacalai).
2. L-Glutamine-penicillin-streptomycin solution (SIGMA).
3. Trypsin-EDTA (SIGMA).
4. 0.01% poly-L-lysine solution(SIGMA).
5. Polyethyleneimine (PEI), linear, MW 25,000 (Polysciences Inc.).
6. HBSS (Hank's balanced salt solution) (SIGMA).
7. 5 mM forskolin (500×) (SIGMA).
8. StemFit® AK02N (Takara).
9. D-PBS(−) (Nacalai).
10. 10 mM Y-27632 (Wako).
11. Easy iMatrix-511 silk 100 ml.
12. Protamine sulfate 10 mg/ml (Mochida).
13. TrypLE Select (Thermo Fisher).
14. 0.5 M EDTA solution (Nacalai).
15. 3 mM CHIR99021 (Tocris).
16. StemPro-34 SFM (Invitrogen).
17. GlutaMAX supplement (Invitrogen).
18. MTG (monothioglycerol) 0.45 M (Nacalai).
19. BMP4 (R&D).
20. VEGF (R&D).
21. bFGF (Wako).
22. Ascorbic acid 2-phosphate (SIGMA).
23. SB431542 (Wako).
24. SCF (R&D).
25. TPO (PeproTech).
26. FLT3L (PeproTech).
27. α-MEM (Invitrogen).

28. Insulin-transferrin-selenium (ITS-G) (100×) (Invitrogen).
29. Human IL-7 (PeproTech).
30. Human IL-15 (PeproTech).
31. PHA-P (Wako).

2.2 Cell

1. 293FT cells (Invitrogen).
2. iPS cells.
3. OP9-DL1.

2.3 Equipment

1. Poly-L-lysine coat dish.
2. Coat a 10 cm dish using 3 ml poly-L-lysine solution diluted to 1/5 with PBS.
3. 0.45 μm filter (Millex-HV 0.45 μm), PVDF, 33 mm (Millipore).
4. CO_2 incubator (5% or 10%).
5. Centrifuge.
6. 24-well plate.
7. 15 ml tube (Nunc).
8. Cell scraper.
9. 60 mm and 100 mm petri dish.
10. 15 ml conical tube.
11. 1.5 ml microcentrifuge tube.
12. Costar ultralow attachment 6-well plate (Corning).
13. Cell counter.
14. Low-oxygen incubator.

3 Methods

3.1 Selection of Promoter for Stable Expression of CAR Throughout Differentiation

Gene transfer with lentiviral vector has been well established for a variety of cell types. Several kinds of promoter are used to drive transgene expression constitutively. When we engineer iPSCs to express transgene, it is important to select appropriate promoter. EF1α, CAG, and PGK promoter were reported to undergo silencing during differentiation to neutrophils [10], and mouse ES cells transduced with the EF1α promoter-containing lentiviral vector lost most of their transgene expression during in vitro differentiation to neural precursors and neuronal cells [11]. On the other hand, hES cells transduced by lentiviral encoding transgene under ubiquitin (UbC) promoter express the transgene throughout prolonged cultivation and in differentiated embryo body (EB) and teratoma [12]. In case of T cell differentiation from iPSCs, we compared EF1α promoter and UbC promoter. Almost all

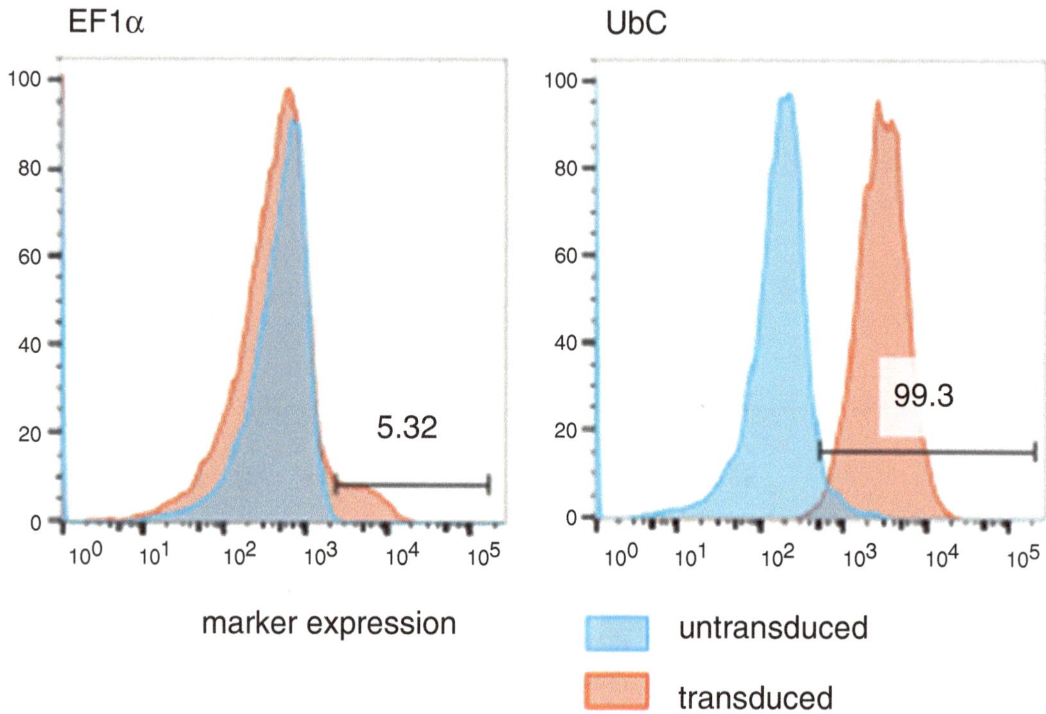

Fig. 1 UbC promoter can drive stable gene expression of CAR during differentiation. iPSC lines were transduced with EF1a promoter lentiviral vector or UbC promoter lentiviral vector. Each iPSC line was differentiated into T-lineage cells and the expression of marker genes was analyzed by flow cytometric analysis

differentiated cells derived from iPSCs transduced by lentivirus of UbC express transgene throughout differentiation, while most of differentiated cells derived from iPSCs transduced by lentivirus of EF1α undergo silencing (Fig. 1). We recommend using UbC promoter to use for stable. Expression of CAR.

3.2 Production of Lentivirus for Transduction of Chimeric Antigen Receptor

1. 293FT cells are seeded in 5×10^6/10 ml DMEM medium/dish on poly-L-lysine treated 10 cm petri dishes.
2. Incubate for about 24 h in 37 °C 10% CO_2 incubator.
3. Prepare HBSS 500 μl/dish for 1.5 ml Eppendorf tube or 5 ml polystyrene tube (Falcon 2058).
4. Add the following plasmid DNA (pCAG-HIVgp 3 μg, pCMV-VSV-G-RSV-Rev3 μg, CAR expression plasmid 6 μg) per dish and mix by vortexing or pipetting.
5. Add 24 μl of PEI (1 mg/ml) per dish and mix by vortexing or pipetting.
6. Incubate at room temperature for 15 to 20 min.
7. Mix by pipetting several times using PIPETMAN (P1000), then add it little by little to the dish and stir the medium thoroughly.

8. Incubate at 37 °C. 5% CO_2 incubator for 12–16 h.
9. Aspirate the culture supernatant and replaced with 7.5 ml of DMEM medium which contains 10 μM forskolin.
10. Incubate at 37 °C. 5% CO_2 incubator for 48 h.
11. Collect virus-containing culture supernatant and remove floating cells by using 0.45 μm filter (*see* **Note 1**).
12. Centrifuge virus-containing culture supernatant at 6000 × *g*, 4 °C for 16 h (*see* **Note 2**).
13. Aspirate the culture supernatant and add α-MEM without FBS (40 μl/dish).
14. Incubate the suspension at 4 °C overnight with shaking.
15. Pipette virus suspension vigorously on ice (*see* **Note 3**).
16. Dispense viral solution into tubes with screw cap and store at −80 °C.

3.3 Transduction of Chimeric Antigen Receptor to iPSCs by Lentiviral Vector

1. Seed 1 × 10^4 iPSCs in several wells of a 24-well plate, and culture media is supplemented with 10 μM Y-27632 (*see* **Note 4**).
2. Culture the plates at 37 °C, CO_2 5%.
3. Set the centrifuge at 32 °C.
4. Aspirate the culture media and replace with 500 μl StemFit AK02N containing 10 μM ROCK inhibitor and 10 μg/ml protamine sulfate.
5. Add various amount of virus solution to iPSCs (*see* **Note 5**).
6. The next day, change to regular culture medium.
7. Medium change is performed every other day. If the color of the medium turns orange or yellow, it should be changed every day.
8. When the cells become more confluent, pass the cells to 6-well plate.
9. After expansion of transacted iPSCs, analyze the transduction efficiency and purify marker positive iPSC by using cell sorter.

3.4 Differentiation of CAR-Engineered iPSCs to T Cell

1. On day 0, CAR-transduced T-iPSC colonies were treated with 0.5× TrypLE Select to harvest as a single-cell suspension. 0.5× TrypLE Select solution was prepared by mixing TrypLE Select and 0.5 mM EDTA solution.
2. Transfer the iPSCs to low-attachment plates to allow for the formation of embryoid bodies (EBs) in StemFit AK02N containing 10 μM Y-27632 and 3 μM CHIR99021.
3. Culture the plates at 37 °C, CO_2 5% O_2 5%.

4. On day 1, after overnight incubation, the formation of embryoid bodies was facilitated. These embryoid bodies were collected and transferred to EB medium (StemPro-34, Invitrogen, with 2 mM L-glutamine, 400 μM monothioglycerol, 50 μg/ml ascorbic acid 2-phosphate, insulin-transferrin-selenium supplements) containing 40 ng/ml of hBMP-4, 50 ng/ml hbFGF, and 50 ng/ml VEGF.
5. Culture the plates at 37 °C, CO_2 5% O_2 5%.
6. On day 2, add 6 μM SB431542 to the culture medium and culture in hypoxia condition.
7. On day 4, embryoid bodies were then cultured in EB medium containing 50 ng/ml hbFGF, 50 ng/ml VEGF, and 50 ng/m hSCF, and culture in hypoxia condition.
8. On day 6, change the medium to EB medium containing 50 ng/ml hbFGF and 50 ng/ml VEGF, 50 ng/m hSCF, 10 ng/ml hFlt3L, and 30 ng/ml TPO in normoxia condition.
9. On day 8–14 of culture, transfer the differentiated cells onto delta-like 1-expressing OP9 (OP9-DL1) feeder cells, and culture them in α-MEM supplemented with 15% FBS in the presence of 10 ng/ml hFlt3L, 5 ng/ml IL-7 50 μg/ml ascorbic acid 2-phosphate, and insulin-transferrin-selenium supplements (100×). Transfer the cells to the new delta-like 1-expressing OP9 (OP9-DL1) feeder cells by 5 days.
10. After 21–28 days of culture, the hematopoietic cells differentiated into CD3+ T-lineage cells.
11. Harvest the differentiated cells and stimulate them by 2 μg/ml PHA-P (phytohemagglutinin) on irradiated PBMC for expansion in α-MEM supplemented with 15% FBS in the presence of 5 ng/ml IL-15, 10 ng/ml IL-7, and 50 μg/ml ascorbic acid 2-phosphate and insulin-transferrin-selenium supplements.

4 Notes

1. If there are many floating cells, centrifuge 500 × *g* 5 min to remove floating cells before filtering.
2. To remove component of medium, concentration of virus solution is recommended.
3. To avoid wasting viral particles, pipette virus suspension vigorously.
4. Depending on virus titer and the degree of cell toxicity, it is desirable to adjust the initial number of iPSCs.
5. When virus titer is not determined, it is recommended to prepare several wells to determine appropriate amount of virus solution.

References

1. Lee DW et al (2015) T cells expressing CD19 chimeric antigen receptors for acute lymphoblastic leukaemia in children and young adults: a phase 1 dose-escalation trial. Lancet 385:517–528
2. Maude SL et al (2014) Chimeric antigen receptor T cells for sustained remissions in leukemia. N Engl J Med 371:1507–1517
3. Davila ML et al (2014) Efficacy and toxicity management of 19-28z CAR T cell therapy in B cell acute lymphoblastic leukemia. Sci Transl Med 6:224ra25–224ra25
4. Themeli M, Rivière I, Sadelain M (2015) New cell sources for T cell engineering and adoptive immunotherapy. Cell Stem Cell 16:357–366
5. Qasim W et al (2017) Molecular remission of infant B-ALL after infusion of universal TALEN gene-edited CAR T cells. Sci Transl Med 9:1–9
6. Nishimura T et al (2013) Generation of rejuvenated antigen-specific T cells by reprogramming to pluripotency and redifferentiation. Cell Stem Cell 12:114–126
7. Vizcardo R et al (2013) Regeneration of human tumor antigen-specific T cells from iPSCs derived from mature CD8+ T cells. Cell Stem Cell 12:31–36
8. Kitayama S et al (2016) Cellular adjuvant properties, direct cytotoxicity of re-differentiated Vα24 invariant NKT-like cells from human induced pluripotent stem cells. Stem Cell Reports 6:213–227
9. Themeli M et al (2013) Generation of tumor-targeted human T lymphocytes from induced pluripotent stem cells for cancer therapy. Nat Biotechnol 31:928–933
10. Zou J et al (2011) Plenary paper oxidase-deficient neutrophils from X-linked chronic granulomatous disease iPS cells: functional correction by zinc finger nuclease – mediated safe harbor targeting. Blood 117:5561–5572
11. Hong S et al (2007) Functional analysis of various promoters in lentiviral vectors at different stages of in vitro differentiation of mouse embryonic stem cells. Mol Ther 15:1630–1639
12. Jiang H et al (2010) Hemato-endothelial differentiation from lentiviral-transduced human embryonic stem cells retains durable reporter gene expression under the control of ubiquitin promoter. Cytotechnology 62:31–42

Chapter 11

In Vitro Differentiation of T Cells: From Nonhuman Primate-Induced Pluripotent Stem Cells

Shoichi Iriguchi and Shin Kaneko

Abstract

In this chapter, we describe a protocol for hematopoietic differentiation of nonhuman primate (NHP)-induced pluripotent stem cells (iPSCs) derived from T cells and generation of T cells. Derivation of T cells from PSCs involves three steps: induction of PSCs to hematopoietic progenitor cells (HPCs), differentiation of HPCs into progenitor T cells, and maturation of progenitor T cells into mature T cells, in particular CD8 single-positive (SP) T cells.

Key words T cells, Nonhuman primate-induced pluripotent stem cells, In vitro differentiation, Hematopoietic progenitor cells, Preclinical model

1 Introduction

Induced pluripotent stem cell (iPSC)-based therapies may become a tractable cell source for regenerative medicine. However, assessment of in vivo long-term safety in the relevant preclinical setting will be crucial to bring the therapies widely acceptable to patients. Although rodent models will play central role to evaluate the safety and potential risks of tumorigenicity of iPSC-derived cells from human iPSCs, the immunogenicity of the cells cannot be measured as human iPS-derived cells are tested in immunodeficient mice. In addition, heterologous immunity between individual humans cannot be recapitulated in mouse models housed in a specific pathogen-free environment. Because humans and nonhuman primates share many physiological and immunological features, development of nonhuman primate preclinical model for iPSC-based medicine will provide information that cannot be answered in the mouse models. To this end, development of nonhuman primate preclinical models have been described and provided important insights into clinical application [1–7]. To date, protocols to generate hematopoietic cells from nonhuman primates iPSC have been demonstrated utilizing protocols similar to human ES/iPS

Shin Kaneko (ed.), *In Vitro Differentiation of T-Cells: Methods and Protocols*, Methods in Molecular Biology, vol. 2048,
https://doi.org/10.1007/978-1-4939-9728-2_11,

differentiation cultures [8–10]. In brief, derivation of T cells from PSCs contains three steps: induction of PSCs to hematopoietic progenitor cells (HPCs), differentiation of HPCs into progenitor T cells, and maturation of progenitor T cells into mature T cells (CD8 single-positive (SP) or CD4 SP). In general, induction can be achieved by coculturing ESCs and iPSCs with stromal feeder cells supporting hematoendothelial differentiation or by means of embryoid body formation accompanied by a stepwise addition of key cytokines for hematoendothelial differentiation. Either way, a $CD34^{+}CD43^{low}$ population, which represents multipotent hematopoietic progenitors, can be detected over a period of 10–14 days or more after initiation via mesoderm intermediates followed by hematoendothelial cells [11–13]. The resulting $CD34^{+}$ cell population was further differentiated into T cells by coculture on a DLL-1 or DLL-4-expressing OP9 cell line in the presence of the lymphopoietic cytokines IL-7 and Flt3L. After differentiating for 1 week in the presence of Notch ligands, the $CD34^{+}$ cells acquired CD7 expression, and a fraction of these $CD7^{+}$ cells acquired CD5 expression and eventually CD4 and CD8 expressions, which gave rise to CD4 and CD8 double-positive (DP) T cells. As in normal development, the T-iPSC-derived DP T cells expressed CD3-TCR complexes on their surfaces, a feature that defines T lymphocyte lineage determination. The final maturation step, where CD4 or CD8 SP T cells emerge via TCR stimulation, appears possible in vitro, but controlling for the desired subset (either CD4 SP only or CD8 SP only) requires further investigation. When iPSCs from T cell clones were induced to redifferentiate, few SP T cells, mostly CD8 SP T cells, can be obtained, but majority of the cells remain as DP T cells during the OP9-DLL-1 cultures. These cells can be induced to differentiate into CD8 SP T cells by TCR stimulation methods with an anti-CD3 monoclonal antibody or PHA-P in the presence of inactivated allogenic peripheral blood mononuclear cell feeders [14, 15]. The protocol described above could be readily adopted to NHP iPSC differentiation protocol, with some discrepancies in cell surface markers. In this chapter of the collection, we will cover protocols to generate T cells from nonhuman primate (NHP) iPSCs derived from peripheral T lymphocytes (T-iPSCs) by on-feeder coculture systems.

2 Materials

2.1 Reagents

1. NHP iPSCs.
2. Mouse embryonic fibroblasts (MEFs) (*see* **Note 1**).
3. OP9-DLL1 mouse bone marrow stromal cell line (RIKEN BRC) (*see* **Note 2**).
4. C3H10T1/2 mouse stromal cell line (RIKEN BRC) (*see* **Note 3**).

5. Human PBMC, cryopreserved.
6. DMEM high glucose (Nacalai).
7. DMEM/nutrient mixture F-12 (DMEM/F-12, SIGMA).
8. α-MEM basal medium powder (Thermo Fisher).
9. Iscove's modified Dulbecco's medium powder (IMDM, Thermo Fisher).
10. Basal Medium Eagle (BME) (Thermo Fisher).
11. KnockOut Serum Replacement (KSR) for ES cells (Thermo Fisher).
12. Insulin-transferrin-selenium solution (ITS), 100× (Thermo Fisher).
13. Ascorbic acid 2-phosphate (SIGMA).
14. L-Glutamine, penicillin, streptomycin (SIGMA).
15. MEM nonessential amino acid (NEAA, 100× solution, GIBCO-Invitrogen).
16. PHA (Wako).
17. 2-Mercaptoethanol (Thermo Fisher).
18. Mitomycin C (Wako).
19. $CaCl_2$.
20. Sodium bicarbonate (Nacalai).
21. NaOH (Nacalai).
22. D-PBS (Ca^{2+} and Mg^{2+} free, Nacalai).
23. Gelatin from porcine skin, Type A (SIGMA).
24. 2.5% Trypsin (GIBCO).
25. EDTA (0.5 M, pH 8.0, Nacalai).
26. Monothioglycerol (MTG, Nacalai).
27. FBS (HyClone, GIBCO, etc.).
28. Trypsin/EDTA (0.05% (wt/vol)/0.5 mM, SIGMA).
29. Human basic FGF (Wako).
30. Human BMP4 (R&D Systems).
31. Human Flt3 ligand (L) (PeproTech).
32. Human SCF (R&D Systems).
33. Human VEGF (R&D Systems).
34. Human IL-7 (PeproTech).
35. Human IL-15 (PeproTech).
36. Trypan blue solution (0.4%, Nacalai).
37. BSA fraction V (SIGMA).

38. Antibodies and sorting reagents (*see* the section 2.4 and **Note 4**).
39. Distilled water.

2.2 Equipment

1. Cell strainer (70 μm, Thermo Fisher Scientific).
2. Nalgene disposable bottle top filter (polyethersulfone membrane with 0.2-μm pore size, Fisher Scientific).
3. T75 tissue culture flask canted neck with 0.2-μm vented plug seal cap (250 ml, nonpyrogenic polystyrene).
4. Tissue culture dishes (polystyrene, TPP).
5. Tissue culture six-well plate (polystyrene flat bottom, TPP).
6. Tissue culture plate, 48-well (polystyrene flat bottom, Costar).
7. Polystyrene round-bottom tube (5 ml, with 35-μm cell strainer cap, BD Biosciences).
8. Polypropylene conical tubes (15, 50 ml).
9. Serological pipette (nonpyrogenic).
10. P1000G pipette tip (MBP).
11. Hemocytometer.
12. Water bath.
13. Microcentrifuge.
14. Sterile biosafety cabinet.
15. CO_2 incubator (37 °C/5%, Thermo Fisher).
16. Inverted microscope (Olympus).
17. Milli-Q water purification system (Millipore).
18. Pipet-Aid (Drummond).
19. Flow cytometer (Becton Dickinson).
20. Microcentrifuge tube (1.5 ml).
21. Bright-field microscope.

2.3 Reagent Setup

1. *Gelatin solution (0.1% (wt/vol))*: Add 500 mg of gelatin to 500 ml of Milli-Q. Solubilize and sterilize by autoclaving for 20 min at 121 °C. Store the solution at 4 °C for up to 6 months.
2. *Mitomycin C solution*: Dissolve 10 mg of lyophilized mitomycin C to 25 ml D-PBS. Mix well and divide into 1.5-ml aliquots. Store at −20 °C.
3. *Human ES/iPS dissociation solution (TK)*: Add 20 ml 2.5% trypsin, 40 ml KSR, and 2 ml 100 mM $CaCl_2$ to 138 ml D-PBS (−). Mix well and divide into 10-ml aliquots and store at −20 °C. After thawing, store at 4 °C for 1 week.

4. *α-MEM basal medium*: Prepare according to the manufacturer's instruction. Divide into 210-ml aliquots and store at 4 °C for up to 2 weeks.
5. *FACS buffer*: FACS buffer contains 2% (vol/vol) FBS in D-PBS. Store at 4 °C for up to 6 months.
6. *MTG solution*: Add 87 μl of MTG to 10 ml of endotoxin-free reagent-grade distilled water. Mix well and divide into 500-μl aliquots. Store for up to 6 months at −20 °C.
7. *Ascorbic acid solution*: Add 50 mg of ascorbic acid to 1 ml of endotoxin-free reagent-grade distilled water. Dissolve completely, divide into 100 μl aliquots, and store for up to 6 months at −20 °C.
8. *BSA/PBS solution (0.1% (wt/vol))*: Dissolve 25 mg of bovine serum albumin fraction V in 25 ml of PBS. Sterilize the solution by filtration using a 0.22-μm membrane filter, and store for up to 6 months at 2–8 °C.
9. *PHA solution*: Dissolve 50 mg of PHA in 50 ml of distilled water. Dissolve completely and divide into 1-ml aliquots and store at −20 °C.
10. *Reconstitution of cytokines*: Reconstitute cytokines according to the product information provided by manufacturer (*see* the section 2.5).

2.4 Antibodies Used to Analyze Differentiation of HPCs and T Cells

1. *Antihuman CD34 (BD Biosciences)*: APC, 563.
2. *Antihuman CD14 (BD Biosciences)*: FITC, M5E2.
3. *Antihuman CD235a (BD Biosciences)*: FITC, GA-R2 (HIR2).
4. *Anti-NHP CD45 (BD Biosciences)*: Brilliant Violet 510, D058-1283.
5. *Antihuman CD7 (BD Biosciences)*: PE/Cy7, M-T701.
6. *Antihuman CD5 (Thermo Fisher)*: APC, CD5-5D7.
7. *Antihuman CD3 (BD Biosciences)*: APC/Cy7, SP34-2.
8. *Antihuman CD4 (BD Biosciences)*: Brilliant Violet 421, L200.
9. *Antihuman CD8α (BioLegend)*: PerCP/Cy5.5, SK1.

2.5 Reconstitution of Cytokines

1. *Human bFGF*: 100 μg/ml, 0.1% (wt/vol) BSA/D-PBS solution, aliquot 50 μl into 1.5-ml tubes, and store at −80 °C.
2. *Human VEGF*: 100 μg/ml, 0.1% (wt/vol) BSA/D-PBS solution, aliquot 50 μl into 1.5-ml tubes, and store at −80 °C.
3. *Human BMP4*: 50 μg/ml, 0.1% (wt/vol) BSA/D-PBS 4 mM HCl, aliquot 50 μl into 1.5-ml tubes, and store at −80 °C.
4. *Human SCF*: 50 μg/ml, 0.1% (wt/vol) BSA/D-PBS solution, aliquot 50 μl into 1.5-ml tubes, and store at −80 °C.

5. *Human FLT3L*: 200 μg/ml, 0.1% (wt/vol) BSA/D-PBS solution, aliquot 50 μl into 1.5-ml tubes, and store at −80 °C.
6. *Human IL-7*: 100 μg/ml, 0.1% (wt/vol) BSA/D-PBS solution, aliquot 50 μl into 1.5-ml tubes, and store at −80 °C.
7. *Human IL-15*: 100 μg/ml, 0.1% (wt/vol) BSA/D-PBS solution, aliquot 50 μl into 1.5-ml tubes, and store at −80 °C.

2.6 Media Composition

1. *MEF growth medium*: DMEM 500 ml, FBS 55.5 ml 10%, PSG 5.5 ml 1%.
2. *10T1/2 medium*: BME 500 ml, FBS 55.5 ml 10%, PSG 5.5 ml 1%.
3. *OP9 medium*: αMEM 210 ml, FBS 37.5 ml 15%, PSG 2.5 ml 1%.
4. *NHP iPSC culture medium*: DMEM/F-12 500 ml, KSR 125 ml 20%, PSG 6.25 ml 1%, NEAA (100×) 6.25 ml 1×, 2ME (55 mM) 1.2 ml, bFGF 50 μl 5 ng/ml.
5. *Hematopoietic differentiation medium A (day 0–6)*: IMDM 500 ml, FBS 90 ml 10%, PSG 6 ml 1%, ITS (100×) 6 ml 1×, MTG 600 μl 450 μM, ascorbic acid 2-phosphate 600 μl 50 μg/ml, VEGF 120 μl 20 ng/ml.
6. *Hematopoietic differentiation medium B (day 7–14)*: Hematopoietic differentiation medium A 50 ml, SCF 50 μl 50 ng/ml, FLT3L 12.5 μl 50 ng/ml.
7. *T cell differentiation medium*: OP9 medium 20 ml, IL-7 1 μl 5 ng/ml, FLT3L 1 μl 10 ng/ml.
8. *T cell stimulation medium*: OP9 medium 100 ml, IL-7 10 μl 10 ng/ml, IL-15 10 μl 10 ng/ml, PHA 20 μl 2 μg/ml.
9. *T cell expansion medium*: OP9 medium 100 ml, IL-7 10 μl 10 ng/ml, IL-15 10 μl 10 ng/ml.

3 Methods

3.1 MEF Preparation for Human ES/iPSC Culture

1. Thaw cryopreserved MEFs and seed into a 100-mm dish at a density of 3×10^6 cells/dish in MEF growth medium. Incubate at least overnight at 37 °C/5% CO_2.
2. Add 300 μl of mitomycin C solution to the overnight culture dish, and inactivate MEFs by incubating 90 min in a 37 °C incubator.
3. While inactivating MEFs, prepare gelatin-coated 60-mm dishes by adding 2 ml 0.1% gelatin solution per dish. Incubate the dishes at a 37 °C incubator for at least 30 min.
4. After the incubation, aspirate medium and wash MEFs twice with 10 ml D-PBS.

5. Add 1 ml of trypsin/EDTA solution and incubate for 5 min at a 37 °C incubator.
6. Add 9 ml of MEF growth medium and collect MEFs by pipetting.
7. Transfer cell suspension into a 15-ml centrifuge tube and centrifuge for 5 min at 300 × *g* at room temperature.
8. Aspirate supernatant and resuspend the pellet in 1 ml of MEF growth medium.
9. Determine a viable cell count by trypan blue.
10. Adjust the volume of cell suspension to the density of 1.5×10^5 cells/ml.
11. Aspirate the gelatin solution from the precoated dishes and add 2 ml cell suspension per dish (3×10^5 cells/dish).
12. Distribute MEFs evenly and incubate MEF dishes in a 37 °C CO_2 incubator. It is important to make dishes with evenly distributed MEFs for efficient differentiation (*see* **Note 5**).

3.2 NHP iPSC Culture

1. Aspirate spent NHP iPSC medium from a dish.
2. Wash cells with 3 ml of D-PBS.
3. Add 1 ml per 60-mm dish of TK solution and incubate at 37 °C in a CO_2 incubator for 5 min.
4. After the incubation, gently tap the edges of the dish to remove MEFs and aspirate TK solution.
5. Add 3 ml of NHP iPSC culture medium and break up the colonies into small cell clumps by pipetting. Use P1000G pipette tip.
6. Transfer cell clumps to a 15-ml centrifuge tube.
7. Centrifuge at 170 × *g* for 5 min at room temperature.
8. Aspirate the supernatant, gently tap the bottom of the tube to loosen the pellet, and resuspend the pellet in 3 ml of NHP iPSC culture medium.
9. Aspirate MEF growth medium from 3.1 and add 3 ml of NHP iPSC medium.
10. Add 0.5 ml of cell suspension to the MEF dishes (a 1:5 split). Note, the split ratio may require optimization depending on PSC lines.
11. Feed NHP iPSCs daily by 3 ml of fresh NHP iPSC medium.
12. Passage NHP iPSCs weekly (*see* **Note 6**).

3.3 Culture of Mouse OP9-DLL1 Cells

1. Aspirate spent medium and wash cells twice with 10 ml D-PBS.
2. Add 1 ml of trypsin/EDTA solution and incubate for 5 min in a 37 °C incubator.

3. Add 9 ml of OP9 medium and collect OP9-DLL1 by pipetting.
4. Transfer cell suspension into a 15-ml centrifuge tube and centrifuge for 5 min at 300 × *g* at room temperature.
5. Aspirate supernatant and resuspend the pellet in 1 ml of OP9 medium.
6. Determine a viable cell count by trypan blue.
7. Add 10 ml of OP9 medium to a 100-mm dish for maintenance.
8. Add 2.0 × 10^5 cells to the 100-mm dish and place it in a CO_2 incubator at 37 °C. Passages should occur every 2–3 days. It is crucial to avoid over confluent during maintenance culture as it greatly compromises T cell differentiation efficiency.
9. For T cell differentiation, add 1.5 × 10^5 cells to 60-mm dishes supplemented with 3 ml of OP9 medium. Dishes are ready for use typically in 2 days (*see* **Note 7**).

3.4 Culture of C3H10T1/2

1. Aspirate spent medium and wash cells twice with 10 ml D-PBS.
2. Add 1 ml of trypsin/EDTA solution and incubate for 5 min in a 37 °C incubator.
3. Add 9 ml of 10T1/2 medium and collect 10T1/2 by pipetting.
4. Transfer cell suspension into a 15-ml centrifuge tube, and centrifuge for 5 min at 300 × *g* at room temperature.
5. Aspirate supernatant and resuspend the pellet in 1 ml of 10T1/2 medium.
6. Determine a viable cell count by trypan blue.
7. Add 10 ml of 10T1/2 medium to a 100-mm dish for maintenance.
8. Add 2.0 × 10^5 cells to the 100-mm dish and place it in a CO_2 incubator at 37 °C. Passages should occur every 3 days. We refresh cell culture every 10 weeks after thawing.
9. For hematopoietic differentiation, add 2.5 × 10^5 cells to 100-mm dishes supplemented with 10 ml of 10T1/2 medium. Dishes are ready for use typically in 2 days.

3.5 Hematopoietic Differentiation from NHP iPSCs

1. Aspirate C3H10T1/2 medium from C3H10T1/2 dishes prepared for coculture.
2. Add 10 ml of hematopoietic differentiation medium A and keep them in a CO_2 incubator.
3. Aspirate spent NHP iPSC medium from a dish prepared for differentiation.
4. Wash cells with 3 ml of D-PBS.
5. Add 1 ml per 60-mm dish of TK solution and incubate at 37 °C in a CO_2 incubator for 5 min.

6. After the incubation, gently tap the edges of the dish to remove MEFs and aspirate TK solution.
7. Add 3 ml of NHP iPSC culture medium and break up the colonies into small cell clumps by pipetting. Use P1000G pipette tip.
8. Transfer cell clumps to a 15-ml centrifuge tube.
9. Centrifuge at 170 × *g* for 5 min at room temperature.
10. Aspirate the supernatant, gently tap the bottom of the tube to loosen the pellet, and resuspend the pellet in 3 ml of hematopoietic differentiation medium A (*see* **Note 8**).
11. Add 300 μl of cell suspension to one C3H10T1/2 dish prepared in **steps 1** and **2** (*see* **Note 9**).
12. Distribute cells evenly with back/forth and right/left movement several times, and incubate dishes in a hypoxia CO_2 incubator (5% CO_2/5% O_2).
13. On day 3, aspirate the old medium and replace with 10 ml of hematopoietic differentiation medium A.
14. On day 7, aspirate the old medium and replace with 10 ml of hematopoietic differentiation medium B. Incubate the dishes in a normoxia CO_2 incubator (5% CO_2/20% O_2).
15. On day 10, replace medium as did on day 7.
16. On day 12, replace medium as did on day 7.
17. On day 14, when round hematopoietic cells become visible within sac-like structure (iPS-sac), harvest the hematopoietic cells by vigorously disturbing iPS-sac with serological pipette.
18. Filtrate the cell suspension through 70-μm nylon mesh filter to remove cell debris.
19. Centrifuge 440 × *g* for 5 min at room temperature.
20. Aspirate the supernatant and resuspend the cells in 1 ml T cell differentiation medium.
21. Determine cell numbers.
22. Cells are ready to be used in downstream applications (flow cytometry and 3.6 T cell differentiation). For flow cytometry, use 3.5×10^5 cells for flow cytometer setup (5×10^4 cells/fluorochrome) and evaluation of hematopoietic differentiation efficacy (1×10^5 cells/sample). Hematopoietic differentiation efficiency can be evaluated by analysis of CD34, CD43, and CD45 expressions.

3.6 T Cell Differentiation on OP9-DLL1

1. After the iPS-sac dissociation, prepare T cell differentiation medium as in 2.6.
2. Aliquot 3 ml of T cell differentiation medium to a 15-ml tube and add 5×10^5 cells/dish.

3. Remove spent medium from a dish of OP9-DLL1 cells at subconfluent prepared for coculture (refer to Subheading 3.3).
4. Add the cell suspension to the OP9-DLL1 dish.
5. Incubate in a 37 °C incubator.
6. On day 3, add 3 ml of T cell differentiation medium. Prepare a new OP9-DLL1 dish for cell transfer.
7. On day 5, harvest the cells by pipetting and centrifuge at 300 × *g* for 5 min at room temperature. Aspirate the supernatant and resuspend the cells in 3 ml of T cell differentiation medium.
8. Aspirate the spent medium from the new OP9-DLL1 dish, and add the differentiating cells.
9. On day 8, add 3 ml of T cell differentiation medium. Prepare a new OP9-DLL1 dish for cell transfer.
10. On day 10, transfer differentiating cells as did on day 5.
11. On day 13, add 3 ml of T cell differentiation medium. Prepare a new OP9-DLL1 dish for cell transfer.
12. On day 15, transfer differentiating cells as did on day 5 (*see* **Note 10**).
13. On day 18, add 3 ml of T cell differentiation medium.
14. On day 20, harvest the cells by pipetting and centrifuge at 300 × *g* for 5 min at room temperature. Aspirate the supernatant and resuspend the cells in 1 ml of FACS buffer.
15. Determine cell number and evaluate T cell differentiation by flow cytometric analysis of CD3, CD4, CD5, CD7, 8α, 8αβ, and CD45 expressions.

3.7 Expansion of NHP iPS-T Cells

3.7.1 Preparation of PBMC Feeder Cells (See Note 10)

1. On the day before start expansion, thaw cryopreserved PBMCs in a 37 °C water bath.
2. Transfer cell suspension to a 15-ml tube containing prewarmed OP9 medium.
3. Centrifuge 8 min at 460 × *g* at room temperature.
4. Aspirate the supernatant and resuspend in 1 ml of OP9 medium.
5. Determine cell numbers.
6. Adjust the cell suspension to the density of 2×10^6 cells/ml with OP9 medium.
7. Add 1 ml/well of 48-well plate and incubate the plate for overnight.

3.7.2 Expansion of iPS-T Cells

1. Irradiate PBMCs prepared for coculture at 100 Gy (*see* **Note 11**).
2. Harvest the irradiated PBMCs from the wells.
3. Centrifuge 8 min at 460 × *g* at room temperature.
4. Aspirate the supernatant and resuspend in 1 ml of OP9 medium.
5. Determine cell numbers.
6. Combine 1×10^5 cells/well of iPS-T cells (from Subheading 3.6) with 2×10^6 cells/well of irradiated PBMCs.
7. Centrifuge 8 min at 460 × *g* at room temperature.
8. Aspirate the supernatant and resuspend in 1 ml/well of T cell stimulation medium.
9. Add 1 ml of cell suspension to a well of 48-well plate.
10. Incubate the plate in a CO_2 incubator for 48 h.
11. On day 2, add 300 μl of T cell expansion medium.
12. On day 3, remove 800 μl of spent medium from the wells by pipette. Add 500 μl of fresh T cell expansion medium to the wells.
13. On day 5–7, observe the culture under the microscope for split. If not splitting, perform half medium change as did on day 3.
14. If culture is ready for split, harvest cells to a 15-ml tube.
15. Centrifuge 300 × *g* for 5 min at room temperature.
16. Aspirate the supernatant and resuspend the cells in 1 ml of T cell expansion medium.
17. Determine cell numbers.
18. Adjust cell suspension to the density of 2.5×10^5 cells/ml with T cell expansion medium.
19. Add 1 ml of cell suspension to new wells and incubate the plate in a CO_2 incubator.
20. From day 5 to 14, perform half medium changes and splitting accordingly (*see* **Note 12**).
21. From day 12 and onward, cells can be used for flow cytometric analysis of T cell surface marker expressions and other functional assays.

4 Notes

1. For efficient PSC differentiation, it is important to maintain PSC in an appropriate condition. This largely depends on the condition of feeder cells used in maintenance culture. Our lab

uses mitomycin C-treated MEFs, freshly isolated from ICR embryos and cryopreserved at P2.

2. Condition of OP9-DLL1 feeder is the most important parameter of successful T cell differentiation. This could be achieved by selection of an appropriate serum lot, careful culture of OP9-DLL1 cells during maintenance, and use of OP9 medium prepared from powder. The most critical composition of the differentiation culture is serum. Our lab spends considerable amount of time to select an appropriate lot of serum. To select an appropriate serum, HPCs are induced to differentiate into T cells in the various test lots and are evaluated for the efficiency of T cell differentiation by analysis of CD7, CD5, CD4, and CD8 expressions after 20 days differentiation. Test lots are also evaluated for its ability to support OP9-DLL1 cells by monitoring cell viability, growth, and morphology. Although OP9-DLL1 feeders gets overconfluent during 5 days differentiation, OP9-DLL1 cells for maintenance culture should not get overconfluent as it reduces the T cell differentiation ability. Thus, it is important to perform passaging before it gets confluent, typically 2–3 days when cultured in an appropriate serum lot.
3. FBS lot for C3H10T1/2 should also be selected based on its ability to support growth of the cells.
4. Antibodies listed here are confirmed to react with rhesus and cynomolgus monkeys; use of the antibodies may require further evaluations.
5. The inactivated MEF feeders should be used by 4 days after preparation.
6. Some hES/iPSC lines may require passaging shorter than 7 days if some colonies undergo spontaneous differentiation. If earlier passages cannot reduce the numbers of differentiated cells, the hES/iPSC line should be subcloned by manual colony picking.
7. If OP9-DLL1 feeder dishes are required the next day, passage cells at 2 × 105 cells/dish instead of 1.5 × 105 cells/dish.
8. Addition of human BMP4 at the final concentration of 40 ng/ml may improve the hematopoietic differentiation efficiency of NHP iPSC lines.
9. The amount of cell suspension may require optimization by NHP iPSC lines you use.
10. Differentiating cells get confluent around day 15 after T cell differentiation. If this happens, split the culture at the ratio of 1: 5–10. Overconfluent culture may result in reduced T cell differentiation efficiency.

11. PBMCs can be purchased from multiple venders or isolated from volunteer donors. Select PBMC lot that support T cell expansion.
12. The optimum dose of irradiation should be empirically determined by each institution.
13. Optimum cell density for medium changes and splitting may be different by iPSC lines and batches of differentiation. Daily observation of the culture is important to effectively expand iPS-T cells.

Acknowledgments

We thank Mr. Yasuyuki Miyake, Shuichi Kitayama, and Dr. Kahoru Taya for help in the preparation. This work is supported in part by KAKENHI to S.I. and S.K.

References

1. Morizane A, Doi D, Kikuchi T et al (2013) Direct comparison of autologous and allogeneic transplantation of iPSC-derived neural cells in the brain of a non-human primate. Stem Cell Reports 1:283–292. https://doi.org/10.1016/j.stemcr.2013.08.007
2. Chong JJH, Yang X, Don CW et al (2014) Human embryonic-stem-cell-derived cardiomyocytes regenerate non-human primate hearts. Nature 510:273–277. https://doi.org/10.1038/nature13233
3. Kawamura T, Miyagawa S, Fukushima S et al (2016) Cardiomyocytes Derived from MHC-homozygous induced pluripotent stem cells exhibit reduced allogeneic immunogenicity in MHC-matched non-human Primates. Stem Cell Reports 6:312–320. https://doi.org/10.1016/j.stemcr.2016.01.012
4. Shiba Y, Gomibuchi T, Seto T et al (2016) Allogeneic transplantation of iPS cell-derived cardiomyocytes regenerates primate hearts. Nature 538:388–391. https://doi.org/10.1038/nature19815
5. Sugita S, Iwasaki Y, Makabe K et al (2016) Successful transplantation of retinal pigment epithelial Cells from MHC homozygote iPSCs in MHC-matched models. Stem Cell Reports 7:635–648. https://doi.org/10.1016/j.stemcr.2016.08.010
6. Kikuchi T, Morizane A, Doi D et al (2017) Human iPS cell-derived dopaminergic neurons function in a primate Parkinson's disease model. Nature 548:592–596. https://doi.org/10.1038/nature23664
7. Morizane A, Kikuchi T, Hayashi T et al (2017) MHC matching improves engraftment of iPSC-derived neurons in non-human primates. Nat Commun 8:385. https://doi.org/10.1038/s41467-017-00926-5
8. Hong SG, Winkler T, Wu C et al (2014) Path to the clinic: assessment of iPSC-BasedCell therapies in vivo in a nonhuman primate model. CellReports 7:1298–1309. https://doi.org/10.1016/j.celrep.2014.04.019
9. Gori JL, Butler JM, Chan Y-Y et al (2015) Vascular niche promotes hematopoietic multipotent progenitor formation from pluripotent stem cells. J Clin Invest. https://doi.org/10.1172/JCI79328DS1
10. D'Souza SS, Maufort J, Kumar A et al (2016) GSK3β inhibition promotes efficient myeloid and lymphoid Hematopoiesis from non-human primate-induced pluripotent stem cells. Stem Cell Reports 6:243–256. https://doi.org/10.1016/j.stemcr.2015.12.010
11. Kardel MD, Eaves CJ (2012) Modeling human hematopoietic cell development from pluripotent stem cells. Exp Hematol 40:601–611. https://doi.org/10.1016/j.exphem.2012.04.001
12. Kennedy M, D'Souza SL, Lynch-Kattman M et al (2007) Development of the hemangioblast defines the onset of hematopoiesis in human ES cell differentiation cultures. Blood 109:2679–2687. https://doi.org/10.1182/blood-2006-09
13. Vodyanik MA, Thomson JA, Slukvin II (2006) Leukosialin (CD43) defines hematopoietic

progenitors in human embryonic stem cell differentiation cultures. Blood 108:2095–2105. https://doi.org/10.1182/blood-2006-02-003327

14. Vizcardo R, Masuda K, Yamada D et al (2013) Regeneration of human tumor antigen-specific T cells from iPSCs derived from mature CD8 (+) T cells. Cell Stem Cell 12:31–36. https://doi.org/10.1016/j.stem.2012.12.006

15. Nishimura T, Kaneko S, Kawana-Tachikawa A et al (2013) Generation of rejuvenated antigen-specific T cells by reprogramming to pluripotency and redifferentiation. Cell Stem Cell 12:114–126. https://doi.org/10.1016/j.stem.2012.11.002

Chapter 12

An Improved Method to Produce Clinical-Scale Natural Killer Cells from Human Pluripotent Stem Cells

Huang Zhu and Dan S. Kaufman

Abstract

Human natural killer (NK) cell-based adoptive anticancer immunotherapy has gained intense interest with many clinical trials actively recruiting patients to treat a variety of both hematological malignancies and solid tumors. Most of these trials use primary NK cells isolated either from peripheral blood (PB-NK cells) or umbilical cord blood (UCB-NK cells), though these sources require NK cell collection for each patient leading to donor variability and heterogeneity in the NK cell populations. In contrast, NK cells derived human embryonic stem cells (hESC-NK cells) or induced pluripotent stem cells (hiPSC-NK cells) provide more homogeneous cell populations that can be grown at clinical scale, and genetically engineered if needed. These characteristics make hESC-/iPSC-derived NK cells an ideal cell population for developing standardized, "off-the-shelf" immunotherapy products. Additionally, production of NK cells from undifferentiated human pluripotent stem cells enables studies to better define pathways that regulate human NK cell development and function. Our group previously has established a stromal-free, two-stage culture system to derive NK cells from hESC/hiPSC in vitro followed by clinical-scale expansion of these cells using interleukin (IL)-21 expressing artificial antigen-presenting cells. However, prior to differentiation, this method requires single-cell adaptation of hESCs/hiPSCs which takes months. Recently we optimized this method by adapting the mouse embryonic fibroblast-dependent hESC/hiPSC to feeder-free culture conditions. These feeder-free hESCs/hiPSCs are directly used to form embryoid body (EB) to generate hemato-endothelial precursor cells. This new method produces mature, functional NK cells with higher efficiency to enable rapid production of an essentially unlimited number of homogenous NK cells that can be used for standardized, targeted immunotherapy for the treatment of refractory cancers and infectious diseases.

Key words Embryonic stem cell, Induced pluripotent stem cells, Embryoid body, Hematopoietic progenitors, Natural killer cells, Cancer immunotherapy, In vitro differentiation

1 Introduction

Human natural killer (NK) cells are an important part of innate immune system with ability to kill malignant and virally infected cells without MHC restriction and without prior sensitization [1]. NK cell cytotoxic activity against tumor cells or infected cells is mediated through a repertoire of germ line encoded activating

Shin Kaneko (ed.), *In Vitro Differentiation of T-Cells: Methods and Protocols*, Methods in Molecular Biology, vol. 2048, https://doi.org/10.1007/978-1-4939-9728-2_12, © Springer Science+Business Media, LLC, part of Springer Nature 2019

and inhibitory cell surface receptors including killer immunoglobulin receptors (KIRs), natural cytotoxicity receptors (NCRs), and the Fc gamma receptor (FcγRIIIa) CD16a that mediates antibody-dependent cellular cytotoxicity (ADCC) [2, 3]. These important characteristics enable NK cells to function as allogeneic effector cells for treatment of refractory cancers and infection disease such as HIV.

Human embryonic stem cells (hESCs) and induced pluripotent stem cells (iPSCs) are ideal starting populations for the development of multiple cell lineages, including NK cells. hESCs/iPSCs provide a model system to study the development of NK cells in vitro [4, 5]. Previous studies from our group have shown that hESC-/iPSC-derived NK cells have potent antitumor and antiviral activities, providing a standardized cell-based treatment for these diseases [6–8].

Multiple clinical studies now demonstrate NK cells can effectively treat acute myeloid leukemia (AML) and other malignancies while not causing serious adverse effects such as graft-versus-host disease (GvHD) or cytokine release syndrome (CRS) [9, 10]. To date, most of NK cell-based adoptive immunotherapy clinical trials have used primary NK cells isolated from donor's peripheral blood (PB-NK cells) [9, 10], umbilical cord blood (UCB-NK cells) [11], or the transformed NK cell line NK-92 [12]. While each have demonstrate clinical efficacy, there are some shortcomings [13]. For example, PB-NK cells and UCB-NK cells are typically a heterogeneous mix of NK cells and other immune cells that can vary from donor to donor [14]. NK-92 cells are aneuploid and need to be irradiated before use which limits the in vivo survival and expansion of these cells—as is known to be a key determinant of antitumor activity [12]. In contrast, hESC-NK cells and hiPSC-NK cells are more homogenous and can generate essentially unlimited cells sources suitable for clinical use [8]. Importantly, hESC/hiPSC-NK cells exhibit similar phenotype, transcriptome, and functions as primary NK cells [6, 8]. Moreover, hESC/iPSC-NK cells can be routinely genetically modified by any of several methods including transposon and viral vectors [15]. Importantly, this genetic modification can be done on the undifferentiated pluripotent stem cells to produce a uniform population of NK cells with the desired effect. For example, the CRISPR/Cas9 or TALEN can be used to delete a gene of interest in hESCs or iPSCs to produce "knockout" human NK cells. Key studies demonstrate the ability to utilize these genetic reporter systems expressed in hESCs and iPSCs to track and regulate development of specific blood cell lineages [4, 5, 16].

The methods to generate NK cells from human pluripotent stem cells have evolved in the last decade [6, 17, 18]. Our initial studies used a stromal cell dependent two-stage culture system [6]. Briefly, to obtain $CD34^{+}CD45^{+}$ hematopoietic progenitor cells, undifferentiated hESCs were culture on stromal cell line (e.g., S17 or M2–10B4 cells) using media containing fetal bovine serum (FBS). Then, hematopoietic progenitor cells were sorted and moved to a second stromal cell line (e.g., AFT024 or EL08-1D2 cells) in media supplemented with SCF, Fms-like tyrosine kinase 3 ligand (FLT3L), IL-3, IL-15, and IL-7 to direct differentiation toward NK cells. Later, our group optimized the protocol by adapting a "spin-embryoid body (EB)" method to derive hematopoietic progenitor cells in defined serum-free, stromal-free conditions [8, 18]. After 11 days culture, hematopoietic progenitor cell-containing EBs were transferred to feeder-free plates in NK differentiation media containing SCF, FLT3L, IL-3, IL-15, and IL-7 for 4 weeks to generate $CD45^{+}CD56^{+}$ NK cells. hESC-NK/iPSC-NK cells obtained using these methods express activating and inhibitory receptors similar to PB-NK cells [8]. More importantly, these NK cells exhibit potent antitumor and antiviral activity both in vitro and in vivo. These functions include effective elimination of tumor cells xenografted in immune-deficient mice and inhibition of HIV-infected targets in an in vivo SCID-hu model [6, 7, 15, 19]. hESC-NK/iPSC-NK cells can be further expanded to clinical scale by using irradiated human stimulator cells, as previously utilized to effectively expand PB-NK or UCB-NK cells [20]. Notably, this method requires an essential single-cell adaptation step to enable hESC/iPSC survival as single cells [18, 21]. This process is not uniform and typically requires 12–15 single-cell passages that take several weeks.

In this chapter, we describe an improved method for the derivation of NK cells from human pluripotent stem cells with higher efficiency and less time than existing protocols [18]. Figure 1

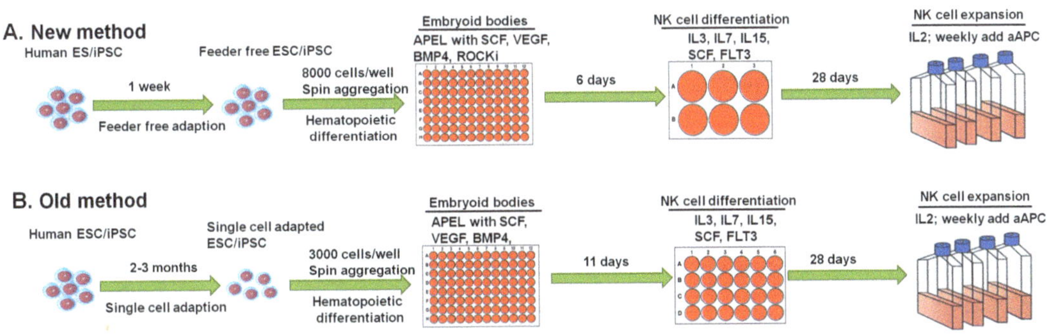

Fig. 1 Schematic diagram of new method (**a**) and old method [18] (**b**) for in vitro NK cell differentiation from human embryonic stem cells (ESC) or induced pluripotent stem cells (iPSC). Briefly, the old method requires a single-cell adaptation process for ESC/iPSC which takes 2–3 months, while the new method starts with feeder-free adapted ESC/iPSC and use ROCKi to help EB formation. Both methods use spin EBs (11 days for old method and 6 days for the new method) to generate hematopoietic progenitors (CD34+ cells). Then EBs are directly transferred into NK cell differentiation conditions. Mature and functional NK cells will develop after 3–5 weeks and can be expanded using IL-2 and aAPCs

provides a schematic diagram to compare the old and new methods. Prior to the generation of EBs, MEF-dependent undifferentiated hESCs/iPSCs are passaged on Matrigel-coated plates in mTeSR media (feeder-free conditions) for 1 week, though this step is not needed for hESC/iPSC already maintained in feeder-free conditions. The feeder-free hESC/iPSC can be directly used to form EB in the presence of Rho-associated protein kinase inhibitor (ROCKi) which decreases the cellular stress response and the apoptotic cell death in stem cell cultures [22]. Indeed, these new conditions produce more hESC-/iPSC-derived $CD34^+$ progenitor cells in only 6 days (Fig. 2g). Following the generation of the hemato-

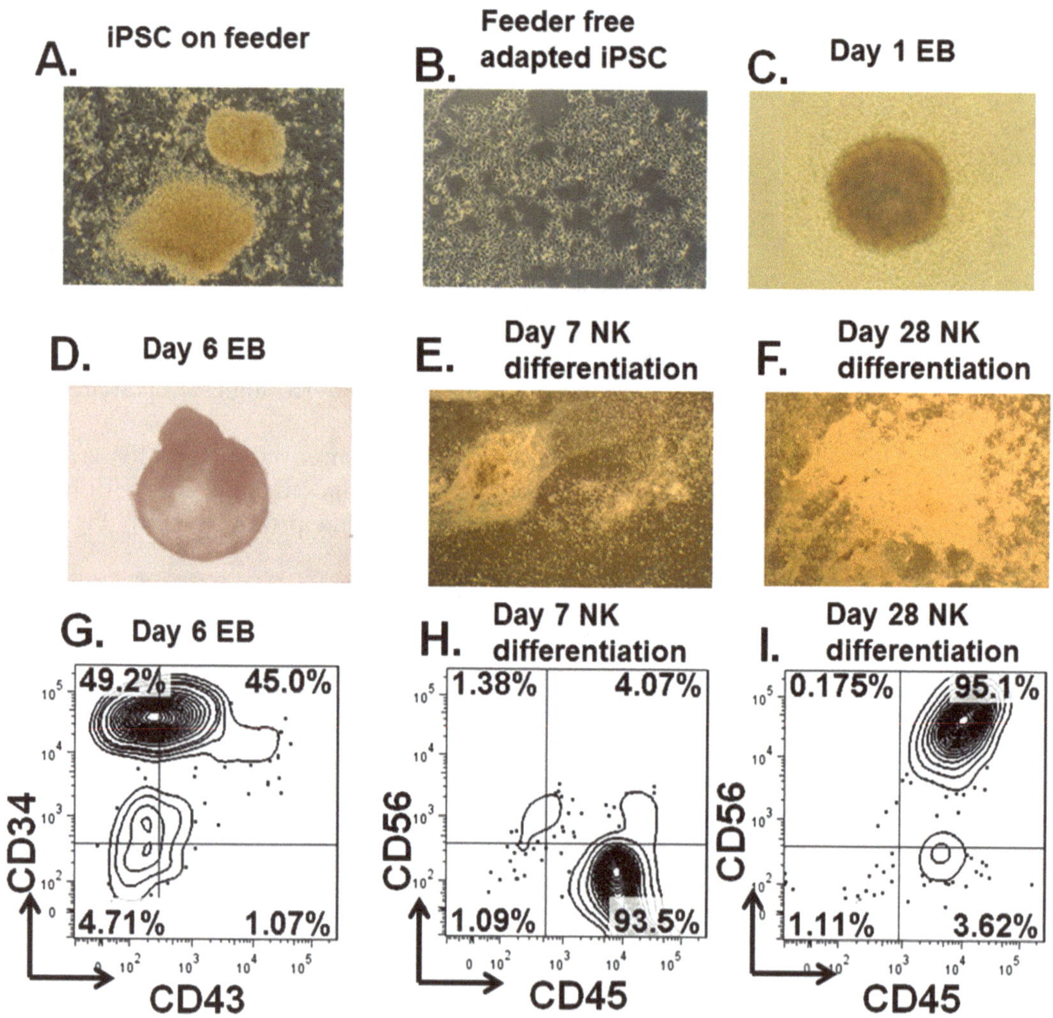

Fig. 2 Phase microscope images and phenotypes of differentiating NK cells from human pluripotent stem cells at different stages. (**a**) Undifferentiated iPSCs on MEFs. (**b**) Feeder-free undifferentiated iPSCs. (**c**) Day 1 embryoid body. (**d** and **g**) Day 6 embryoid body. (**e** and **h**) Day 7 NK cell differentiation. (**f** and **i**) Day 28 NK cell differentiation. Original magnification × 100 for **a–d** and ×20 for **e** and **f**

endothelial precursor cells in the EBs, NK cell differentiation is supported using a stromal cell-free protocol similar to the existing protocol [18]. NK cells derived using this method have similar growth rate, phenotype, and activity compared to PB-NK cells and UCB-NK cells generated by existing methods. In summary, this improved method allows efficient, rapid, and reproducible production of hESC-/iPSC-NK cells from different pluripotent stem cell lines without the requirement of time-consuming single-cell adaptation process thus facilitate clinical-scale production and translation of hESC-/iPSC-NK cells.

2 Materials

2.1 Cell Lines

We have successfully produced NK cells from several different iPSC lines, including those derived from human fibroblasts (FiPSC), human peripheral blood mononuclear cells (PBiPSCs) [23], or human CD34+ cells isolated from umbilical cord blood (UiPSC) [8]. Human ES cell line H9 was obtained from Wicell, Madison WI. K562 aAPC cells expressing membrane-bound IL-21 (Clone 9. mbIL-21) [20] were kindly provided by Dr. Dean Lee and Dr. Lawrence Cooper, MD Anderson Cancer Center, Houston, TX.

2.2 Cell Culture Media and Reagents

1. Undifferentiated hESC and hiPSCs: mTsER™1 (STEMCELL Technologies, Vancouver, Canada).
2. K562 aAPC cells and MOLM13 cells: RPMI-1640, 2 mM L-glutamine, 10% FBS, and 1% penicillin/streptomycin.
3. Spin EB differentiation medium: STEMdiff™ APEL™2 Medium (STEMCELL Technologies, Vancouver, Canada), 40 ng/mL SCF (PeproTech), 20 ng/mL BMP4 (R&D systems), 20 ng/mL VEGF (R&D systems).
4. NK cell differentiation medium [18]: 56.6% DMEM+ GlutaMAX™-I (Life Technologies), 28.3% F12+ GlutaMAX™-I (Life Technologies), 15% heat-inactivated human AB serum (Valley Biomedical), 1% P/S, 2 mM L-glutamine, 1 μM β-mercaptoethanol, 5 ng/mL sodium selenite (Sigma-Aldrich), 50 μM ethanolamine (MP Biomedicals), 20 mg/L ascorbic acid (Sigma-Aldrich), 5 ng/mL IL-3 (PeproTech), 20 ng/mL SCF (Peprotech), 20 ng/mL IL-7 (PeproTech), 10 ng/mL IL-15 (PeproTech), and 10 ng/mL FLT3 ligand (FLT3L) (PeproTech), Store at 4 °C in the dark.
5. NK cell expansion medium: RPMI-1640, 10% FBS, 2 mM L-glutamine, 1% P/S, and 50 U/mL IL-2 (PeproTech).
6. Antibodies [8]: NKp46-PE, NKp44-PE, CD56-APC, NKG2D-PE, TRAIL-PE, and FAS ligand-PE are from Becton,

Dickinson and Company (Franklin Lakes, NJ, http://www.bd.com). CD16-APC is obtained from eBioscience Inc. (San Diego, CA, http://www.ebioscience.com). CD158a/h-PE, CD158j-PE, CD158i-PE, CD158e1/e2, and CD159a-PE are from Beckman Coulter (Fullerton, CA, http://www.beckmancoulter.com).

7. Other material: Matrigel matrix (Corning), TrypLE Select (Life Technologies), Collagenase Type IV (STEMCELL Technologies), ROCKi: Y-27632 (Sigma), 96-well round-bottom plates (for EB formation, NUNC), 24-well or 6-well tissue culture plates (Corning), CellEvent® Caspase-3/7 Green Detection Reagent (Thermal Fisher Scientific), CellTrace™ Violet (Thermo Fisher Scientific), SYTOX™ AADvanced™ dead cell stain solution (Thermal Fisher Scientific).

3 Method

3.1 Feeder-Free Adaptation of Feeder-Dependent hESC/iPSC

Before generating EBs, undifferentiated feeder-dependent hESCs/iPSCs need to be transferred to feeder-free conditions with the use of mTeSR™1 on Matrigel (Corning)-coated plates, as previously described [24]. If hESCs/iPSCs are already maintained on feeder-free conditions, skip this step. It is critical to make sure the hESC/iPSC are not differentiated before and after adaptation. This step will change the hESC/iPSC from clump culture to monolayer culture and increase efficiency of making single-cell suspension. Cells should be able to adapt to feeder-free culture within 1–2 passages and then exhibit morphology consistent with feeder-free human pluripotent stem cells (Fig. 2a, b). The following instructions are used for passaging cells from one well of a 6-well plate.

1. Before passaging, coat new plate with 1× Matrigel (*see* instruction below), and incubate in 37 °C incubator for 2–4 h before use. Preparing 1× Matrigel:
   (a) Thaw 5 mL stock vial of Matrigel at 4 °C overnight.
   (b) Quickly aliquot Matrigel using pre-chilled tips into pre-chilled Eppendorf tubes.
   (c) Store at −80 °C for up to 6 months.
   (d) To prepare 1× Matrigel working solution, resuspend one Matrigel aliquot at 1:100 ratio into cold DMEM/F12. 1× Matrigel working solution can be stored at −4 °C for up to 2 weeks.
   (e) Coat plate with sufficient volume to cover surface (e.g., use 1 mL per well of 6-well plate).
   (f) Incubate in 37 °C incubator for 2 to 4 h before use.

2. After making sure the feeder-dependent hESC/iPSC is free of differentiated colonies and at right confluence (70–80%), aspirate media and pass cells by incubating cells for 15 min at 37 °C in pre-warmed Collagenase Type IV (1 mg/mL).
3. Aspirate the collagenase and wash cells once with 3 mL PBS. Then add 1 mL of mTeSR™1 medium. Then gently detach the colonies by scraping with a serological glass pipette.
4. Collect detached aggregates and add 3 mL of mTeSR™1 medium. To break up the cell aggregates to appropriate size (50–200 μM), gently pipette up and down 2–3 times using a 5 mL serological pipette.
5. Plate cell aggregate mixture 1:2 onto Matrigel pre-coated plate.
6. Change media daily and repeat procedure once when cells become 70–80% confluent. Most of the feeder cells should be eliminated after two passages.
7. After two passages with collagenase, passage cells by incubating cells for 5 min at 37 °C in pre-warmed TrypLE Select.
8. Carefully dissociate cells aggregates to single cell using 1 mL micropipette and transfer cells to a 15 mL conical tube. Add additional 5 mL medium (DMEM/F12 + 10% FBS).
9. Pass cell mixture through a 70 μM cell strainer to remove aggregates.
10. Centrifuge cells at 300 × *g* for 5 min.
11. Aspire supernatant and wash cells once with 5 mL PBS.
12. Resuspend cells with mTeSR™1 medium and plate cells 1:2 onto Matrigel pre-coated plate. Cells can be used for spin EB formation when become confluent.

3.2 Generation of Hematopoietic Progenitor Cells from hES/iPS Cells by Spin EB Formation

Our group adapted a "spin EB" protocol to generate hematopoietic progenitors [8, 21, 25]. Cells inside EBs can form self-stromal cells to support following lymphocyte development, eliminating the use of xeno-derived stromal cells such as OP9 [18]. While there is variability between different hESC and iPSC lines, we can consistently obtain >50% CD34+ cells for most of lines using this method (Fig. 2) (*see* **Note 1**). The following instructions are used for collecting cells from one well of a 6-well plate for EB formation.

1. Two days before spin EB formation, pass 200,000 feeder-free hESCs/iPSCs onto 1 well of Matrigel-coated 6-well plate which should reach 70–80% confluence on the day of spin EBs setup. Generally, 1 well of a 6-well plate are sufficient for two plates of spin EB.

2. Aspirate culture medium and detach cells by incubating with 1 mL incubating cells for 5 min at 37 °C in pre-warmed TrypLE Select.
3. Carefully dissociate cells aggregates to single cell using 1 mL micropipette and transfer cells to a 15 mL conical tube. Add additional 5 mL medium (DMEM/F12 + 10% FBS). Pass cell mixture through a 70 μM cell strainer to remove aggregates.
4. Centrifuge cells, remove supernatant, and wash cells once with 5 mL PBS. Then resuspend the cells in 1 mL APEL media.
5. Count cells and dilute cells to appropriate density using APEL media containing cytokines (SCF: 40 ng/mL BMP4: 20 ng/mL, VEGF: 20 ng/mL) plus 10 μM ROCKi (Y-27632) to be used for plating. Typically, cells are seeded at 8000 cells/well in 100 μL media (80,000 cells/mL) in the 96-well plates (*see* **Note 2**).
6. Pipet 100 μL of the cell suspension into 96-well plates using a multichannel pipet. Centrifuge 96-well plates at 300 × *g* for 5 min, and incubate the plates at 37 °C, 5% CO_2 for 6 days. We have found 6 days are enough to give rise to >50% CD34+ cells for most of hESC and iPSC lines.

3.3 Derivation of Human NK Cells from Spin EB

Spin EBs can be transferred into either 24-well plates or 6-well plates coated with 2% gelatin or without coating. We have found that 6-well plate is more suitable for medium change and 2% gelatin can help EBs attachment.

1. Before transferring EBs, coat new 6-well plates with 2% gelatin (*see* instruction below) and incubate in 37 °C incubator for 2–4 h before use.

   Preparing 2% gelatin-coated plate:

   (a) Prepare a 2% (w/v) solution by dissolving gelatin (SIGMA #G1890) in tissue culture grade water.

   (b) Sterilize by autoclaving at 121 °C, 15 psi for 30 min.

   (c) Coat culture surface with 5–10 μL gelatin solution/cm2 (about 1 mL per well of 6-well plate).

   (d) Allow to incubate 37 °C for 1 h, then aspirate remaining gelatin solution before introducing cells and medium.
2. Add 3 mL NK differentiation medium containing all of the cytokines (IL-3: 5 ng/mL SCF: 20 ng/mL, IL-7: 20 ng/mL, IL-15: 10 ng/mL) to each well of gelatin-coated 6-well plate.
3. Spin EBs are directly transferred into 6-well plates on day 6. Carefully transfer EBs from 96-well plates to a 10 cm dish using 10 mL serological pipette, and remove most of medium (*see* **Note 3**). Then add 2 mL of NK differentiation medium containing all of the cytokines. Distribute 14–16 EBs into each

well of the 6-well plate using 1 mL micropipette. Typically, EBs from one 96-well plate can make one 6-well NK differentiation plate.

4. Medium changes are done every 5–7 days (*see* **Note 4**). Following the first week of the NK cell differentiation, IL-3 is no longer added to the media.
5. Continue medium changes for 3–4 weeks and check expression of NK cell makers (CD45+CD56+) by flow cytometry on suspension cells (Fig. 2e–i). Cells can be collected passing them through a 70 μm filter to remove any clumps when >80% of suspension cells are CD45+CD56+ (*see* **Note 5**).

3.4 Expansion of hESC-/iPSC-Derived NK Cells to Clinical Scale

Typically, we can obtain 2–20 × 10^6 NK cells from one 6-well plate. To further expand the NK cells for downstream applications, artificial antigen-presenting cells (aAPCs) are used to generate >10^9 NK cells (Fig. 3).

1. We use membrane-bound IL-21 expressing (mbIL-21) K562 cells [8, 20] as aAPCs to stimulate NK cell expansion. Before adding to NK cells, aAPCs are irradiated with 10,000 rads and made as frozen stocks.
2. After collecting from 6-well plate, NK cells are maintained in NK expansion medium containing 50 U/mL IL-2 (add freshly) at a density of 3 × 10^5 cells/mL. Irradiated aAPCs

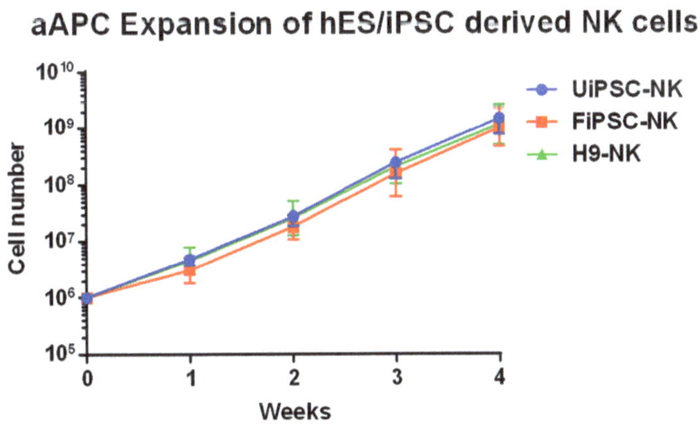

Fig. 3 Growth curve of NK cells derived from hES/iPSC using method presented in this chapter. 2 iPSC lines (iPSC reprogrammed from umbilical CD34+ cells (UiPSC) and from fibroblast (FiPSC)) and 1 hESC line H9-derived NK cells were transferred from NK differentiation conditions and placed in aAPC for expansion. Cells were stimulated with aAPCs weekly and cell number reached >10^9 from 10^6 starting cell number in 4 weeks. Each line represents mean of three independent experiments

are thawed and added to NK cells at a ratio of 1:1. Media are changed every 3–4 days containing 50 U/mL of freshly added IL-2.

3. hESC-derived or iPSC-derived NK cells can be expanded for >3 months without a decrease in cell viability or cytolytic activity.

3.5 Phenotypic and Functional Characterization of hESC-/iPSC-Derived NK Cells

hESC-/iPSC-derived NK cells can be phenotyped by measuring surface marker expression such as CD16, NKG2D, NKp44, NKp46, TRAIL, FasL, etc. We normally characterize phenotypes and killing activity of hESC-/iPSC-derived NK cells after 3 weeks of expansion. Functions of NK cells can be accessed by measurement of direct cytolytic activity tumor cells (such as K562) by Caspase-3/7 flow cytometry assay or immunological assays for cytotoxic granule or cytokine release, or test antitumor activity in vivo xenograft models [6, 19]. NK cells developed using this method show a mature NK cell phenotype and cytotoxicity (Fig. 4).

1. We typically test the expression of following receptors by flow cytometry: CD56, CD16, CD94, NKG2D, NKp44, NKp46, KIR, FasL, and TRAIL.
2. After analyzing phenotypes, functional of hESC-/iPSC-derived NK cells are evaluated using flow cytometry-based Caspase-3/7 assay against tumor targets:
   (a) Target cells were counted and pre-stained with CellTrace™ Violet at a final concentration of 5 μM in PBS for 15 min at 37 °C.
   (b) After staining, target cells were washed in complete culture medium prior to being mixed with NK cell cultures at the indicated effector to target (E:T) ratios.
   (c) Incubate NK cells and target cells mix at 37° for 3.5 h.
   (d) Add CellEvent® Caspase-3/7 Green Detection Reagent for an additional 30 min of culture for a total incubation time of 4 h.
   (e) During the final 5 min of staining, add 1 μL of SYTOX™ AADvanced™ dead cell stain solution and mixed gently.
   (f) Cells were then analyzed by flow cytometry.

4 Notes

1. We are able to obtain >50% CD34+ cells for most of iPSC and ES lines tested using this method. We also found that EBs containing >10% CD34+ cells can be successfully differentiated to mature NK cells.

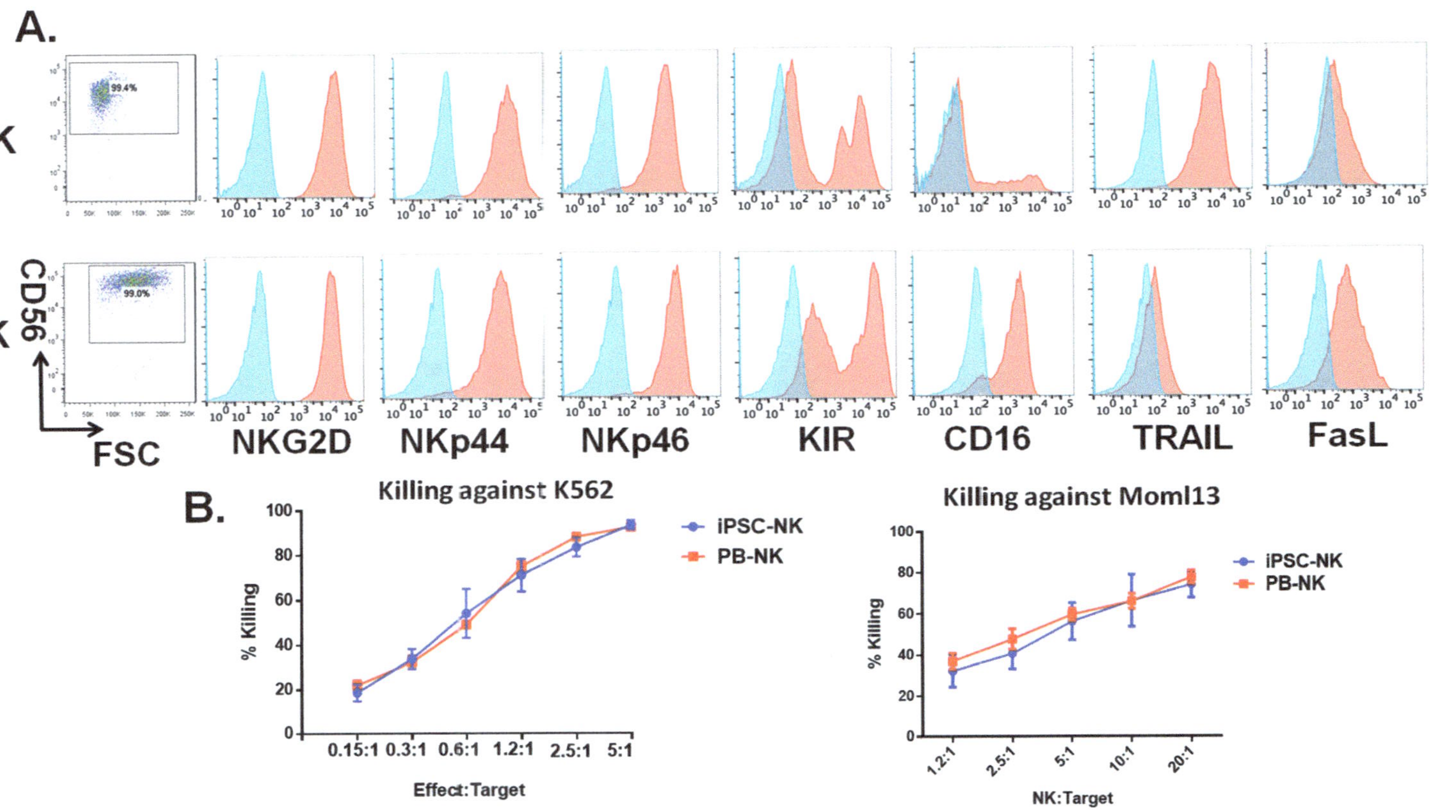

Fig. 4 iPSC-NK cells are phenotypically mature and have similar killing activities against tumor targets as PB-NK cells. (**a**) iPSC-NK cells and PB-NK cells were stained for a panel of NK cell receptors. Expression of each marker is shown by representative flow cytometry plots. (**b**) In vitro killing against K562 cells and MOLM13 cells was determined using CellEvent™ Caspase-3/7 Green Flow Cytometry Assay. Different amounts of target cells were incubated with either iPSC-NK cells or PB-NK cells (effector–target ratio from 0.15:1 to 5:1 for K562, 1.2:1 to 20:1 for MOLM13 cells) for 4 h. Data are represented as mean ± SEM

2. 8000 cells/well works for most of cell lines. But the optimize cell number should be optimized depending on the growth rate of the cell line. We seed more cells than our previous protocol [18] (3000 cells/well) as we reduced the EB formation time from 11 days to 6 days.
3. It is critical to remove most of the medium and resuspend with NK cell differentiation medium.
4. After 14 days, medium needs to be changed every 3 days.
5. It usually takes 4 weeks to obtain >90% CD45+CD56+ cells from spin EBs NK cells for most of ES or iPSC lines (Fig. 2). Some cell line might only require 3 weeks and some line may take longer (about 5 weeks).

References

1. Vivier E, Raulet DH, Moretta A, Caligiuri MA, Zitvogel L, Lanier LL, Yokoyama WM, Ugolini S (2011) Innate or adaptive immunity? The example of natural killer cells. Science 331:44–49
2. Morvan MG, Lanier LL (2016) NK cells and cancer: you can teach innate cells new tricks. Nat Rev Cancer 16(1):7–19
3. Jing Y, Ni Z, Wu J, Higgins L, Markowski TW, Kaufman DS, Walcheck B (2015) Identification of an ADAM17 cleavage region in human CD16 (FcgammaRIII) and the engineering of a non-cleavable version of the receptor in NK cells. PLoS One 10(3):e0121788
4. Angelos MG, Ruh PN, Webber BR, Blum RH, Ryan CD, Bendzick L, Shim S, Yingst AM, Tufa DM, Verneris MR, Kaufman DS (2017) Aryl hydrocarbon receptor inhibition promotes hematolymphoid development from human pluripotent stem cells. Blood 129 (26):3428–3439
5. Ferrell PI, Xi J, Ma C, Adlakha M, Kaufman DS (2015) The RUNX1 +24 enhancer and P1 promoter identify a unique subpopulation of hematopoietic progenitor cells derived from human pluripotent stem cells. Stem Cells 33 (4):1130–1141
6. Woll PS, Grzywacz B, Tian X, Marcus RK, Knorr DA, Verneris MR, Kaufman DS (2009) Human embryonic stem cells differentiate into a homogeneous population of natural killer cells with potent in vivo antitumor activity. Blood 113(24):6094–6101
7. Ni ZY, Knorr DA, Clouser CL, Hexum MK, Southern P, Mansky LM, Park IH, Kaufman DS (2011) Human pluripotent stem cells produce natural killer cells that mediate anti-HIV-1 activity by utilizing diverse cellular mechanisms. J Virol 85(1):43–50
8. Knorr DA, Ni Z, Hermanson D, Hexum MK, Bendzick L, Cooper LJ, Lee DA, Kaufman DS (2013) Clinical-scale derivation of natural killer cells from human pluripotent stem cells for cancer therapy. Stem Cells Transl Med 2 (4):274–283
9. Miller JS, Soignier Y, Panoskaltsis-Mortari A, McNearney SA, Yun GH, Fautsch SK, McKenna D, Le C, Defor TE, Burns LJ, Orchard PJ, Blazar BR, Wagner JE, Slungaard A, Weisdorf DJ, Okazaki IJ, McGlave PB (2005) Successful adoptive transfer and in vivo expansion of human haploidentical NK cells in patients with cancer. Blood 105(8):3051–3057
10. Romee R, Rosario M, Berrien-Elliott MM, Wagner JA, Jewell BA, Schappe T, Leong JW, Abdel-Latif S, Schneider SE, Willey S, Neal CC, Yu L, Oh ST, Lee YS, Mulder A, Claas F, Cooper MA, Fehniger TA (2016) Cytokine-induced memory-like natural killer cells exhibit enhanced responses against myeloid leukemia. Sci Transl Med 8(357):357ra123
11. Verneris MR, Miller JS (2009) The phenotypic and functional characteristics of umbilical cord blood and peripheral blood natural killer cells. Br J Haematol 147(2):185–191
12. Tonn T, Schwabe D, Klingemann HG, Becker S, Esser R, Koehl U, Suttorp M, Seifried E, Ottmann OG, Bug G (2013) Treatment of patients with advanced cancer with the natural killer cell line NK-92. Cytotherapy 15 (12):1563–1570
13. Zhu H, Lai YS, Li Y, Blum RH, Kaufman DS (2018) Concise review: human pluripotent

stem cells to produce cell-based cancer immunotherapy. Stem Cells 36(2):134–145

14. Passweg JR, Tichelli A, Meyer-Monard S, Heim D, Stern M, Kuhne T, Favre G, Gratwohl A (2004) Purified donor NK-lymphocyte infusion to consolidate engraftment after haploidentical stem cell transplantation. Leukemia 18(11):1835–1838
15. Ni ZY, Knorr DA, Bendzick L, Allred J, Kaufman DS (2014) Expression of chimeric receptor CD4 zeta by natural killer cells derived from human pluripotent stem cells improves in vitro activity but does not enhance suppression of HIV infection in vivo. Stem Cells 32 (4):1021–1031
16. Sugimura R, Jha DK, Han A, Soria-Valles C, da Rocha EL, Lu YF, Goettel JA, Serrao E, Rowe RG, Malleshaiah M, Wong I, Sousa P, Zhu TN, Ditadi A, Keller G, Engelman AN, Snapper SB, Doulatov S, Daley GQ (2017) Haematopoietic stem and progenitor cells from human pluripotent stem cells. Nature 545(7655):432–438
17. Woll PS, Martin CH, Miller JS, Kaufman DS (2005) Human embryonic stem cell-derived NK cells acquire functional receptors and cytolytic activity. J Immunol 175(8):5095–5103
18. Hermanson DL, Ni Z, Kaufman DS (2015) Human pluripotent stem cells as a renewable source of natural killer cells. In: Cheng T (ed) Hematopoietic differentiation of human pluripotent stem cells. Springer, Dordrecht, pp 69–79
19. Hermanson DL, Bendzick L, Pribyl L, McCullar V, Vogel RI, Miller JS, Geller MA, Kaufman DS (2016) Induced pluripotent stem cell-derived natural killer cells for treatment of ovarian cancer. Stem Cells 34(1):93–101
20. Denman CJ, Senyukov VV, Somanchi SS, Phatarpekar PV, Kopp LM, Johnson JL, Singh H, Hurton L, Maiti SN, Huls MH, Champlin RE, Cooper LJ, Lee DA (2012) Membrane-bound IL-21 promotes sustained ex vivo proliferation of human natural killer cells. PLoS One 7(1): e30264
21. Ng ES, Davis R, Stanley EG, Elefanty AG (2008) A protocol describing the use of a recombinant protein-based, animal product-free medium (APEL) for human embryonic stem cell differentiation as spin embryoid bodies. Nat Protoc 3(5):768–776
22. Claassen DA, Desler MM, Rizzino A (2009) ROCK inhibition enhances the recovery and growth of cryopreserved human embryonic stem cells and human induced pluripotent stem cells. Mol Reprod Dev 76(8):722–732
23. Zou L, Chen QS, Quanbeck Z, Bechtold JE, Kaufman DS (2016) Angiogenic activity mediates bone repair from human pluripotent stem cell-derived osteogenic cells. Sci Rep-Uk 6
24. Ludwig TE, Bergendahl V, Levenstein ME, Yu J, Probasco MD, Thomson JA (2006) Feeder-independent culture of human embryonic stem cells. Nat Methods 3(8):637–646
25. Hexum MK, Tian X, Kaufman DS (2011) In vivo evaluation of putative hematopoietic stem cells derived from human pluripotent stem cells. Methods Mol Biol 767:433–447

Chapter 13

In Vitro Detection of Cellular Adjuvant Properties of Human Invariant Natural Killer T Cells

Rong Zhang, Shuichi Kitayama, Tianyi Liu, Norihiro Ueda, Yumi Tokumitsu, Hiroaki Mashima, Hideki Ohdan, Shin Kaneko, and Yasushi Uemura

Abstract

Invariant natural killer T (iNKT) cells are a subset of T lymphocytes that play a crucial role in the tumor surveillance. The activation of iNKT cells by their specific ligand α-galactosylceramide (α-GalCer) induces the activation of dendritic cells (DCs) via reciprocal interaction, which results in the generation of cellular immunity against cancer. Here we describe a method to detect DC-mediated cellular adjuvant properties of human iNKT cells in vitro.

Key words Invariant natural killer T cells, Dendritic cells, T lymphocytes, Cellular adjuvant, Cancer immunotherapy

1 Introduction

Invariant NKT (iNKT) cells are characterized by the expression of a semi-invariant T cell receptor (TCR) composed of a canonical invariant TCR α-chain (Vα14-Jα18 for mice and Vα24-Jα18 for humans) and a TCR β-chain using limited Vβ segments (Vβ8.2, Vβ7, and Vβ2 in mice and a Vβ11 in humans) [1, 2]. Using this TCR, they recognize a limited number of glycosphingolipids presented by CD1d on antigen-presenting cells. The in vivo activation of iNKT cells with α-galactosylceramide (α-GalCer), a synthetic glycosphingolipid originally derived from a marine sponge, leads to the activation of dendritic cells (DCs) through a CD40 ligand (CD40L)-CD40 interaction, followed by the induction of cellular immunity [3, 4]. This early event and the massive release of

Rong Zhang and Shuichi Kitayama are co-first authors, and Shin Kaneko and Yasushi Uemura are co-senior authors.

Shin Kaneko (ed.), *In Vitro Differentiation of T-Cells: Methods and Protocols*, Methods in Molecular Biology, vol. 2048, https://doi.org/10.1007/978-1-4939-9728-2_13,

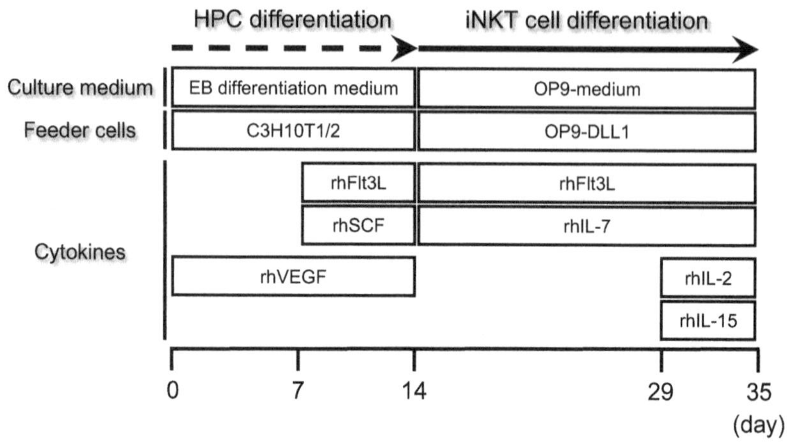

Fig. 1 Schematic representation of the differentiation protocol of re-iNKT cells. The addition of rhIL-2 and rhIL-15 at the final stage of T cell differentiation is quite different from that of conventional T cell differentiation

cytokines underlie the powerful adjuvant properties for antitumor immune responses [5–10].

Because of its important role in shaping of the immune response, the cross talk between iNKT cells and DCs has been a field of extensive study [11–15]. As there is few in vivo models that allow for analyzing the generation of tumor antigen-specific T cell responses via human iNKT cell/DC interaction, the development of efficient methods has been expected. In vitro serial co-culture method devised by us allows for detection of not only the DC activation but also the subsequent $CD8^+$ T cell activation. This system enabled us to detect the enhanced expression of CD86 and the IL-12p70 production in α-GalCer-DCs that were co-cultured with induced pluripotent stem cell (iPSC)-derived redifferentiated iNKT (re-iNKT) cells (Figs. 1 and 2) [16–18]. Moreover, the addition of a tumor antigen-derived peptide in the co-culture efficiently stimulated $CD8^+$ T cells and which could be detected as HLA tetramer-positive cells by flow cytometry (Figs. 3 and 4).

2 Materials

All the materials added in the cell culture should be low endotoxin grade and preservative-free.

2.1 Cells

1. Peripheral blood mononuclear cells (PBMCs), obtained from peripheral blood of healthy donors.
2. Human iNKT cells.
3. C3H10T1/2.

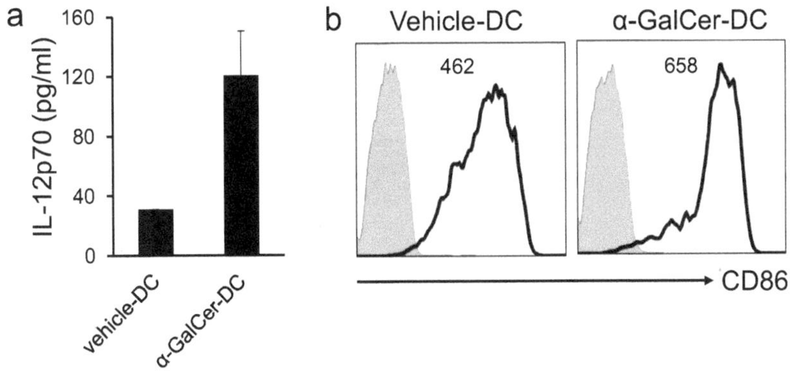

Fig. 2 Detection of DC activation (**a**) IL-12p70 production by DCs Vehicle- or α-GalCer-DCs were co-cultured with re-iNKT cells for 16 h. Levels of IL-12p70 in the culture supernatant were evaluated with ELISA. (**b**) Staining histograms of the surface CD86 (white) and isotype-matched controls (gray) are shown Vehicle- or α-GalCer-DCs were co-cultured with re-iNKT cells for 16 h. CD86 expression of DCs was analyzed by flow cytometry. Filled histogram: isotype control. Open histogram: anti-CD86

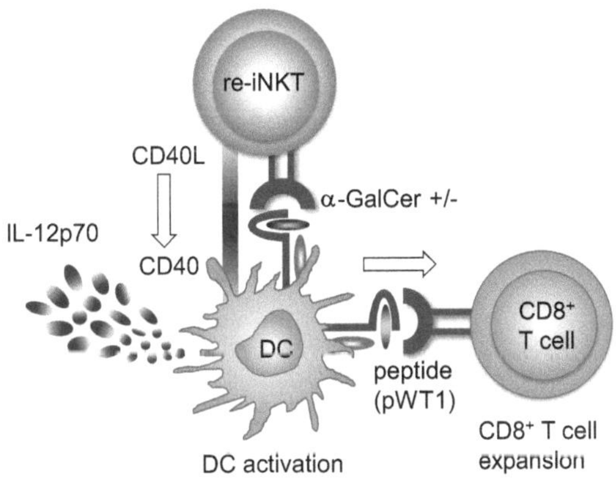

Fig. 3 Schematic representation of the antigen-specific T cell expansion assay. iNKT cells and DCs ± α-GalCer were initially cultured for 12 h to activate the DCs, after which the DCs and iNKT cells were irradiated and cultured with autologous $CD8^+$ T cells in the presence of antigenic peptide (HLA-A24-restricted WT1 peptide)

4. OP9-DLL1 (Delta 1-expressing OP9).
5. Redifferentiated iNKT cells from iPSCs (re-iNKT cells).
6. Human monocyte-derived DCs.
7. Human $CD8^+$ T cells.

2.2 Buffer

1. Dulbecco's Phosphate-Buffered Saline with neither calcium nor magnesium (D-PBS (−)).
2. Wash buffer for flow cytometry:
   D-PBS (−) containing 2% fetal bovine serum (FBS).

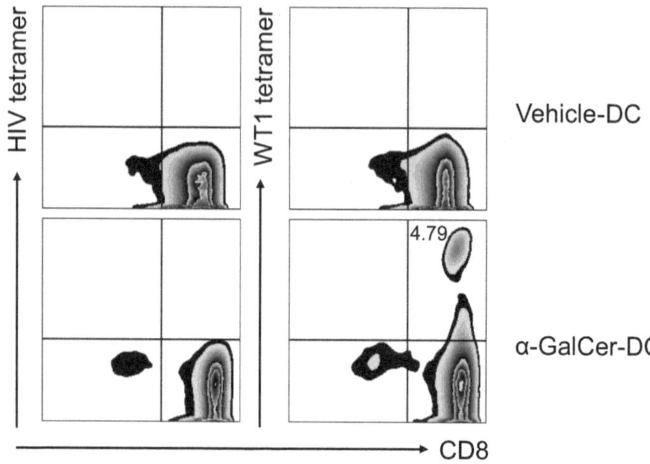

Fig. 4 $CD8^+$ T cell expansion stimulated by re-iNKT cell/α-GalCer-DC interaction. Frequency of WT1/HLA-A24 tetramer-positive $CD8^+$ T cells at day 10 was analyzed by flow cytometry. HIV-env/HLA-A24 tetramer was used as a control

2.3 Reagent for Culture Medium

1. Heat-inactivated (at 56 °C, 30 min) human serum (HS).
2. Fetal bovine serum (FBS).
3. Roswell Park Memorial Institute 1640 medium (RPMI 1640).
4. Iscove's Modified Dulbecco's Medium (IMDM).
5. Alpha Modified Eagle Minimum Essential Medium (α-MEM).
6. Penicillin G (final concentration, 100 units/mL).
7. Streptomycin (final concentration, 100 μg/mL).
8. L-Glutamine (final concentration, 2 mM).
9. Human insulin (final concentration, 10 μg/mL).
10. Human transferrin (final concentration, 5.5 μg/mL).
11. Sodium selenite (final concentration, 5 ng/mL).
12. Monothioglycerol (0.45 mM).
13. Ascorbic acid (50 μg/mL).
14. EB differentiation medium:

    IMDM supplemented with 15% FBS, 10 μg/ML human insulin, 5.5 μg/mL human transferrin, 5 ng/mL sodium selenite, 2 mM L-glutamine, 0.45 mM monothioglycerol, and 50 ng/mL ascorbic acid.
15. OP9 medium:

    α-MEM supplemented with 15% FBS, 2 mM L-glutamine, 100 U/mL penicillin, and 100 μg/mL streptomycin.

2.4 Cytokines

1. Recombinant human (rh)-interleukin (IL)-2.
2. rh-IL-4.
3. rhIL-7.
4. rh-IL-15.
5. rh-granulocyte-macrophage colony-stimulating factor (rhGM-CSF).
6. rh-FMS-like tyrosine kinase 3 ligand (rhFlt3L).
7. rh-stem cell factor (rhSCF).
8. rh-vascular endothelial growth factor (rhVEGF).

2.5 Antigens

1. α-Galactosylceramide (α-GalCer, KRN7000).
2. Human leukocyte antigen (HLA) class I-restricted 8- or 9-mer synthetic peptide (> 90% purity).

2.6 Antibodies, HLA Tetramer, and Reagents

1. Magnetic bead-conjugated anti-human CD14 monoclonal antibody (mAb) (for positive magnetic cell sorting) (Miltenyi Biotec; Bergisch Gladbach, Germany).
2. $CD8^+$ T cell isolation kit (for negative magnetic cell sorting) (Miltenyi Biotec).
3. Anti-human CD3 mAb (low endotoxin/azide-free, agonistic antibody such as clone HIT3a and OKT3).
4. Fluorescence-conjugated anti-human Vα24 mAb (for flow cytometry).
5. Fluorescence-conjugated anti-human invariant NKT cell mAb (specific for invariant Vα24JαQ complementary determining region (CDR) 3) (for flow cytometry).
6. Fluorescence-conjugated anti-human CD86 mAb (for flow cytometry).
7. Fluorescence-conjugated anti-human CD8 mAb (clone T8) (for tetramer staining).
8. Human γ-globulin (2.5 mg/mL dissolved with D-PBS (−)).
9. Fluorescence-conjugated HLA class I/peptide complex tetramer (detection tetramer combined with a relevant peptide and control tetramer combined with an irrelevant peptide).
10. 7-Aminoactinomycin D (7-AAD) or propidium iodide (PI) for detection of dead cells.
11. Ficoll-Paque (used to separate blood to mononuclear cells).

2.7 Culture Plates

1. 6-well tissue culture plate treated by vacuum gas plasma.
2. 24-well tissue culture plate treated by vacuum gas plasma.

3 Methods

3.1 Generation of Human iNKT Cells

1. PBMCs obtained from peripheral blood by Ficoll-Paque density gradient centrifugation are seeded on 24-well plate at 1.0×10^7 cells/mL/well in RPMI1640 supplemented with 5% HS, penicillin G (100 units/mL), streptomycin (100 μg/mL), L-glutamine (2 mM), and α-GalCer (10 ng/mL).
2. At day 13, wells of new 24-well tissue culture plate are pre-coated with anti-CD3 mAb (10 μg/mL D-PBS) overnight at 4 °C.
3. After expansion for 14 days, the cells are blocked Fc receptor (FcR) by incubating with human γ-globulin and stained with anti-human Vα24 mAb and anti-human invariant NKT cell (clone 6B11) mAb.
4. $V\alpha24^{+}6B11^{+}$ cells are isolated with an automated cell sorter.
5. The isolated cells are stimulated with plate-bound anti-CD3 mAb (Subheading 3.1, **step 2**) in the presence of rhIL-2 (100 U/mL), rhIL-15 (10 ng/mL), and irradiated (4500 cGy) autologous PBMCs seeded in 24-well plate (5.0×10^6 cells/well) (*see* **Notes 1** and **2**).
6. After expansion for 14 days, iNKT cells ($1.0\text{~}2.0 \times 10^6$ cells/well) are stimulated with irradiated allogeneic PBMCs (5.0×10^6 cells/well) pre-pulsed for 5 h with α-GalCer (10 ng/mL) in the presence of rhIL-2 (100 U/mL) and rhIL-15 (10 ng/mL) (*see* **Notes 1** and **2**).
7. After expansion for 9~10 days, resting state iNKT cells can be used for co-culture.

3.2 Generation of Human re-iNKT Cells

The differentiation of iNKT cells from iNKT cell-derived iPSCs is basically same as conventional T cell differentiation method described in our previous reports [16, 17].

1. Clumps of iPSC are transferred onto C3H10T1/2 feeder cells and cultured in EB medium containing rhVEGF (20 ng/mL) (Fig. 1).
2. On day 7, rhSCF and rhFlt3L (50 ng/mL each) are added to the culture.
3. On day 14, hematopoietic progenitor cells are collected and transferred onto OP9-DLL1 cells and co-cultured in OP9 medium in the presence of rhIL-7 (1 ng/mL) and rhFlt3L (10 ng/mL).
4. On day 29, rhIL-2 (100 U/mL) and rhIL-15 (10 ng/mL) are added to the culture.
5. On day 35, re-iNKT cells can be recovered and stimulated for expansion same as iNKT cells (Subheading 3.1, **step 6**) (*see* **Notes 1** and **2**).

3.3 Generation of Immature Dendritic Cells

1. $CD14^+$ monocytes are isolated from PBMCs by positive magnetic cell sorting.
2. $CD14^+$ monocytes are cultured in 6-well plate at 2.0×10^6 cells/2 mL/well in RPMI1640 supplemented with 5% HS in the presence of rhGM-CSF and rhIL-4 (50 ng/mL each) (*see* **Note 3**).
3. On days 2 and 3, the DC cultures receive an additional dose of rhGM-CSF and rhIL-4 (50 ng/mL each) (*see* **Note 4**).
4. On day 6, nonadherent DCs are harvested and used as immature DCs (*see* **Note 5**).

3.4 Analysis of DC Activation

1. Vehicle or α-GalCer is added to the immature DC culture (Subheading 3.3, **step 4**) and cultured for 5 h.
2. Vehicle or α-GalCer pre-pulsed (5 h) immature DCs are recovered and washed twice with RPMI1640 supplemented with 5% HS without containing cytokines (*see* **Note 5**).
3. Vehicle or α-GalCer pre-pulsed immature DCs (Subheading 3.4, **step 2**) are seeded on 24-well plate at 5.0×10^5 cells/500 μL/well in RPMI1640 supplemented with 5% HS without containing cytokines (*see* **Note 5**).
4. iNKT cells (5.0×10^4 cells/500 μL in RPMI1640 supplemented with 5% HS/well) are added to the well of DC culture (Subheading 3.4, **step 3**) and cultured for 16 h (*see* **Note 6**).
5. Culture supernatants are harvested and stored at −80 °C.
6. The levels of IL-12p70 are measured using enzyme-linked immunosorbent assay (ELISA) (Fig. 2a).
7. After harvesting the culture supernatants, the cells are recovered and blocked FcR with human γ-globulin for 5 min.
8. The cells are stained with anti-human Vα24 mAb, anti-CD86 mAb, and 7-AAD.
9. After Ab staining, the cells are washed twice with D-PBS (−) containing 2% FBS.
10. Levels of CD86 expression of 7-AAD- and Vα24-negative cells are analyzed by flow cytometry (Fig. 2b).

3.5 Detection of Antigen-Specific $CD8^+$ T Cell Expansion

1. Vehicle or α-GalCer (5 h) pre-pulsed immature DCs are seeded on 24-well plate at 5.0×10^5 cells/500 μL RPMI1640 supplemented with 5% HS without cytokines/well.
2. iNKT cells (5.0×10^4 cells/500 μL RPMI1640 supplemented with 5% HS) are added to the DC culture (Subheading 3.5, **step 1**) and cultured for 12 h (*see* **Note 6**).
3. $CD8^+$ T cells are isolated from PBMCs by negative magnetic cell sorting (*see* **Note 7**).

4. After 12 h of iNKT cell/DC co-culture (Subheading 3.5, **step 2**), differentially conditioned DCs are irradiated (3000 cGy) and the $CD8^+$ T cells (2.0×10^6 cells/500 μL RPMI1640 supplemented with 5% HS) are added to the co-culture in the presence of antigenic peptide (5 μM) (e.g., HLA-A24-restricted Wilms tumor 1 (WT1) peptide) (Fig. 3).
5. After 10 days of culture, the cells are recovered, FcR blocked, and stained with HLA class I tetramer (e.g., HLA-A24-WT1 tetramer) followed by staining with anti-human CD8 mAb, anti-human Vα24 mAb, and 7-AAD.
6. After staining, the cells are washed twice with D-PBS (−) containing 2% FBS.
7. The frequencies of CD8 and tetramer double-positive cells are determined by flow cytometry (Fig. 4). The presence of dead cells and Vα24 iNKT cells was excluded by running parallel 7-AAD- and Vα24-stained samples.

4 Notes

1. The iNKT cell culture is kept as it is at least for 2~3 days after stimulation.
2. The iNKT cell expansion can be confirmed by staining with anti-human Vα24 mAb, anti-human Vβ11 mAB, or anti-human invariant NKT cell mAb (clone 6B11).
3. DCs are sensitive to microbial products such as endotoxin. It is necessary to avoid their exposure to microbial products.
4. The robust treatment of DCs induces DC activation characterized by the enhanced CD86 expression. Therefore, it should be avoided pipetting as much as possible on adding cytokines.
5. The robust pipetting of DCs induces DC activation. Therefore, DCs are gently treated on recovery of cells.
6. It is important to use resting state iNKT cells. Activated iNKT cells express high levels of CD40 ligand and induce α-GalCer-independent activation of DCs.
7. DCs and $CD8^+$ T cells used in the antigen-specific $CD8^+$ T cell expansion analysis need to be obtained from the same donor from whom the iNKT cells are established in order to avoid alloreactive responses.

Acknowledgments

This study was supported by grants from the Grants-in-Aid 17K10717 and 17K10806 from the Ministry of Education, Culture, Sports, Science and Technology (MEXT) of Japan; the

National Cancer Center Research and Development Fund (28-A-8); and the Practical Research for Innovative Cancer Control from Japan Agency for Medical Research and Development (AMED 15ck0196002h0103). We thank Mrs. M. Ozaki and K. Kobayashi (National Cancer Center) for secretarial support.

References

1. Bendelac A, Savage PB, Teyton L (2007) The biology of NKT cells. Annu Rev Immunol 25:297–336. https://doi.org/10.1146/annurev.immunol.25.022106.141711
2. Molling JW, Moreno M, van der Vliet HJ, van den Eertwegh AJ, Scheper RJ, von Blomberg BM, Bontkes HJ (2008) Invariant natural killer T cells and immunotherapy of cancer. Clin Immunol 129:182–194. https://doi.org/10.1016/j.clim.2008.07.025
3. Fujii S, Shimizu K, Kronenberg M, Steinman RM (2002) Prolonged IFN-gamma-producing NKT response induced with alpha-galactosylceramide-loaded DCs. Nat Immunol 3(9):867–874. https://doi.org/10.1038/ni827
4. Kitamura H, Iwakabe K, Yahata T, Nishimura S, Ohta A, Ohmi Y, Sato M, Takeda K, Okumura K, Van Kaer L, Kawano T, Taniguchi M, Nishimura T (1999) The natural killer T (NKT) cell ligand alpha-galactosylceramide demonstrates its immunopotentiating effect by inducing interleukin (IL)-12 production by dendritic cells and IL-12 receptor expression on NKT cells. J Exp Med 189(7):1121–1128
5. Motohashi S, Nakayama T (2009) Invariant natural killer T cell-based immunotherapy for cancer. Immunotherapy 1(1):73–82. https://doi.org/10.2217/1750743x.1.1.73
6. Fujii SI, Shimizu K (2017) Exploiting antitumor immunotherapeutic novel strategies by deciphering the cross talk between invariant NKT cells and dendritic cells. Front Immunol 8:886. https://doi.org/10.3389/fimmu.2017.00886
7. Liu H, Chen L, Liu J, Meng H, Zhang R, Ma L, Wu L, Yu S, Shi F, Li Y, Zhang L, Wang L, Feng S, Zhang Q, Peng Y, Wu Q, Liu C, Chang X, Yang L, Uemura Y, Yu X, Liu T (2017) Co-delivery of tumor-derived exosomes with alpha-galactosylceramide on dendritic cell-based immunotherapy for glioblastoma. Cancer Lett 411:182–190. https://doi.org/10.1016/j.canlet.2017.09.022
8. Nair S, Dhodapkar MV (2017) Natural killer T cells in Cancer immunotherapy. Front Immunol 8:1178. https://doi.org/10.3389/fimmu.2017.01178
9. Krijgsman D, Hokland M, Kuppen PJK (2018) The role of natural killer T cells in cancer—a phenotypical and functional approach. Front Immunol 9:367. https://doi.org/10.3389/fimmu.2018.00367
10. Gasser O, Sharples KJ, Barrow C, Williams GM, Bauer E, Wood CE, Mester B, Dzhelali M, Caygill G, Jones J, Hayman CM, Hinder VA, Macapagal J, McCusker M, Weinkove R, Painter GF, Brimble MA, Findlay MP, Dunbar PR, Hermans IF (2018) A phase I vaccination study with dendritic cells loaded with NY-ESO-1 and alpha-galactosylceramide: induction of polyfunctional T cells in high-risk melanoma patients. Cancer Immunol Immunother 67(2):285–298. https://doi.org/10.1007/s00262-017-2085-9
11. Lameris R, Schneiders FL, de Gruijl TD, van der Vliet HJ (2014) Exploiting the CD1d-iNKT cell axis for potentiation of DC-based cancer vaccines. Methods Mol Biol 1139:155–165. https://doi.org/10.1007/978-1-4939-0345-0_14
12. Gottschalk C, Mettke E, Kurts C (2015) The role of invariant natural killer T cells in dendritic cell licensing, cross-priming, and memory CD8(+) T cell generation. Front Immunol 6:379. https://doi.org/10.3389/fimmu.2015.00379
13. Keller CW, Freigang S, Lunemann JD (2017) Reciprocal crosstalk between dendritic cells and natural killer T cells: mechanisms and therapeutic potential. Front Immunol 8:570. https://doi.org/10.3389/fimmu.2017.00570
14. Liu TY, Uemura Y, Suzuki M, Narita Y, Hirata S, Ohyama H, Ishihara O, Matsushita S (2008) Distinct subsets of human invariant NKT cells differentially regulate T helper responses via dendritic cells. Eur J Immunol 38(4):1012–1023. https://doi.org/10.1002/eji.200737838
15. Uemura Y, Liu TY, Narita Y, Suzuki M, Nakatsuka R, Araki T, Matsumoto M, Iwai LK, Hirosawa N, Matsuoka Y, Murakami M, Kimura T, Hase M, Kohno H, Sasaki Y,

Ichihara Y, Ishihara O, Kikuchi H, Sakamoto Y, Jiao SC, Senju S, Sonoda Y (2009) Cytokine-dependent modification of IL-12p70 and IL-23 balance in dendritic cells by ligand activation of Valpha24 invariant NKT cells. J Immunol 183(1):201–208. https://doi.org/10.4049/jimmunol.0900873

16. Nishimura T, Kaneko S, Kawana-Tachikawa A, Tajima Y, Goto H, Zhu D, Nakayama-Hosoya-K, Iriguchi S, Uemura Y, Shimizu T, Takayama N, Yamada D, Nishimura K, Ohtaka M, Watanabe N, Takahashi S, Iwamoto A, Koseki H, Nakanishi M, Eto K, Nakauchi H (2013) Generation of rejuvenated antigen-specific T cells by reprogramming to pluripotency and redifferentiation. Cell Stem Cell 12(1):114–126. https://doi.org/10.1016/j.stem.2012.11.002

17. Kitayama S, Zhang R, Liu TY, Ueda N, Iriguchi S, Yasui Y, Kawai Y, Tatsumi M, Hirai N, Mizoro Y, Iwama T, Watanabe A, Nakanishi M, Kuzushima K, Uemura Y, Kaneko S (2016) Cellular adjuvant properties, direct cytotoxicity of re-differentiated Valpha24 invariant NKT-like cells from human induced pluripotent stem cells. Stem cell reports 6(2):213–227. https://doi.org/10.1016/j.stemcr.2016.01.005

18. Yamada D, Iyoda T, Vizcardo R, Shimizu K, Sato Y, Endo TA, Kitahara G, Okoshi M, Kobayashi M, Sakurai M, Ohara O, Taniguchi M, Koseki H, Fujii SI (2016) Efficient regeneration of human Valpha24(+) invariant natural killer T cells and their anti-tumor activity in vivo. Stem Cells 34(12):2852–2860. https://doi.org/10.1002/stem.2465

Chapter 14

In Vitro Differentiation of T Cells from Murine Pluripotent Stem Cells

Xiaoniao Chen, Fengyang Lei, Liqiang Wang, Xiaofang Xiong, and Jianxun Song

Abstract

In recent years cancer immunotherapy, especially the cell-based immunotherapy, has reached several milestones and achieved a lot of cancer remission in the clinics. Obtaining a more potent and effective cytotoxic T lymphocytes (CTLs) for cancer immunotherapy is always the ultimate goal for the researchers. However, the difficulty in harvesting a large number of tumor antigen-specific CTLs from the tumor patient is still a major obstacle we need to overcome. In our previous studies, it is shown that pluripotent stem cell-derived CTL—especially the genetically engineered antigen-specific CTLs—may serve as a good source of unlimited number of highly reactive and antigen-specific CTLs. Here we present a two-step method for the generation of antigen-specific T lymphocytes from iPS cells by in vitro priming and in vivo maturation.

Key words Stem cell biology, T cells, Induced pluripotent stem cells, Differentiation, Notch signaling

1 Introduction

T cell-based cancer immunotherapy has achieved a lot of promising results in recent years [1]. Cytotoxic T lymphocytes (CTLs), especially the tumor antigen-specific CTLs, can recognize malignant tumor cells by precisely interacting tumor antigens with the T cell receptors (TCR) and release cytotoxins (perforin and granzyme B) as well as cytokines (IFN-γ, TNF-α) to kill malignant cells. It is known that a less-differentiated and central-memory-like CTL is the desiring arsenal for T cell-based immunotherapy, because less-differentiated CTLs have a higher potential for proliferation, are less susceptible to apoptosis than terminally differentiated CTL, and have a more sensitive response to homeostatic cytokines [2–6]. However, it is impossible to obtain this rare population of CTLs from patients, especially the tumor patients whose immune

Xiaoniao Chen and Fengyang Lei contributed equally to this work.

Shin Kaneko (ed.), *In Vitro Differentiation of T-Cells: Methods and Protocols*, Methods in Molecular Biology, vol. 2048, https://doi.org/10.1007/978-1-4939-9728-2_14, © Springer Science+Business Media, LLC, part of Springer Nature 2019

system is significantly suppressed. We have taken a different approach to generate this highly reactive and antigen (Ag)-specific CTLs for the T cell cancer immunotherapy.

Overexpressing a unique TCR to the self-renewable stem cells for immune reconstitution builds up a therapeutic potential for treating a variety of diseases [7–9]. However, there is another impossibility to obtain embryonic tissue from tumor patients. Although hematopoietic stem cells (HSCs) are good candidates for therapeutic purposes, which have been widely applied in clinic [10–12], HSCs have reduced differentiation and proliferative capacities following ex vivo isolation, and HSCs are difficult to expand in in vitro [13, 14]. Recently the introduction of iPS cell technology has given us a third path to obtain pluripotent cells that could be applied for many therapeutic purposes. Furthermore, iPS bears an indefinite capability of in vitro proliferation and directed differentiation under certain conditions. Therefore, we believe that iPS cells have a greater potential to be used in the cellular immunotherapy for cancer.

Here, we present a two-step method for the generation of antigen-specific T lymphocytes from iPS cells by in vitro priming and in vivo maturation. In vitro stimulation with a Notch ligand Delta-1 drives early stage T cell differentiation from iPS cells, and TCR gene transduction results in iPS cells maturating into antigen-specific T cells in vivo, which shows a strong antigen reactivity. In general, we demonstrate antigen-specific CTL is able to differentiate from iPS cell. Our studies provide a potentially efficient and feasible approach for generating antigen-specific CTLs for T cell-based cancer immunotherapies and facilitate the development of therapeutic strategies for diseases [15–17].

2 Materials

2.1 Mice/Cell Lines

1. iPS-MEF-Ng-20D-17 cell line is kind gift from Dr. Shinya Yamanaka at Kyoto University [18]. It was generated from mouse embryonic fibroblasts by retroviral transduction of a combination of *Oct3/4*, *Sox2*, *Klf4*, and *c-Myc* genes. iPS cells are routinely cocultured with irradiated SNL76/7 feeder cells supplemented with 15% FBS-DMEM medium.
2. OP9 and OP9-DL1 cell lines are generous gifts from Dr. Juan Carlos Zuniga-Pflucker at the University of Toronto [19]. OP9 and OP9-DL1 cell lines are routinely maintained in 20% FBS-α-MEM medium.
3. SNL76/7 cell line (ATCC) is maintained in 10% FBS-DMEM medium. Prior to coculture with iPS cells, SNL76/7 cells will undergo a 5000 Rads irradiation in the ^{60}Co irradiator, and the irradiated cells are designated irSNL76/7.

4. Mice: 6–8 weeks of age female C57/BL6J and B6.127S7-*Rag1*^{tm1Mom}/J mice are purchased from the Jackson Laboratory.
5. Platinum-E (Plat-E) retroviral packaging cells (*see* **Note 4**).

2.2 Cell Culture

1. 6- and 24-well culture plate (BD).
2. 100 mm culture dish (BD).
3. 70 μm cell strainer (BD).
4. 1 mL and 10 mL syringes (BD).
5. 27 G$^{1/2}$ and 18 G1$^{1/2}$ needles (BD).
6. Various plastic pipettes (5 mL, 10 mL, 25 mL, 50 mL; BD).
7. 0.22 μm bottle top filter (Corning).
8. 0.4 μm filter (Millipore).
9. Dulbecco's Modified Eagle Medium (DMEM) (Invitrogen).
10. α-Minimal Essential Medium (α-MEM) (Invitrogen): with one pack of α-MEM powder, 2.2 g $NaHCO_3$ and 1 L of $_{dd}H_2O$ are added. Filtration of prepared medium through a 0.22 μm bottle top filter is mandatory for sterilization. FBS, antibiotics, and cytokines are added into the medium accordingly.
11. Phosphate-buffered saline (PBS): 137 mmol/L NaCl, 2.7 mmol/L KCl, 10 mmol/L $Na_2HPO_4{\cdot}2H_2O$, 2 mmol/L KH_2PO_4, pH adjusted to 7.4, and autoclaved before use.
12. 100× penicillin and streptomycin mix (10,000 U/mL, Invitrogen).
13. Fetal bovine serum (FBS, heat-inactivated, HyClone).
14. Flt-3 ligand (Flt-3 L, PeproTech).
15. Interleukin 7 (IL-7, PeproTech).
16. Gelatin (Sigma-Aldrich).
17. Brefeldin A (Invitrogen).
18. Fc blocker 24G2 (BD).
19. Permeabilizing kit (BioLegend).

2.3 Retroviral Transduction

1. GeneJammer transfection reagent (Stratagene).
2. 5 μg/mL polybrene (Sigma-Aldrich).
3. FACSAria SORP cell sorter (BD).

2.4 Peptide Stimulation

1. $CD8^+$ MACS separation system (Miltenyi Biotec).
2. $OVA_{257–264}$ peptide (GenScript).

2.5 PCR Assay

1. DNeasy Blood and Tissue DNA isolation kit (Qiagen).
2. PCR Master Mix Kit (Qiagen).

3. *OT-I* forward primer: 5′-ACGTGTATTCCCATCTCTGGA CAT-3′.
4. *OT-I* reserve primer: 5′-TGTTCATAATTGGCCCGAGAG CTG-3′.

2.6 Antibodies

1. CD4: GK1.5 (BioLegend).
2. CD8: 6A242 (Santa Cruz).
3. TCR Vβ5: MR9–4 (BioLegend).
4. TCR Vβ: H57–597 (BioLegend).
5. CD3: 17A2 (BioLegend).
6. CD45: 30-F11 (BioLegend).
7. CD117: 2B8 (BioLegend).
8. IL-2: JES6-5H4 (BD Pharmingen).
9. IFN-γ: XMG1.2 (BD Pharmingen).

3 Methods

3.1 Generation of Retroviral Vector

1. MSCV-IRES-DsRED (MiDR) vector is designed and constructed based on MSCV-IRES-GFP vector by replacing the GFP module with a DsRED reporter module.
2. OT-I T cell receptor Vα2 chain and Vβ5 chain are individually subcloned and linked by a 2A sequence before inserting into the MiDR vector to make OT-I/MiDR construct.

3.2 Retroviral Transduction and Cell Sorting

1. Plat-E packaging cells are used to generate pseudovirus that will be used for the transduction. On Day −3, 3×10^6 Plat-E cells are seeded on a 100 mm petri dish one day prior to transfection.
2. On Day −2, Plat-E cells will be transfected with OT-I MiDR vector by using GeneJammer transfection reagent.
3. On Day −1, 1×10^6 iPS cells will be seeded into one well of a 0.1% gelatin precoated 24-well plate (*see* **Note 1**).
4. Transduction will be performed by spinning at $1000 \times g$ for 1 h at 32 °C in the presence of 5 μg/mL polybrene on Day 0.
5. After transduction, plates will be placed in 32 °C, 5% CO_2 incubator overnight to further promote the retroviral integration.
6. On Day 1, repeat the transduction as described above. A 6-well plate will be precoated with irSNL76/7 feeder cells for future use.

7. On day 2, transduced iPS cells will be harvested by trypsin, centrifuged at 800 rpm for 5 min, and seeded on the precoated irSNL76/7 feeder cells.
8. At confluency, cells will be harvested, centrifuged at 800 rpm for 5 min, and processed for cell sorting. GFP and DsRED double positive cells will be sorted by FACSAria cell sorter. Sorted cells will be continuously cultured on irSNL76/7 feeder cells for future use.

3.3 In Vitro Priming with OP9-DL1 Cells

1. On Day 0, 5×10^4 iPS cells will be seeded on a 100 mm culture dish containing confluent OP9-DL1 cell monolayer in 20% FBS α-MEM media (*see* **Note 2**).
2. On Day 3, change culture medium.
3. On Day 5, cells will be harvested and centrifuged at 800 rpm for 5 min before incubating on a fresh 100 mm culture dish for 30 min in 37 °C incubator (*see* **Note 3**). Floating cells will be collected and counted; 5×10^5 cells will be transferred to a fresh culture dish containing confluent OP9-DL1 cell monolayer in 20% FBS α-MEM media. Cytokine mFlt-3 L (final concentration, 5 ng/mL) will be added in the culture.
4. On Day 8, loosely attached cells will be gently pipette off (*see* **Note 4**).
5. After harvesting cells from the coculture, cells will be centrifuged at 800 rpm for 5 min and resuspended in 20% FBS α-MEM media supplemented with mFlt-3 L (5 ng/mL) and mIL-7 (1 ng/mL).
6. Cells will be transferred into a 6-well culture plate coated with confluent OP9-DL1 cells (*see* **Note 5**).
7. From day 10, culture media will be changed in every other day (20% FBS α-MEM media supplemented with mFlt-3 L (5 ng/mL) and mIL-7 (1 ng/mL)).
8. Culture plates coated with feeder OP9-DL1 cells will be changed in 4–6 days depending on the growth of the feeder cells.

3.4 In Vitro Analysis

1. Morphological changes of differentiated cells and cell recovery rates (Fig. 1).
2. Gene integration and expression are analyzed by both PCR and RT-PCR analyses.
3. At different days of coculture with OP9-DL1 cells, live cell images are taken by phase-contrast microscope.
4. Cell recovery rates are calculated based on the number of cells that harvested from culture (Fig. 2a).

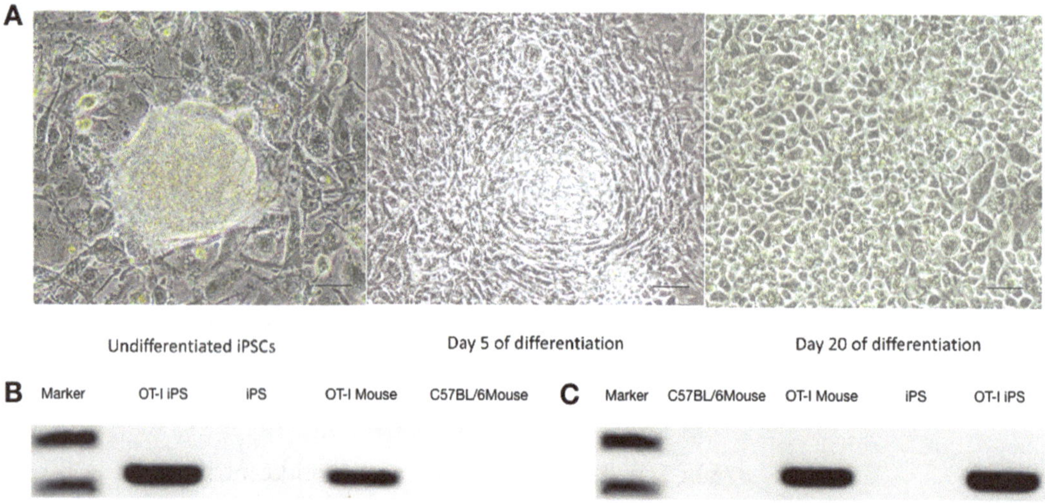

Fig. 1 Morphology of iPS cell under differentiation. (**a**) At various days, mouse iPS cells are imaged (scale bars, 50 μm). (**b**) RT-PCR and (**c**) PCR characterization of OT-I TCRVβ5 gene integration and expression

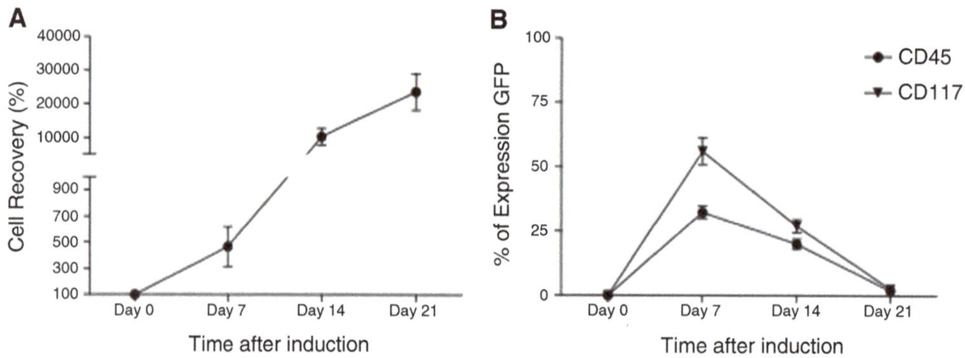

Fig. 2 iPS cell differentiation in the presence of Notch signaling. (**a**) Cell recovery of iPS cell-derived cells. Numbers of iPS cell-derived cells present on day 0 are assigned a value of 100%, and numbers of cells on days 7, 14, and 21 are used to calculate the percentage recovery relative to day 0. Data represent the mean ± SEM. (**b**) Changes in hematopoietic marker CD117 and CD45 expression in iPS cells (GFP+). Expression of each marker was determined at the indicated days. Data represent the mean ± SEM

5. At different days of coculture, cells will be removed from culture by trypsinization and washed with cold PBS before proceeding to cell surface staining.
6. Before staining with different fluorochrome-conjugated antibodies, cells are blocked by Fc blocker 24G2 in 4 °C for 20 min.
7. After staining in 4 °C for 20 min, cells will be washed three times in cold PBS before loading on flow cytometer.
8. Flow cytometric analysis of surface marker changes (Fig. 3a).
9. Activation of in vitro differentiated iPS cells (Fig. 3b).

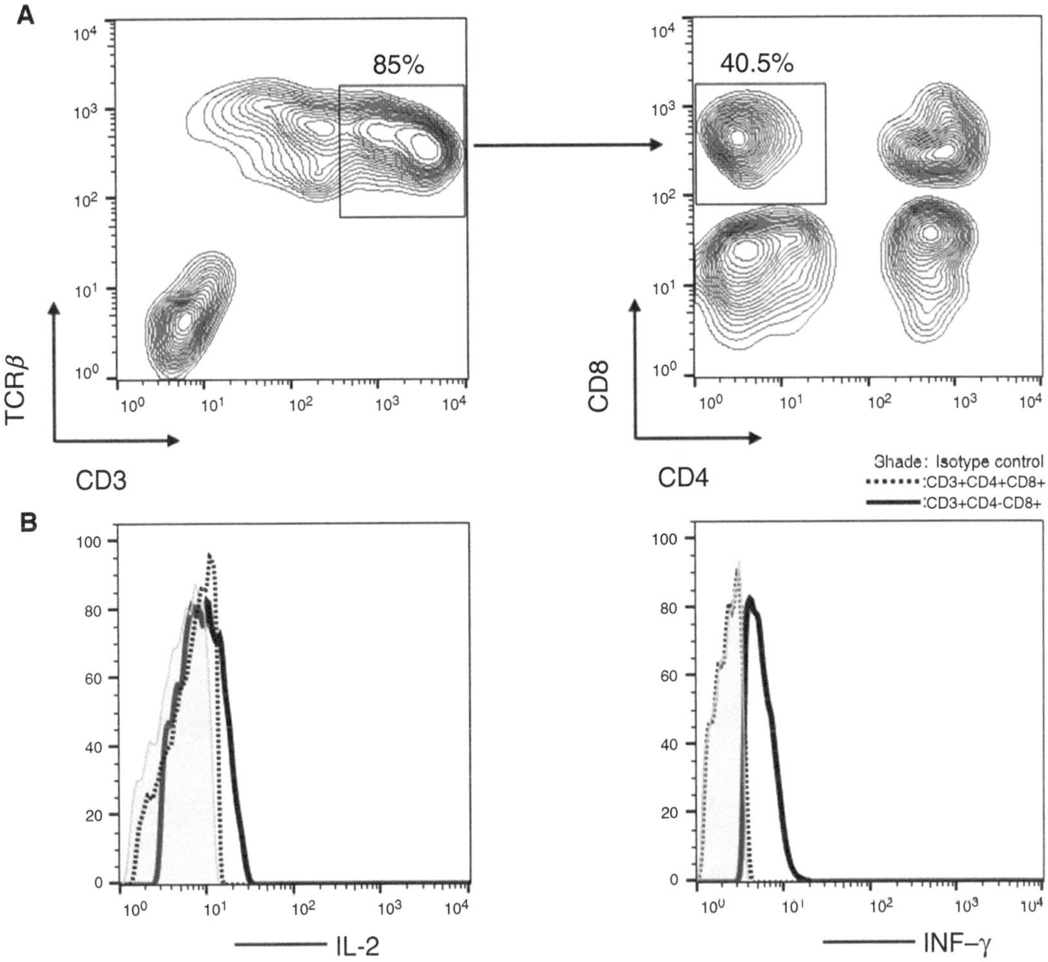

Fig. 3 T cell differentiation from iPS cells. Mouse iPS cells are cocultured with OP9-DL1 cells as described in Fig. 1. On day 22, iPS cell-derived cells were isolated and analyzed. (**a**) CD4+ CD8− or CD4− CD8+ cells after gating on CD3+ and TCRβ+ populations. (**b**) Cells are stimulated with plate-coated anti-CD3 and soluble anti-CD28 antibodies for 40 h at 37 °C at 5% CO_2. IL-2 and IFN-γ are analyzed by intracellular staining, after gating on live CD4−CD8+ T cells (dark lines; shaded areas indicate isotype controls)

10. One day before activation assay, precoat a 24-well plate with anti-CD3 (final concentration, 4 μg/mL, in PBS) at 4 °C overnight.
11. On day 22 of coculture with OP9-DL1 cells, iPS cell-derived T cells are harvested from culture and washed with cold PBS before stimulating with plate-coated anti-CD3 and soluble anti-CD28 antibodies (final concentration, 4 μg/mL).
12. Incubating plates in 37 °C, 5% CO_2 incubator for 40 h and then Brefeldin A is added into culture for another 4 h (*see* **Note 6**).
13. At the end of stimulation, cells are harvested, washed, and blocked by Fc blocker as described above. Prepared cells are

stained for surface markers as CD8 and TCR Vβ chain by using fluorochrome-conjugated antibodies.

14. After surface staining, cells are fixed by using 4% formaldehyde and permeabilized by using BioLegend's Permeabilizing kit.
15. After permeabilization, intracellular molecules IL-2 and IFN-γ are stained by using different fluorochrome-conjugated antibodies.
16. Before final flow cytometric examination, cells are washed three times in cold PBS to exclude excessive antibodies.

3.5 In Vivo Maturation

1. On Day 22 of coculture, partially differentiated iPS cells are trypsinized off, centrifuged at 800 rpm for 5 min, and incubated on a fresh culture dish in 37 °C for 30 min (*see* **Note 7**).
2. Floating cells will be collected, passed through a 70 μm nylon strainer to exclude cell clumps, and washed three times with cold PBS (*see* **Note 8**).
3. Cells will be resuspended maintained in cold PBS with a concentration of 1.5×10^7 cells/mL before injection.
4. 200 μL cell suspension or 3×10^6 cells are adoptively transferred into 4-week-old B6.129S7-$Rag1^{tm1Mom}/J$ mice through the tail vein. Three weeks are allowed for in vivo maturation of partially differentiated iPS cells (*see* **Note 9**).
5. Maturation in $Rag1^{-/-}$ mice. After three weeks of in vivo development, $Rag1^{-/-}$ mice are sacrificed; spleen and lymph nodes are removed from mice (Fig. 4a).
6. Single cells are processed through mechanical breakdown. Red blood cells are lysed by ACK lysis buffer and mononucleocytes are collected and washed twice in cold PBS.
7. After washing, cells are blocked with Fc blocker 24G2 in 4 °C for 20 min and before staining with different fluorochrome-conjugated anti-CD3, anti-CD4, anti-CD8, and anti-TCRβ antibodies in 4 °C for 20 min.
8. At the end of staining, cells are washed three times in cold PBS before flow cytometric analysis.

3.6 Ex Vivo Analysis

1. Three weeks after in vivo development, a large population of CD8+TCRVβ5+ cells were present in the lymphoid organs of $Rag1^{-/-}$ mice received in vitro primed OT I-iPSCs as compared with other controls.
2. Sacrifice the mice and collect lymphoid organs such as spleen and lymph nodes from the mice.
3. Harvest CD8+ T cells by negative selection and stimulate the CD8+ T cells with $OVA_{257–264}$ peptide.

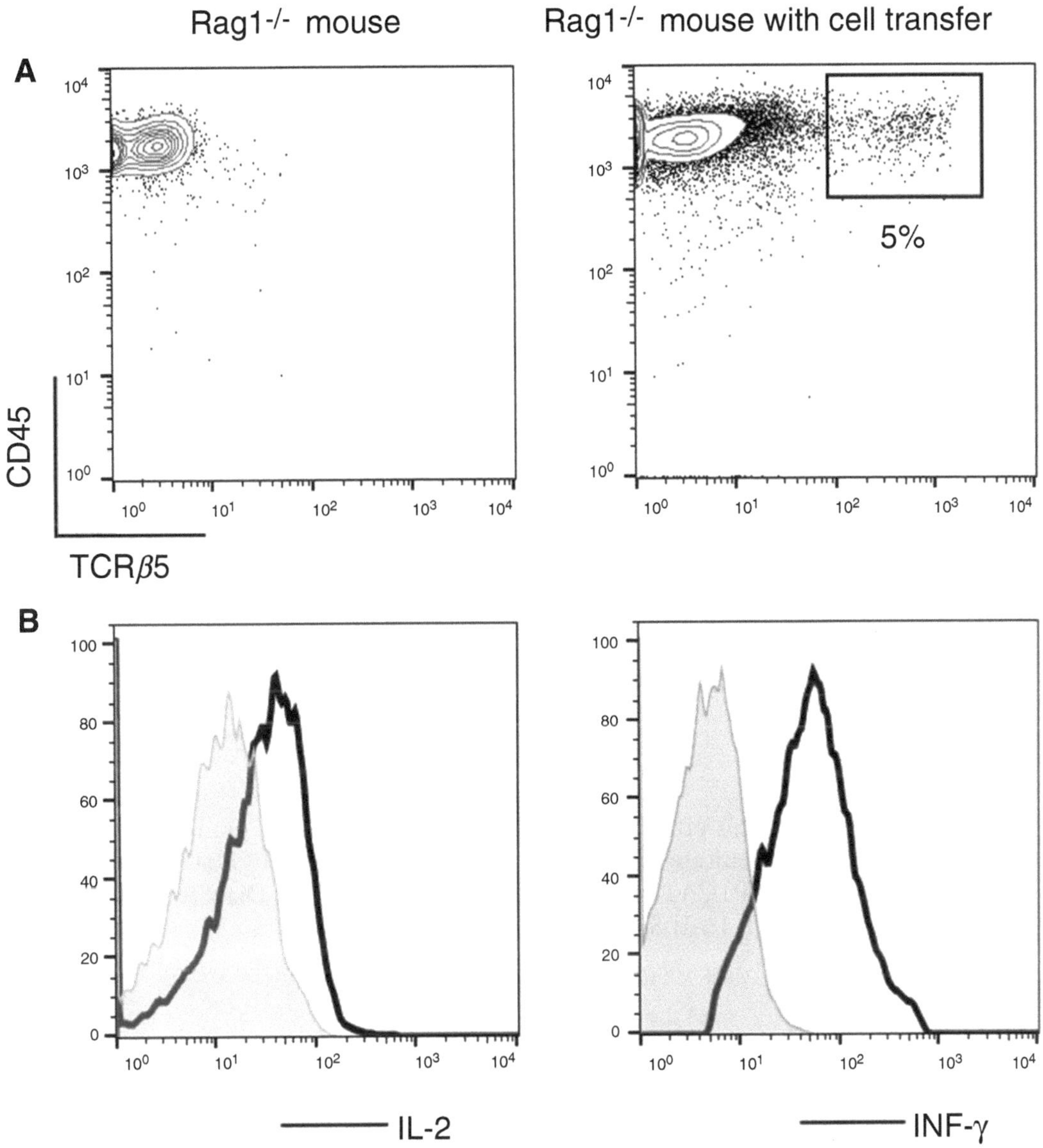

Fig. 4 Antigen-specific CD8+ T cell development from iPS cells matured in Rag1$^{-/-}$ mice. In vitro primed OT-I TCR gene-transduced iPS cells are injected IV into Rag1$^{-/-}$ mice. After 3 weeks, OVA-specific CD8+Vβ5+ T cell development is determined. (**a**) CD8+Vβ5+ T cells from pooled lymph nodes and spleen are analyzed by flow cytometry, after gating on CD8+ populations. (**b**) IL-2 and IFN-γ production (dark lines; shaded areas indicate isotype controls) are determined by intracellular cytokine staining, after gating on the CD8+Vβ5+ populations

4. After stimulation, intracellular staining of IL-2 and IFN-γ is performed as described above.
5. At the end of the staining, wash cells in cold PBS twice and analyze the cytokine profile by flow cytometry (Fig. 4b).

4 Notes

1. Precoat the plate with 0.1% gelatin at 4 °C overnight.
2. To successfully induce the T cell differentiation, a confluent monolayer of OP9-DL1 cells is required.
3. Sitting at 37 °C for 30 min is to remove the excessive OP9-DL1 cells.
4. Wash the OP9-DL1 feeding layer with 10 mL PBS one more time to get the maximal recovery of partially differentiated iPS cells.
5. Usually iPS cells recovered from one 100 mm culture dish will be transferred into one well of the 6-well plate.
6. Brefeldin A is commonly used in intracellular staining to retain the synthesized cytokine molecules.
7. 37 °C, 30 min incubation is to exclude OP9-DL1 cells.
8. Large cell clump will cause pulmonary embolism to mice.
9. Before IV injection, mice will be placed under an infrared light to dilate their tail vein which will facilitate the injection.

Acknowledgments

This project was funded, in part, under the grants from National Institute of Health Grant R01AI121180, R21AI109239, and R01CA221867 and the American Diabetes Association (1-16-IBS-281) to J.S.

References

1. Schuster SJ, Svoboda J, Chong EA, Nasta SD, Mato AR, Anak O, Brogdon JL, Pruteanu-Malinici I, Bhoj V, Landsburg D, Wasik M, Levine BL, Lacey SF, Melenhorst JJ, Porter DL, June CH (2017) Chimeric antigen receptor T cells in refractory B-cell lymphomas. N Engl J Med 377(26):2545–2554. https://doi.org/10.1056/NEJMoa1708566
2. Hataye J, Moon JJ, Khoruts A, Reilly C, Jenkins MK (2006) Naive and memory CD4+ T cell survival controlled by clonal abundance. Science 312(5770):114–116. https://doi.org/10.1126/science.1124228
3. Seki Y, Yang J, Okamoto M, Tanaka S, Goitsuka R, Farrar MA, Kubo M (2007) IL-7/STAT5 cytokine signaling pathway is essential but insufficient for maintenance of naive CD4 T cell survival in peripheral lymphoid organs. J Immunol 178(1):262–270
4. Stemberger C, Huster KM, Koffler M, Anderl F, Schiemann M, Wagner H, Busch DH (2007) A single naive CD8+ T cell precursor can develop into diverse effector and memory subsets. Immunity 27(6):985–997. https://doi.org/10.1016/j.immuni.2007.10.012
5. Siewert C, Lauer U, Cording S, Bopp T, Schmitt E, Hamann A, Huehn J (2008) Experience-driven development: effector/memory-like alphaE+Foxp3+ regulatory T cells originate from both naive T cells and naturally occurring naive-like regulatory T cells. J Immunol 180(1):146–155
6. Hinrichs CS, Borman ZA, Gattinoni L, Yu Z, Burns WR, Huang J, Klebanoff CA, Johnson LA, Kerkar SP, Yang S, Muranski P, Palmer DC, Scott CD, Morgan RA, Robbins PF, Rosenberg SA, Restifo NP (2011) Human

effector CD8+ T cells derived from naive rather than memory subsets possess superior traits for adoptive immunotherapy. Blood 117 (3):808–814. https://doi.org/10.1182/blood-2010-05-286286

7. Alajez NM, Schmielau J, Alter MD, Cascio M, Finn OJ (2005) Therapeutic potential of a tumor-specific, MHC-unrestricted T-cell receptor expressed on effector cells of the innate and the adaptive immune system through bone marrow transduction and immune reconstitution. Blood 105 (12):4583–4589. https://doi.org/10.1182/blood-2004-10-3848
8. Yang L, Baltimore D (2005) Long-term in vivo provision of antigen-specific T cell immunity by programming hematopoietic stem cells. Proc Natl Acad Sci U S A 102(12):4518–4523. https://doi.org/10.1073/pnas.0500600102
9. Zhao Y, Parkhurst MR, Zheng Z, Cohen CJ, Riley JP, Gattinoni L, Restifo NP, Rosenberg SA, Morgan RA (2007) Extrathymic generation of tumor-specific T cells from genetically engineered human hematopoietic stem cells via notch signaling. Cancer Res 67 (6):2425–2429. https://doi.org/10.1158/0008-5472.CAN-06-3977
10. Boztug K, Schmidt M, Schwarzer A, Banerjee PP, Diez IA, Dewey RA, Bohm M, Nowrouzi A, Ball CR, Glimm H, Naundorf S, Kuhlcke K, Blasczyk R, Kondratenko I, Marodi L, Orange JS, von Kalle C, Klein C (2010) Stem-cell gene therapy for the Wiskott-Aldrich syndrome. N Engl J Med 363(20):1918–1927. https://doi.org/10.1056/NEJMoa1003548
11. Mendez-Ferrer S, Michurina TV, Ferraro F, Mazloom AR, Macarthur BD, Lira SA, Scadden DT, Ma'ayan A, Enikolopov GN, Frenette PS (2010) Mesenchymal and haematopoietic stem cells form a unique bone marrow niche. Nature 466(7308):829–834. https://doi.org/10.1038/nature09262
12. Peerani R, Zandstra PW (2010) Enabling stem cell therapies through synthetic stem cell-niche engineering. J Clin Invest 120(1):60–70. https://doi.org/10.1172/JCI41158
13. Boitano AE, Wang J, Romeo R, Bouchez LC, Parker AE, Sutton SE, Walker JR, Flaveny CA, Perdew GH, Denison MS, Schultz PG, Cooke MP (2010) Aryl hydrocarbon receptor antagonists promote the expansion of human hematopoietic stem cells. Science 329 (5997):1345–1348. https://doi.org/10.1126/science.1191536
14. Himburg HA, Muramoto GG, Daher P, Meadows SK, Russell JL, Doan P, Chi JT, Salter AB, Lento WE, Reya T, Chao NJ, Chute JP (2010) Pleiotrophin regulates the expansion and regeneration of hematopoietic stem cells. Nat Med 16(4):475–482. https://doi.org/10.1038/nm.2119
15. Lei F, Zhao B, Haque R, Xiong X, Budgeon L, Christensen ND, Wu Y, Song J (2011) In vivo programming of tumor antigen-specific T lymphocytes from pluripotent stem cells to promote cancer immunosurveillance. Cancer Res 71(14):4742–4747. https://doi.org/10.1158/0008-5472.CAN-11-0359
16. Lei F, Haque M, Sandhu P, Ravi S, Song J, Ni B, Zheng S, Fang D, Jia H, Yang JM, Song J (2017) Development and characterization of naive single-type tumor antigen-specific CD8+ T lymphocytes from murine pluripotent stem cells. Oncoimmunology 6(7):e1334027. https://doi.org/10.1080/2162402X.2017.1334027
17. Lei F, Haque R, Weiler L, Vrana KE, Song J (2009) T lineage differentiation from induced pluripotent stem cells. Cell Immunol 260 (1):1–5. https://doi.org/10.1016/j.cellimm.2009.09.005
18. Okita K, Ichisaka T, Yamanaka S (2007) Generation of germline-competent induced pluripotent stem cells. Nature 448(7151):313–317. https://doi.org/10.1038/nature05934
19. Schmitt TM, Zuniga-Pflucker JC (2002) Induction of T cell development from hematopoietic progenitor cells by delta-like-1 in vitro. Immunity 17(6):749–756

Chapter 15

Clonogenic Culture of Mouse Thymic Epithelial Cells

Miho Sekai, Jianwei Wang, and Yoko Hamazaki

Abstract

The thymus plays an essential role in the development and selection of T cells by providing a unique microenvironment that is mainly composed of thymic epithelial cells (TECs). We previously identified stem cells of medullary TECs (mTECs) that are crucial for central tolerance induction using a novel clonogenic culture system. We also found that medullary thymic epithelial stem cells (mTESCs) maintain life-long mTECs regeneration and central T cell self-tolerance in mouse models. The clonogenic efficiency of TECs in vitro is highly correlated to the TEC reconstitution activity in vivo. Here, we describe the clonogenic culture system to evaluate the self-renewing activity of TESCs. The colonies are derived from TESCs, are visualized and quantified by rhodamine-B staining on a feeder layer, and can be passaged in vitro. Thus, our system enables quantitative evaluation of TESC activity and is useful for dissecting the mechanisms that regulate TESC activity in physiological aging as well as in various clinical settings.

Key words Thymus, Thymic epithelial cells, Thymic epithelial stem cells, Clonogenic assay

1 Introduction

The thymus is the primary organ responsible for generating new T cells that possess diverse T cell receptor repertoires. The T cell generation process requires thymic stromal components, especially thymic epithelial cells (TECs), which form a three-dimensional (3D) meshwork structure for construction of the tissue framework. Previous studies have shown that the 2D culture of TECs fails to support thymocyte development due to the reduced expression of components essential for T cell commitment and differentiation, such as FoxN1, MHC class II, and Notch ligands (Delta-like 1 and Delta-like 4) [1, 2], suggesting the 3D meshwork architecture is necessary for maintaining functional TECs. Two types of TECs, cortical TECs (cTECs) and medullary TECs (mTECs), play distinctive roles in thymocyte proliferation, differentiation, and selection: cTECs are involved in the diverse repertoire formation, CD4- and CD8-lineage decision, and MHC restriction; mTECs regulate central T cell self-tolerance including the deletion of autoreactive T cells and the generation of T regulatory cells [3, 4]. *Foxn1* is a

Shin Kaneko (ed.), *In Vitro Differentiation of T-Cells: Methods and Protocols*, Methods in Molecular Biology, vol. 2048, https://doi.org/10.1007/978-1-4939-9728-2_15,

transcriptional activator crucial for the differentiation of both types of TECs, and *Foxn1* deficiency causes severe immunodeficiency in both mice and humans due to developmental defects in TECs and consequent absence of T cells in the periphery [5, 6].

Despite the crucial importance of TECs in T cell generation, their developmental mechanisms are not well understood, partly because of their rare composition in the thymus and the lack of appropriate molecular markers. In the past decade, however, methods for isolating and analyzing mouse primary TECs have been established [7–9]; the enzymatic digestion of thymic tissues yields single cell suspensions, which enable analysis of the molecular characteristics of TECs by flow cytometry [7–11]. These technical advancements have allowed for the precise characterization of functionally different TEC subsets, driving further studies on TECs as immune regulators.

More recently, several groups, including us, have identified mouse TEC progenitors and stem cells using TEC clonogenic culture systems [12–14], providing important insights on the mechanisms of TEC generation during ontogeny as well as TEC maintenance in the adult thymus [12–16]. Importantly, the clonogenic activity of TECs detected in our 2D culture system reflects well the in vivo reconstitution potential of TECs; the potential is quite high in embryo, decreases in newborn stages, and remains low in adults [14]. Furthermore, using our system, single cell-derived colonies can be passaged and differentiated into mature TECs in vivo upon implantation into the 3D thymic microenvironment. Thus, our 2D clonogenic culture system can be used to assess TEC self-renewal ability, providing a useful tool for analyzing the mechanisms that decrease TEC stem cell (TESC) activity, which corresponds to thymic involution, as well as for screening molecules that modulate TESC activity.

2 Materials

2.1 Isolation and Enrichment of Thymic Epithelial Cells

1. Fetal calf serum (FCS, HyClone).
2. Phosphate-buffered saline without Ca^{2+} or Mg^{2+} (PBS).
3. Ethylenediaminetetraacetate (EDTA): 500 mM in distilled water.
4. Roswell Park Memorial Institute (RPMI)-1640 medium (Wako, Japan).
5. 1% FCS/RPMI: 1% (v/v) FCS in RPMI.
6. 70-μm cell strainer.
7. 0.22-μm filter (Millipore).
8. Liberase TM (Roche): 10 U/mL in PBS; store at −20 °C.

9. DNase I (Worthington): 10 mg/mL in PBS; store at −20 °C.
10. FACS buffer: 1% FCS and 5 mM EDTA in PBS.
11. 1-mL-wide orifice pipette tips.
12. 1-mL syringe plungers.
13. Anti-mouse CD16/CD32 antibody (clone 2.4G2).
14. Mouse CD45 MicroBeads (Miltenyi Biotec).
15. autoMACS Pro Separator (Miltenyi Biotec).
16. 70-μm filter.

2.2 STO Cultivation and Feeder Layer Preparation

1. STO cell line (Riken BRC, RCB0536).
2. Dulbecco's Modified Eagle's Medium (DMEM), high glucose (Wako, Japan).
3. 10,000 U/mL penicillin/10,000 μg/mL streptomycin.
4. STO medium: DMEM supplemented with 5% FCS and 100 U/mL penicillin/100 μg/mL streptomycin.
5. Trypsin-EDTA (0.25%), phenol red (Thermo Fisher Scientific): 0.25% Trypsin and 1 mM EDTA in phosphate buffer.
6. Mitomycin-C (Kyowa Hakko Kirin, Japan): 0.4 mg/mL in PBS; store at −20 °C.
7. 0.1% gelatin (Sigma): 0.1% (w/v) gelatin in PBS; store at room temperature.

2.3 Clonogenic Thymic Epithelial Cell Culture

1. KnockOut Serum Replacement (KSR, Thermo Fisher Scientific).
2. Ham's F-12 (Thermo Fisher Scientific).
3. Hydrocortisone (Calbiochem): 4 mg/mL in PBS; store at −20 °C.
4. 1 mg/mL cholera toxin (Wako, Japan).
5. 10 mg/mL insulin (Sigma).
6. 3,3,5-Triiodo-L-thyronine (Sigma): 10^{-4} M in PBS; store at −20 °C.
7. TEC culture medium: 3:1 mixture of DMEM and Ham's F-12 supplemented with 10% (v/v) KSR, 0.4 μg/mL hydrocortisone, 10^{-10} M cholera toxin, 5 μg/mL insulin, 2×10^{-9} M 3,3,5-triido-L-thyronine, and 100 U/mL penicillin/100 μg/mL streptomycin.
8. 0.1% BSA/PBS: 0.1% (w/v) bovine serum albumin (BSA) in PBS.
9. Recombinant mouse epidermal growth factor (EGF, Thermo Fisher Scientific): 100 μg/mL in 0.1% BSA/PBS; store at −20 °C.

10. 10^7 U/mL leukemia inhibitory factor (LIF).
11. Defined trypsin inhibitor (DTI, Thermo Fisher Scientific).
12. Cloning cylinders.
13. Vaseline.
14. 0.05% Trypsin-EDTA: 0.05% Trypsin and 5 mM EDTA prepared from Trypsin-EDTA (0.25%) (Subheading 2.2, **item 5**) and EDTA (Subheading 2.1, **item 3**) in PBS.

2.4 Rhodamine-B Staining

1. Formalin: 3.7% formalin in PBS.
2. Rhodamine-B: 1% (w/v) rhodamine-B in 70% ethanol.

3 Methods

Procedures 3.1 to 3.5 are performed under a laminar flow hood. All equipment and reagents are sterilized by spraying with 70% ethanol and filtration through a filter with a 0.22-μm pore or autoclave.

3.1 Isolation of Thymic Epithelial Cells by Enzymatic Digestion

This enzymatic digestion method is modified from the protocol reported by Seach et al. [8]. Unless otherwise stated, all procedures after dissection are carried out on ice.

1. Dissect the thymus from a mouse using scissors and forceps.
2. Remove any fat and excess connective tissue using fine forceps in a 60-mm dish containing 10 mL of 1% FCS/RPMI.
3. Place the thymus into a 70-μm cell strainer in a new 60-mm dish containing 5 mL of 1% FCS/RPMI and mince the tissue into small pieces using sharp scissors.
4. Stir the medium using a 1-mL syringe plunger to release thymocytes.
5. Transfer the thymic fragments inside the cell strainer to a 15-mL tube containing 2 mL of 1% FCS/RPMI (*see* **Note 1**).
6. Gently pipette the fragments up and down (~10 times) with a 1-mL pipette attached to a wide orifice pipette tip.
7. Settle the fragments for at least 2 min and discard the supernatant. The supernatant mostly contains thymocytes.
8. Add 2 mL of 1% FCS/RPMI to the tube and repeat **steps 6** and 7.
9. Add 1 mL enzyme mixture (1% FCS/RPMI containing 0.5 U/mL Liberase TM and 0.2 mg/mL DNase I).
10. Incubate for 12 min in a 37 °C water bath. While incubating, gently pipette the fragments every 4 min with a 1-mL pipette attached to a wide orifice pipette tip (*see* **Note 2**).

11. Settle the fragments for at least 2 min and transfer the supernatant to a new 15-mL tube containing 10 mL of 1% FCS/RPMI.
12. Add 1 mL enzyme mixture to the fragments and repeat **step 10**.
13. After the digestion, there should be no visible fragments. If there are, repeat **steps 11** and **12** until the fragments are completely disassociated (*see* **Note 3**). Pool the cell suspension and the supernatant(s) (**steps 8–13**) and pass through a 70-μm filter.
14. Centrifuge at 400 × *g* for 5 min at 4 °C.
15. Discard the supernatant and resuspend the cell pellet in an appropriate volume of 1% FCS/RPMI.
16. Determine the cell count.

3.2 Enrichment of Thymic Epithelial Cells by Magnetic Separation

1. Resuspend the cells in 95 μL of FACS buffer per 1×10^7 cells.
2. Add 5 μL of mouse CD45 MicroBeads per 1×10^7 cells.
3. Mix well and incubate for 30 min on ice. During the incubation, suspend every 10 min.
4. Wash the cells by adding 10 mL of FACS buffer and centrifuge at 400 × *g*, 4 °C for 10 min.
5. Discard the supernatant, resuspend the cell pellet in 1 mL of FACS buffer per 1×10^8 cells, and pass through a 70-μm filter into a new tube.
6. Use the "Deplete" program on an autoMACS Pro Separator to enrich $CD45^-$ cells. The positive fraction after enrichment is discarded, and the negative fraction is kept on ice.
7. Centrifuge the CD45 negative fraction at 400 × *g*, 4 °C for 10 min, and discard the supernatant.
8. Resuspend in TEC culture medium.

3.3 Preparation of Feeder Layer Cultures

All culture dishes are pre-coated with 0.1% gelatin prior to use, and the cells are cultured in a humidified incubator at 37 °C, 5% CO_2 (hereafter, 37 °C incubator).

3.3.1 Gelatin Coating

1. Add 2 or 3 mL of 0.1% gelatin to 60-mm or 100-mm culture dishes, respectively. Note that the surface of the dishes should be covered with gelatin.
2. Incubate at least 30 min in the 37 °C incubator.
3. Remove the gelatin solution as much as possible and discard (*see* **Note 4**).

3.3.2 Thawing Cells from a Frozen Stock

1. Thaw a vial of frozen STO cells (1×10^6 cells) by gentle agitation in 37 °C water bath (*see* **Note 5**).
2. Remove the vial from the water bath as soon as the frozen contents are thawed.
3. Transfer the cell suspension to a new 15-mL tube.
4. Add 10 mL STO medium dropwise to the cell suspension and gently swirl to mix.
5. Pipette up and down to mix the cell suspension.
6. Centrifuge at $300 \times g$, 4 °C for 10 min.
7. Discard the supernatant and resuspend in 10 mL STO medium.
8. Transfer all the suspension to a 100-mm culture dish or two 60-mm culture dishes.
9. Incubate the cultures in the 37 °C incubator.

3.3.3 Subculturing Cells

STO cells are routinely cultured in 100-mm culture dishes and passaged prior to reaching confluence (~80% confluent) as follows:

1. Remove the culture medium from the cell culture dishes.
2. Rinse the cell layer with 5 mL PBS to remove all components of the serum to prevent trypsin activity.
3. Add 1 mL of 0.25% Trypsin-EDTA to the culture dishes and incubate in the 37 °C incubator. Confirm that the cells dispersed within 5 min by observing under a microscope.
4. Add 9 mL STO medium and gently pipette up and down to mix.
5. Transfer the cell suspension to a new 15-mL tube.
6. Centrifuge at $300 \times g$ for 5 min.
7. Discard the supernatant and resuspend in STO medium.
8. Add appropriate aliquots of the cell suspension to 100-mm dishes (*see* **Note 6**).
9. Incubate the cultures in the 37 °C incubator.

3.3.4 Mitomycin-C Treatment

Mitomycin-C treatment is used to inactivate the replication of STO cells and should be performed no more than two days before the use of feeder layers.

1. Add 0.4 mg/mL mitomycin-C solution to the cultures at a final concentration of 10 μg/mL.
2. Incubate the cultures in the 37 °C incubator for 2 h (*see* **Note 7**).
3. Remove the medium from the cell culture dishes.

4. Rinse the cells with 10 mL PBS three times to completely remove all components containing mitomycin-C, as these could affect cell growth in later steps.
5. Add 1 mL of 0.25% Trypsin-EDTA to the culture dishes and incubate in the 37 °C incubator. Confirm that the cells dispersed within 5 min by observing under a microscope.
6. Add 9 mL of STO medium and gently pipette up and down to mix.
7. Transfer the cell suspension to a new 15-mL tube.
8. Centrifuge at 300 × *g* for 5 min.
9. Discard the supernatant and resuspend in an appropriate volume of STO medium.
10. Determine the cell count.
11. Add replication-inactivated STO cells at a density of 2.5×10^4 cells/cm^2 in 60-mm culture dishes and jiggle the culture dishes in every direction to evenly distribute the cells.
12. Incubate the cultures in the 37 °C incubator.
13. After more than 4 h, ensure the cells are evenly distributed by observing under a microscope.

3.4 Clonogenic Thymic Epithelial Cell Culture

1. Prior to use, check the STO feeder condition, which should be 80–90% confluence in the even distribution.
2. Remove the culture medium from the STO feeder layers.
3. Add 5 mL of TEC culture medium to the feeder cultures.
4. Add the TEC-enriched cell suspension (*see* Subheading 3.2, CD45 negative fraction in **step 8**) onto the feeder layers.
5. Incubate the cultures in the 37 °C incubator.
6. Refresh the medium every 3 days. After 3 days of cultivation, the medium is supplemented with 10 ng/mL EGF and 10^3 U/mL LIF.
7. The clonogenic TEC cultures are maintained until analysis.

3.5 Passaging Cells for Serial Cultivation

TEC colonies should be visible around 3 days after cultivation (*see* Fig. 1) and are passaged after 7 to 10 days in culture.

3.5.1 Passaging Individual Colonies

1. Mark the positions of the colonies in the bottom of the culture dish by observing under a microscope.
2. Remove the culture medium from the cell culture dishes.
3. Rinse the cells with 5 mL PBS to remove all components of serum so as to prevent trypsin activity.
4. Dip the bottom of the cloning cylinder in sterile Vaseline (petroleum jelly).

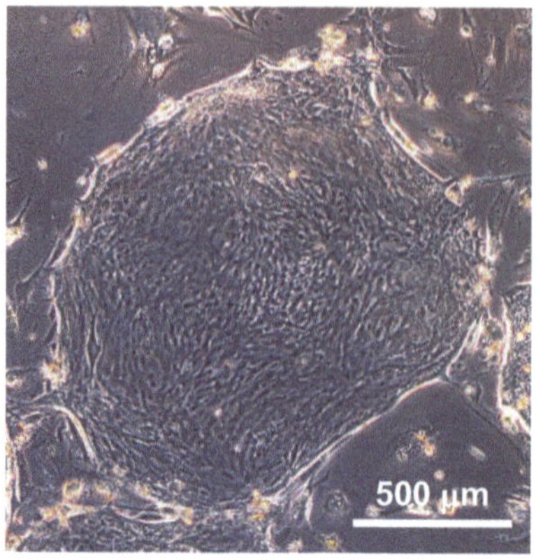

Fig. 1 Morphology of a TEC colony on day 7

5. Place the cloning cylinder in each marked colony (*see* **Note 8**).
6. Add 20 μL of 0.25% Trypsin-EDTA to each cloning cylinder.
7. Incubate in the 37 °C incubator. Confirm that the cells dispersed within 5 min by observing under a microscope.
8. Add 200 μL DTI to each cloning cylinder and pipette up and down to mix.
9. Transfer the cell suspension from a single cloning cylinder to a 60-mm culture dish with a fresh STO feeder layer (*see* Subheading 3.3.4 for feeder preparation).
10. Incubate cultures in the 37 °C incubator.

3.5.2 Passaging Bulk Colonies of a Culture Dish

1. Remove the culture medium from the cell culture dishes.
2. Rinse the cells with 5 mL PBS to remove all components of serum so as to prevent trypsin activity.
3. Add 1 mL of 0.05% Trypsin-EDTA and incubate in the 37 °C incubator. Most of the feeder layers should be detached within 2 min. The detachment should be confirmed by observation under a microscope (*see* **Note 9**).
4. Remove the feeder layer cell suspension and rinse the attached colonies with 5 mL PBS.
5. Add 1 mL of 0.25% Trypsin-EDTA and incubate in the 37 °C incubator for 5 min.
6. Jiggle the culture dish to check trypsinization. Colonies should be detached from the dishes and dissociated within 5 to 10 min. If the dissociation is incomplete, incubate in the 37 °C incubator for another 2 min (*see* **Note 10**).
7. Add 2 mL DTI to the cultures and pipette up and down to mix.
8. Transfer the cell suspension to a new 5-mL tube.

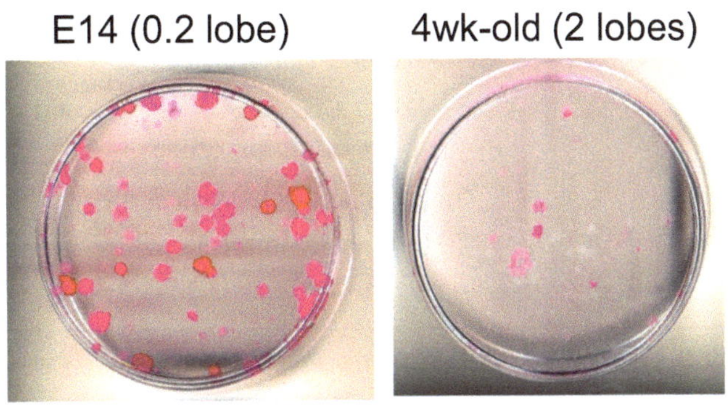

Fig. 2 Embryonic and adult colonies. TECs isolated from embryonic day 14 and 4 weeks post birth are cultured on STO feeders, and colonies are fixed and stained with rhodamine-B 10 days later

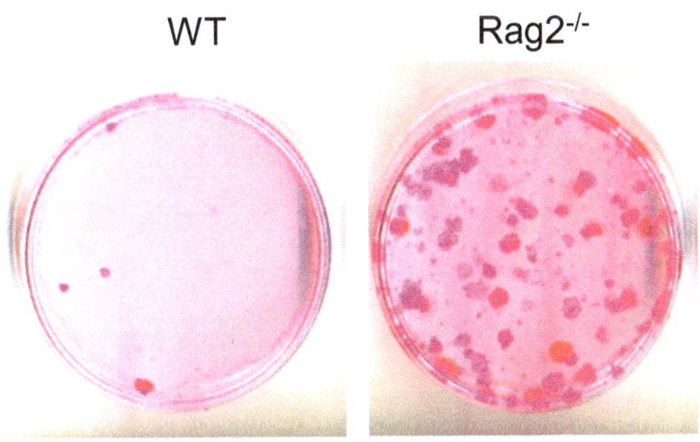

Fig. 3 TEC colonies from 4-week-old wild-type and Rag2$^{-/-}$ thymi (1 thymus/60-mm dish). Rag2$^{-/-}$ TECs show higher clonogenic activity

9. Centrifuge at 300 × *g* for 5 min.
10. Discard the supernatant and resuspend in an appropriate volume of TEC culture medium.
11. Determine the cell count.
12. Add an appropriate cell number to a 60-mm culture dish with a fresh STO feeder layer.
13. Incubate the cultures in the 37 °C incubator.

3.6 Rhodamine-B Staining of Thymic Epithelial Cell Colonies

TEC colonies can be visualized by rhodamine-B staining (*see* Figs. 2 and 3).

1. Remove the culture medium from the cell culture dishes.
2. Rinse the cells with 5 mL PBS twice.
3. Add 2 mL of 3.7% formalin to fix the cells for 15 min at room temperature.

4. Discard the fixative solution and rinse the cells with 5 mL distilled water twice.
5. Add 1% Rhodamine-B to stain the TEC colonies for around 2 h at room temperature.
6. Rinse the cells with 5 mL of distilled water five times.
7. Dry at room temperature overnight (*see* **Note 11**).

4 Notes

1. Coat evenly the inside of the tube with 1% FCS/RPMI.
2. After 12 min incubation, the fragments should be passed through a normal orifice 1-mL pipette tip.
3. The total enzyme reaction time should not exceed 30 min.
4. Do not rinse the dish with PBS or culture medium.
5. Thawing should be rapid (less than 2 min).
6. Aliquot the cells at a dilution ratio of 1:5 to 1:10.
7. Do not exceed 2.5 h.
8. Vaseline enables the cloning cylinder to adhere to the cell culture dish.
9. More than 2 min incubation will also detach the colony cells.
10. Exceeding 12 min will reduce cell survival.
11. Dry until the dish can be turned upside down without the content falling off.

Acknowledgments

We thank Dr. P. Karagiannis for proofreading. This work was supported by grants from the Japanese Ministry of Education, Culture, Science, Sports, and Technology (24590580, 25111505, 15H01154, 17H05641, 18K19442, and 18H02640 to Y.H. and 15K19125 and 19K16689 to M.S.), iPS Cell Research Fund, AMED under Grant Number (JP19gm5010001 and JP19bm0104001) to Y.H., and the Takeda Science Foundation to Y.H. and M.S.

References

1. Anderson KL, Moore NC, McLoughlin DE, Jenkinson EJ, Owen JJ (1998) Studies on thymic epithelial cells in vitro. Dev Comp Immunol 22(3):367–377
2. Mohtashami M, Zuniga-Pflucker JC (2006) Three-dimensional architecture of the thymus is required to maintain delta-like expression necessary for inducing T cell development. J Immunol 176(2):730–734
3. Petrie HT, Zuniga-Pflucker JC (2007) Zoned out: functional mapping of stromal signaling microenvironments in the thymus. Annu Rev

Immunol 25:649–679. https://doi.org/10.1146/annurev.immunol.23.021704.115715

4. Takahama Y, Ohigashi I, Baik S, Anderson G (2017) Generation of diversity in thymic epithelial cells. Nat Rev Immunol 17(5):295–305. https://doi.org/10.1038/nri.2017.12
5. Boehm T (2008) Thymus development and function. Curr Opin Immunol 20 (2):178–184. https://doi.org/10.1016/j.coi.2008.03.001
6. Vaidya HJ, Briones Leon A, Blackburn CC (2016) FOXN1 in thymus organogenesis and development. Eur J Immunol 46 (8):1826–1837. https://doi.org/10.1002/eji.201545814
7. Gray DH, Chidgey AP, Boyd RL (2002) Analysis of thymic stromal cell populations using flow cytometry. J Immunol Methods 260 (1–2):15–28
8. Seach N, Wong K, Hammett M, Boyd RL, Chidgey AP (2012) Purified enzymes improve isolation and characterization of the adult thymic epithelium. J Immunol Methods 385 (1–2):23–34. https://doi.org/10.1016/j.jim.2012.07.023
9. Jain R, Gray DH (2014) Isolation of thymic epithelial cells and analysis by flow cytometry. Curr Protoc Immunol 107(3):26 21–26 15. https://doi.org/10.1002/0471142735.im0326s107
10. Gray D, Abramson J, Benoist C, Mathis D (2007) Proliferative arrest and rapid turnover of thymic epithelial cells expressing Aire. J Exp Med 204(11):2521–2528. https://doi.org/10.1084/jem.20070795
11. Gray DH, Seach N, Ueno T, Milton MK, Liston A, Lew AM, Goodnow CC, Boyd RL (2006) Developmental kinetics, turnover, and stimulatory capacity of thymic epithelial cells. Blood 108(12):3777–3785. https://doi.org/10.1182/blood-2006-02-004531
12. Ucar A, Ucar O, Klug P, Matt S, Brunk F, Hofmann TG, Kyewski B (2014) Adult thymus contains FoxN1(−) epithelial stem cells that are bipotent for medullary and cortical thymic epithelial lineages. Immunity 41(2):257–269. https://doi.org/10.1016/j.immuni.2014.07.005
13. Wong K, Lister NL, Barsanti M, Lim JM, Hammett MV, Khong DM, Siatskas C, Gray DH, Boyd RL, Chidgey AP (2014) Multilineage potential and self-renewal define an epithelial progenitor cell population in the adult thymus. Cell Rep 8(4):1198–1209. https://doi.org/10.1016/j.celrep.2014.07.029
14. Sekai M, Hamazaki Y, Minato N (2014) Medullary thymic epithelial stem cells maintain a functional thymus to ensure lifelong central T cell tolerance. Immunity 41(5):753–761. https://doi.org/10.1016/j.immuni.2014.10.011
15. Kim MJ, Miller CM, Shadrach JL, Wagers AJ, Serwold T (2015) Young, proliferative thymic epithelial cells engraft and function in aging thymuses. J Immunol 194(10):4784–4795. https://doi.org/10.4049/jimmunol.1403158
16. Meireles C, Ribeiro AR, Pinto RD, Leitao C, Rodrigues PM, Alves NL (2017) Thymic crosstalk restrains the pool of cortical thymic epithelial cells with progenitor properties. Eur J Immunol 47(6):958–969. https://doi.org/10.1002/eji.201746922

Chapter 16

Single-Cell Transcriptome Analysis of T Cells

Willem Van Der Byl, Simone Rizzetto, Jerome Samir, Curtis Cai, Auda A. Eltahla, and Fabio Luciani

Abstract

Single-cell RNA-seq (scRNA-seq) has provided novel routes to investigate the heterogeneous populations of T cells and is rapidly becoming a common tool for molecular profiling and identification of novel subsets and functions. This chapter offers an experimental and computational workflow for scRNA-seq analysis of T cells. We focus on the analyses of scRNA-seq data derived from plate-based sorted T cells using flow cytometry and full-length transcriptome protocols such as Smart-Seq2. However, the proposed pipeline can be applied to other high-throughput approaches such as UMI-based methods. We describe a detailed bioinformatics pipeline that can be easily reproduced and discuss future directions and current limitations of these methods in the context of T cell biology.

Key words Single-cell RNA sequencing, scRNA-seq, T cells, Alignment, T cell receptor reconstruction, Gene expression matrix, Clustering, Differential gene expression

1 Introduction

Current methods to analyze single-cell (sc-)RNA-seq data from T cells encompass two general approaches. The first approach is based on the generation of low-throughput (<500 cells per sample) full-length transcripts, often using a plate-based sorting technique. The second approach is high-throughput (typically >1000 cells per sample) but generates only partial-length transcripts (3′- or 5′-based) (*see* previous reviews with detailed descriptions of these approaches [1–4]).

The full-length scRNA-seq approach is usually performed via the Smart-Seq2 protocol [5, 6] or modifications thereof. This approach allows for detection of complex genetic rearrangements, such as the reconstruction of the full-length TCR [7].

The high-throughput approach is often performed via protocols that utilize 3′-based cell barcodes and unique molecular identifiers (UMI) to characterize individual cells and genes from a bulk population. Common protocols are the 10× Chromium

Shin Kaneko (ed.), *In Vitro Differentiation of T-Cells: Methods and Protocols*, Methods in Molecular Biology, vol. 2048, https://doi.org/10.1007/978-1-4939-9728-2_16, © Springer Science+Business Media, LLC, part of Springer Nature 2019

microfluidics system [8], the Drop-seq method [9], and other flow cytometry-based protocols, e.g., MARS-seq [10] or Cell-Seq [11]. An exhaustive overview of the state of the art of scaling up technologies has been recently published [12]. The advantage of these high-throughput methods is in the capacity to sequence thousands of cells with costs comparable to those required to analyze 200–500 cells with a full-length transcriptome approach. The reduction in cost is largely driven by sequencing only fragments of transcripts.

In this chapter, we shall focus on the full-length approach. In particular, we will detail a bioinformatics pipeline that includes both gene expression analyses and full-length T cell receptor (TCR) reconstruction. We will briefly summarize the experimental component for the generation of the scRNA-seq data as these have been recently detailed elsewhere [3, 13, 14].

The bioinformatics pipeline outlined below is largely based on the VDJPuzzle pipeline [15] and provides flexibility in the tools utilized throughout the various steps based on the specific needs (Table 1). The proposed bioinformatics workflow (Fig. 1) brings together in house scripts and freely available software packages that have been tested on scRNA-seq data derived from human and animal T and B cells. We also provide a list of alternative software packages that can be utilized at each step (*see* Table 1). We will then discuss the advantages and limitations of full-length transcript methods over the high-throughput methods and discuss some future directions.

Table 1
Software packages used in the bioinformatics pipeline presented in this chapter and alternatives. The software packages used in the pipeline are in bold

Step in the pipeline	Software packages	Description	References
Quality control (QC) of raw reads	**Trimmomatic** FastQC	Perform a series of analyses to test quality score, coverage, and read length	[16] [17]
Alignment of scRNA-seq data	**STAR** TopHat2	Mapping sequences derived from contiguous as well as noncontiguous (splice) genomic segments	[18] [19]
Quantification of gene expression	*Alignment based:* Cufflinks **RSEM** Salmon align StringTie *Alignment-free:* Kallisto Salmon Sailfish	Quantify the abundance of each transcript from the reads aligned. Methods can be alignment based (time consuming) and alignment-free (time efficient)	[20] [21] [22] [23] [24] [22] [25]

(continued)

Table 1
(continued)

Step in the pipeline	Software packages	Description	References
QC of expression matrix	**Scater** Seurat	Software packages including tools for analysis and quality control of gene expression matrices	[26] [27]
Normalization	**Scran** Scater SCNorm	Software packages including methods to account for variations in gene expression due to amplification and read depth	[26] [28] [29]
Imputation/ dropouts	DropLasso BC-Seq Sconc Seurat	Statistical models to infer expression of dropout genes and infer rate of dropout	[30] [31] [32] [27]
Batch effects	Scran Seurat Limma ComBat	Software packages to remove or account for sources of unwanted variation	[26] [27] [33] [34]
Dimensionality reduction	Destiny Monocle **Scater** Seurat	Statistical models that provide a set of variable genes that best represent the data in a low-dimensional space These include PCA, tSNE, and independent component analysis	[35] [36] [28] [27]
Clustering[a]	**SC3** CIDR Monocle TSCAN	Statistical models that identify subsets of cells based on the quantification of the distance between cells as a function of the gene expression	[37] [38] [36] [39]
Differential expression	**MAST** SCDE Monocle EdgeR	Use statistical inference (typically univariable, gene by gene) to identify the genes with greatest variation in expression between subsets	[40] [41] [36] [42]
Inference of trajectories from gene expression data	Monocle DPT (diffusion pseudo time) CellTree TSCAN	Pseudo-time and diffusion pseudo-time are methods to infer gene expression dynamics from snapshot single-cell data	[36] [43] [44] [39]
Gene set enrichment analysis	**MAST** DAVID GSEA (Broad I nstitute)	Statistical methods to identify categories of genes that may be enriched in a specific gene subset or to identify functional profiles and disease association	[40] [45] [46]
TCR reconstruction	**VDJPuzzle** TraCER Trapes	Methods to assemble the full-length TCR from scRNA-seq data	[7] [47] [48]

[a]This is a very common topic and >50 software packages are available. Please refer to **Note 1**

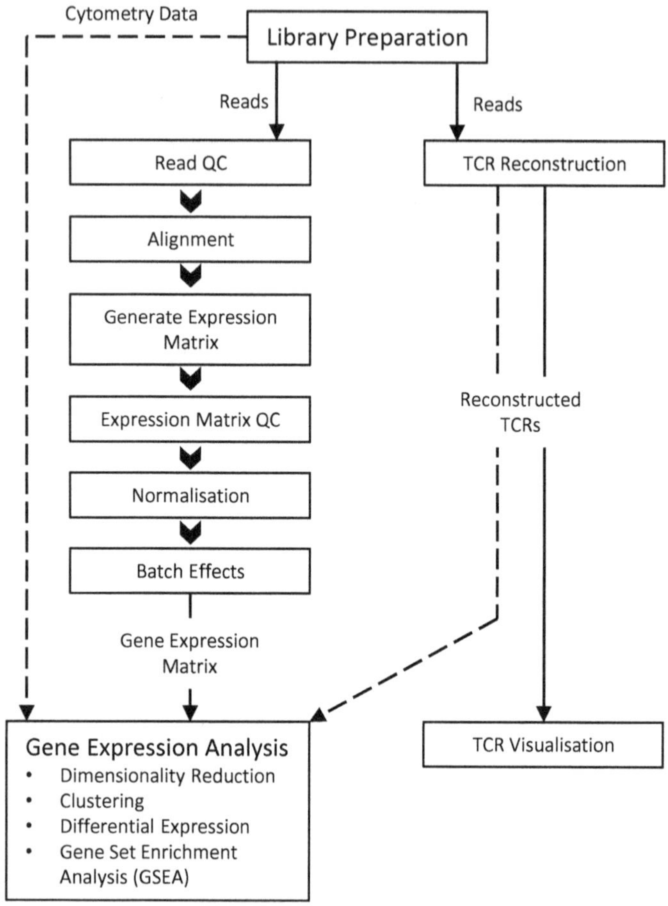

Fig. 1 Outline of the scRNA-seq bioinformatics pipeline presented in this chapter. Arrows represent data flow. Dashed arrows are optional data flows

2 Materials

2.1 Hardware

This process requires two different computers, a "server" and a "workstation" (*see* **Note 2**), each with different system requirements detailed below:

1. Server.
   (a) OS: Linux.
   (b) Memory: 64 GB.
   (c) Hard disk space: 45 GB for software, the reference genome, and indexes plus approximately three times the size of the input data.
   (d) No graphical user interface required as all tools run from the command line.
   (e) A Python interpreter is required (preferably Python 3).

(f) Used for read QC, alignment, quantification, and receptor reconstruction (Subheadings 3.3, 3.4, and 3.13).

2. Workstation.

   (a) OS: MacOS, Linux, or Windows.

   (b) Memory: 4 GB.

   (c) Hard disk space: 5 GB.

   (d) Graphical user interface required for visualizing data.

   (e) Used for normalization, clustering, differential expression, and TCR reconstruction (Subheadings 3.5–3.12 and 3.14).

2.2 Software and Knowledge Requirements

These instructions assume a basic understanding of the Unix command line and the R programming language. If the server is a Linux cluster, the commands need to be scheduled as tasks to execute on the cluster. We provide the basic Linux commands. See the cluster's documentation for information about how to schedule tasks to execute these commands. We have designed the instructions to allow the reader to copy and paste and execute the commands without modifications while allowing the reader to easily modify the commands for individual circumstances (e.g., using single-end instead of paired-end reads).

2.3 Input Data

Sequence data from single T cells in ".fastq.gz" format (*see* **Note 3**) are required as input. These should be paired-end reads with two files per cell. The instructions should work with single-ended reads with some minor modifications to the alignment and quantification steps.

As a practical example of the analysis pipeline, we will utilize a dataset already available from a previous publication [7] and available at https://www.ebi.ac.uk/arrayexpress/experiments/E-MTAB-4850/. These data are from hepatitis C virus-specific CD8 + T cells obtained via flow cytometric tetramer staining for HLA-A1-restricted epitope. Libraries were generated using Smart-Seq2 protocol. There were three experimental conditions: CD8 T cells directly sorted from peripheral blood mononuclear cells (PBMCs) and cells derived from cell lines originating from the same patient which were previously expanded ex vivo with cognate antigen. For these expanded cells two subsets were considered: those that were directly sorted after thawing and staining of a vial (unstimulated condition) and those that were restimulated with antigen for 24 h prior to sorting (stimulated condition). Twenty samples were sequenced per treatment, with two samples in each treatment being bulk samples consisting of ten cells per sample. The original paper included a total of 72 samples, 60 sequenced using NextSeq and 150 bp paired-end reads and 12 sequenced using MiSeq with 250 bp paired-end reads. For this chapter we restricted the analyses to the 60 NextSeq samples for the sake of clarity and simplicity.

3 Methods

3.1 Library Preparation

Here we briefly describe the steps for library preparation following sorting. It cannot be understated that quality experimental design and execution prior to generation of scRNA-seq data pays dividends to the bioinformatician tasked with analyzing the final sequencing output. Due to the nature of scRNA-seq, the initial very small amount of mRNA present in each cell impacts significantly on the successful preparation of cDNA libraries. Thus, we recommend a thorough understanding of experiment-specific techniques and biases prior to any attempted analyses.

The most widely used protocol for full-length transcriptome library preparation is the Smart-Seq2 protocol [6], which relies on the physical separation of single cells into individual wells on 96- or 384-well plates. The key steps for generating sequencing ready libraries are reverse transcription, template switching, PCR amplification, and barcoding through a tagmentation reaction. Reverse transcription from polyadenylated transcripts is performed using the reverse transcriptase from the Moloney murine leukemia virus (MMLV) and oligo-dT primers that carry an additional sequence to provide a recognition site during PCR amplification. Upon reaching the 5′ end of its template, the MMLV reverse transcriptase introduces 2–5 nucleotide overhangs onto the 3′ end of the cDNA molecule [49]. A template switch oligonucleotide (TSO) facilitates template switching of the MMLV reverse transcriptase and continued extension of the cDNA molecule. Importantly the TSO primer also includes a sequence identical to that carried by the initial oligo-dT primer. Thus, each cDNA molecule has recognition sites for PCR amplification at its 5′ and 3′ ends that were introduced during reverse transcription and template switching, respectively. PCR amplification may then be performed with a single primer specific for these recognition sites. Researchers should minimize amplification biases by reducing the number of cycles as much as practical while generating sufficient product for subsequent steps. Finally, each library undergoes cell-specific barcoding and fragmentation by reaction with Tn5 transposase according to the Illumina dual indexing workflow and becomes ready for sequencing. The main distinction of the Smart-Seq2 protocol is that cell-specific barcodes are introduced following PCR amplification rather than during reverse transcription. Consequently, reads from all regions of a transcript carry cell-specific barcodes in comparison to high-throughput approaches, which typically introduce barcodes during reverse transcription at 5′ or 3′ ends only but facilitate the pooling of cells at earlier steps.

In this protocol ERCC controls can be also added. These are synthetic RNA molecules of different length and concentration that can be used as spike-ins to control for RNA capture efficiency and PCR amplification bias. These methods are not always utilized as it is difficult to validate the necessary assumption that each cell has the same amount of spike-in, cell-to-cell fluctuations may affect the capture efficiency, and the use of synthetic spike-in transcripts may not behave in the same way as endogenous RNA molecules [50]. If the quantity of endogenous RNA per cell is unknown, as often is the case, there is a risk to generate insufficient or overabundant spike-in coverage for normalization. In the pipeline below we will point out where additional steps are required for ERCC spike-ins.

3.2 Software Installation

The workflow requires the installation of R (or RStudio) on the workstation and of various bioinformatics tools and a reference genome on the server.

On the server we use conda to install all the software packages required. Conda simplifies the installation of software and allows the creation of multiple environments containing different sets of software.

1. Go to the conda website (https://conda.io) and follow the instructions to download and install Miniconda. (There is also a distribution called Anaconda that includes a large number of preinstalled software packages, but the install is large and we don't require most of the software packages.)
2. Conda needs to be configured with a set of channels to download software from. Use the following shell commands to set up the channels:

```
conda config --add channels conda-forge
conda config --add channels bioconda
```

3. Create and activate a conda environment. This environment will contain all the software required for alignment and quantification.

```
conda create -n sc_rna_seq
source activate sc_rna_seq
```

4. Install the software tools required for alignment and quantification into the conda environment. Several tools are required:

```
conda install bowtie2
conda install trimmomatic
conda install star
conda install rsem
```

5. Download VDJPuzzle from https://bitbucket.org/kirbyvisp/vdjpuzzle2.

6. Create a conda environment for VDJPuzzle using the environment-explicit.txt in the folder you extracted VDJPuzzle to:

```
conda create --name vdjpuzzle --file environment-
explicit.txt
```

7. Download the reference genome for the relevant species (human or mouse) from the Ensembl website, taking note of which Ensembl release version was used (*see* **Note 4**). The full genome sequence, in FASTA format, and the associated gene set, in GTF format, are required (*see* **Note 3**). The version number is important to ensure the expression data is matched with the correct gene metadata (*see* Subheading 3.5, **step 5**). Save these files to a central location on the server and uncompress them. To simplify the instructions, we assume that these files have been placed in the same directory.
8. Add links to genome sequence and gene set files in the directory where these files are located. Call these links "genome.fa" and "annotation.gtf". The purpose of this step is to simplify the instruction below by providing short aliases for the genome reference files. For example, to create the links for the Ensembl version 92 files, run these commands:

```
ln -s Homo_sapiens.GRCh38.dna.primary_assembly.fa gen-
ome.fa
ln -s Homo_sapiens.GRCh38.92.gtf annotation.gtf
```

9. Set up some useful environment variables. These variables will be referenced in the commands in these instructions. For example, the THREADS and GENOME_DIR environment variables are used when creating a genome index in **step 11**.

   The following environment variables are required:

   THREADS: the number of threads that programs should use. Set this to the number of cores on the server's CPU.

   GENOME_DIR: the directory where the reference genome files were placed in the previous steps. This directory should contain the "genome.fa" and "annotation.gtf" links.

   VDJPUZZLE_DIR: the folder where vdjpuzzle2 was extracted to in **step 5**.

   Make sure these variables are in the environment before running any of the shell commands below (*see* **Note 5**). For example, the environment variables could be created using commands similar to the following:

```
export THREADS=8
export GENOME_DIR=/srv/genomes/human
export VDJPUZZLE_DIR=~/vdjpuzzle2
```

10. Build a Bowtie [51] index for the genome by running this command in the genome directory:

```
bowtie2-build --threads ${THREADS} genome.fa bt_index
```

This index is used for receptor reconstruction (*see* Subheading 3.13).

11. Create an index for the STAR [18] alignment software and a transcript references for the RSEM [21] quantification software using the command:

```
rsem-prepare-reference -p ${THREADS} --star --star-sjdboverhang 100 \
    --gtf ${GENOME_DIR}/annotation.gtf ${GENOME_DIR}/genome.fa \
    ${GENOME_DIR}/rsem_ref
```

This is an RSEM command that calls out to STAR to create an index and creates its own transcript reference. If ERCC spike-ins are used, the ERCC sequences need to be appended to the reference genome before creating the index and transcript references (*see* **Note 6**). As our example data doesn't have spike-ins we create the index on the original reference genome files.

That completes the software installation on the server. The server now has the software and data it requires to process scRNA-seq datasets.

The following step needs to be performed on the server when beginning work on a new dataset.

12. Create a directory to contain the sequencing data and results. We will call this directory the "working directory." This directory will initially contain the reads generated during sequencing, stored in FASTQ files. FASTQ files normally have a ". fastq" extension or ".fastq.gz" if they are compressed. A FASTQ file may contain reads for many samples, one sample, or part of a sample, depending on the technology and process used for sequencing. Here we assume paired end reads stored in two compressed files (".fastq.gz") per sample, one file for each end of the pair.

The experimental data should be organized on the hard drive in the following directory hierarchy:

```
<root folder>
    samples/
        sample1/
            reads_R1.fastq.gz
            reads_R2.fastq.gz
        sample2/
            reads_R1.fastq.gz
            reads_R2.fastq.gz
        ...
```

```
results/
    expr/
    vdj/
vdjworking/
```

Where <root folder> is the working directory. The "samples" directory contains one subdirectory for each sample. Replace "sample1," "sample2," etc., with the names of the samples. The reads_R1.fastq.gz and reads_R2.fastq.gz files in each sample directory are the two paired-end reads files. This directory structure is designed primarily to make the following instructions easier to write and understand. VDJPuzzle requires the data for each sample to have its own subdirectory, as well as requiring the FASTQ files to end in "_R1" and "_R2". Taking that into consideration, this directory structure, and the process presented in this chapter, may be treated as a starting point for developing a procedure suited for individual requirements.

Lastly, on the workstation the following software installation steps need to be performed.

13. Download and install R and RStudio. These instructions require R version 3.5 or greater.
14. Open RStudio and execute the following commands to install the required R packages:

```
install.packages("tidyverse")
install.packages("Rtsne")
install.packages("tcR")
install.packages("plotly")
source("https://bioconductor.org/biocLite.R")
biocLite("tximport")
biocLite("SingleCellExperiment")
biocLite("biomaRt")
biocLite("scater")
biocLite("scran")
biocLite("SC3")
biocLite("MAST")
```

3.3 Quality Control on Reads

The following steps need to be repeated once per sample. The scope of this step is to check the quality of the original reads and prepare for alignment. To reduce tedium, these steps can be automated using scripts (*see* **Note 7**) or Snakemake (*see* **Note 8**).

The sample name is represented by "${SAMPLE}" in the commands. This needs to be replaced with the name of the current sample when executing the commands, or the commands can be executed unchanged by setting up an environment variable:

```
export SAMPLE="sample_name"
```

Run all the commands within the working directory (*see* **step 12** in Software Installation above).

1. Activate the conda environment for alignment and quantification:

```
source activate sc_rna_seq
```

2. Reads from FASTQ files have associated quality scores. We use Trimmomatic [16] to trim low-quality bases from each read and filter out low-quality reads. This improves the performance of the alignment steps as the reads will better match the target genome. Trimmomatic can also remove any adaptor sequences appended to the read during library preparation. The options passed to Trimmomatic will vary according to the sequencing protocol used. The following command works with our example data, which has paired-end reads and adaptor sequences from Nextera library preparation:

```
(ADAPTORS=$CONDA_PREFIX/share/trimmomatic/adapters && \
trimmomatic PE -threads ${THREADS} -phred33 \
  samples/${SAMPLE}/reads_R1.fastq.gz \
  samples/${SAMPLE}/reads_R2.fastq.gz \
  samples/${SAMPLE}/R1_PAIRED.fastq.gz \
  samples/${SAMPLE}/R1_UNPAIRED.fastq.gz \
  samples/${SAMPLE}/R2_PAIRED.fastq.gz \
  samples/${SAMPLE}/R2_UNPAIRED.fastq.gz \
  ILLUMINACLIP:${ADAPTORS}/NexteraPE-PE.fa:2:30:10 \
  LEADING:3 TRAILING:3 SLIDINGWINDOW:4:15 MINLEN:36)
```

3.4 Alignment

To align scRNA-seq data we use STAR [18]. Following alignment the number of reads aligned to each gene is quantified using RSEM

Table 2
Common file types used in the analysis of RNA-seq data

File type	Extension(s)	Description
FASTA format	.fasta, .fa	Stores nucleotide or protein sequences
FASTQ format	.fastq	An extension of FASTA that includes a quality score for each nucleotide or amino acid. Used to store the output of high-throughput sequencing
Gene transfer format (GTF)	.gtf	A tab-separated file that describes the location of genes within a DNA sequence
Sequence alignment map (SAM)	.sam	Stores the alignment position of sequences to a reference genome, along with the sequences and quality scores
Binary alignment map (BAM)	.bam	A compressed SAM file

[21]. To simplify the instructions, we have used a facility in RSEM which calls the STAR command to align the reads before applying the RSEM quantification algorithm (*see* below). The alignment and quantification steps can be split into two commands to allow greater control over the parameters passed to STAR or to use alternative alignment or quantification software (*see* Table 1). STAR (and most other aligners) creates a BAM file (*see* Table 2) which is used by RSEM (and other quantification software). Be aware that the BAM file often needs to be sorted before passing it to the software package for gene expression quantification (e.g., RSEM). This can be achieved using the samtools sort command.

As with the quality control on reads section, this step needs to be repeated for each sample. It should be executed on the server, in the sc_rna_seq Conda environment, and can also be automated (*see* **Notes 7** and **8**). Alignment and quantification is typically time consuming and requires a significant amount of computing power. An alternative is to use a pseudo-aligner, which is fast enough to use on a workstation computer (*see* **Note 9**).

1. Run RSEM to align the reads and quantify gene expression using these commands:

```
mkdir results/expr/${SAMPLE}
rsem-calculate-expression -p ${THREADS} --paired-end --seed 12345 \
  --star --star-gzipped-read-file --no-bam-output \
  samples/${SAMPLE}/R1_PAIRED.fastq.gz \
  samples/${SAMPLE}/R2_PAIRED.fastq.gz \
  ${GENOME_DIR}/rsem_ref results/expr/${SAMPLE}/rsem
```

Here we are executing the alignment with STAR directly from RSEM. RSEM then takes the aligned reads and estimates expression levels for each gene and saves the results to a file called "rsem.genes.results" in the "results/expr/${SAMPLE}" directory.

3.5 Creation of Expression Matrix

After completing the alignment steps, the "results/expr" directory will contain a subdirectory for each sample containing RSEM result files. These files contain expression levels for each gene, measured in raw counts and transcripts per million (TPM) (*see* **Note 10** for more details on quantification measures). The data need to be combined into expression matrices with one row per gene and one column per sample. Most tools used for downstream analysis require this format. We will create two matrices: one containing raw counts and the other containing TPM. We use the TPM values for our data analysis, but the counts are useful for calculating quality control metrics (*see* Subheading 3.6).

Execute the following steps on the workstation using the R programming environment.

1. Copy the "results" directory from the working directory on the server to the workstation. (Alternatively, these steps could be executed in place on the server—*see* **Note 11**.)
2. Set the current working directory in R to the directory where the "results" directory was copied to (i.e., the parent directory of the "results" directory).
3. We use the *tximport* library [52] to load the TPM and count values from the "results" files into gene by sample matrices. Execute the following R code to create the matrices:

```
library(tximport)

# Get sample names and input files
sample_names <- list.dirs(path="results/expr",
                          full.names=F, recursive=F)
infiles <- paste0("results/expr/", sample_names, "/rsem.genes.results")

# Load data
txi <- tximport(infiles, type="rsem", txIn=F, txOut=F)
raw_counts <- txi$counts
raw_tpm <- txi$abundance
colnames(raw_counts) <- colnames(raw_tpm) <- sample_names
```

4. When analyzing expression data, it can be useful to know some details about each sample such as experimental treatment. We assume that a file called "sample_info.txt" exists containing sample information and has the following attributes:
   - The file is tab separated.
   - The first line of the file contains column names.
   - One of the columns is called "name" and the values in this column match the sample names used during alignment.

   For example, here are some lines from the sample information file for our example dataset:

Name	Number	Treatment	Bulk
Stim-single1_S1	1	Stim	FALSE
Stim-single2_S2	2	Stim	FALSE
Stim-ten2_S20	20	Stim	TRUE
Unstim-single1_S21	21	Unstim	FALSE

   Use the following code to load the sample information and sort it into the same order as the columns of the expression matrices. This code can be adapted to different file formats with minor modifications. For example, the CSV format can be accommodated by using the `read_csv` function in place of the `read_tsv` function.

```
# Load sample information and convert to data frame
library(tidyverse)
sample_info <- read_tsv("sample_info.txt", col_names=T)
sample_info <- as.data.frame(sample_info)
rownames(sample_info) <- sample_info$name

# Sort sample_info in same order as samples in raw_counts
sample_info <- sample_info[colnames(raw_counts),]
```

5. Download gene annotations using the following code. We download the gene annotations provided by Ensembl. Replace the version number in the code with the Ensembl version that was downloaded during Software Installation. The dataset parameter should also be changed if nonhuman genes are used; for example, use the "mmusculus_gene_ensembl" dataset for mouse genes.

```
library(biomaRt)
version <- 92
if (grepl(paste0(" ",version,"$"), listMarts()[1,"version"])) {
  # Required version is the latest version
  ensembl <- useMart("ensembl", dataset="hsapiens_gene_ensembl")
} else {
  # Required version is an archived version
  arc <- listEnsemblArchives()
  ensembl <- useMart("ensembl",
                     host=arc[arc[,1]==paste0("Ensembl ",version),3],
                     dataset="hsapiens_gene_ensembl")
}
gene_annotation <- getBM(attributes=
    c('ensembl_gene_id','external_gene_name','description',
      'chromosome_name','start_position','end_position',
      'gene_biotype'),
      mart=ensembl)
```

If ERCC spike-ins are being used, the ERCC genes should also be included in the gene annotations (*see* **Note 12**).

6. Next, we match the gene IDs in the annotations to the row names from the count matrix and make sure the annotations are sorted in the same order.

```
rownames(gene_annotation) <- gene_annotation$ensembl_gene_id
gene_annotation <- gene_annotation[rownames(raw_counts),]
```

7. Genes in Ensembl are identified by an ID such as "ENSG00000139187," but each gene also has a human readable name such as "KLRG1." Here we switch the gene names in our dataset from the IDs to the human readable versions. Ensembl provides human readable names, but unfortunately some of these names are duplicated, so we need to make the names unique (using the `make.unique` function in R). We also create a "feature_symbol" column containing the gene names as this is required by the *SC3* package, which we use for clustering. Execute this code to create the gene names:

```
gene_annotation$feature_symbol <-
  make.unique(gene_annotation$external_gene_name)
rownames(gene_annotation) <- gene_annotation$feature_symbol
rownames(raw_counts) <- gene_annotation$feature_symbol
rownames(raw_tpm) <- gene_annotation$feature_symbol
```

8. We use the `SingleCellExperiment` class to encapsulate all the expression values and annotations we need. A `SingleCellExperiment` object keeps the expression matrix and annotations synchronized and provides many methods to make working with the data more convenient [53]. Use the following code to create the `SingleCellExperiment` object:

```
library(SingleCellExperiment)
sce <- SingleCellExperiment(
  assays = list(counts = as.matrix(raw_counts),
                tpm = as.matrix(raw_tpm)),
  rowData = gene_annotation,
  colData = sample_info
)
```

9. Lastly, save the experiment data to a file for later use:

```
saveRDS(sce, "raw_data.rds")
```

3.6 Quality Control on Expression Matrix

Single-cell RNA-seq data may be affected by poor quality of the sorted cells, issues with sorting, and poor RNA extraction and amplification. These and other factors can impact on the quality of the gene expression estimated from scRNA-seq data. Poor-quality samples can be removed from the dataset before further analysis. In this section we filter out the poor-quality samples and filter out low expression genes.

We use a Bioconductor package called *scater*, which provides some useful functions for quality control and data visualization [28]. *Scater* provides a series of metrics, such as total counts per cell or mean expression per gene, that are useful for quality control.

1. Load the raw data that was saved to disk in the previous section.

```
library(SingleCellExperiment)
sce.orig <- readRDS("raw_data.rds")
```

2. Execute the following code to prepare the data for quality control:

```
library(scater)
# Create log counts
assay(sce.orig, "logcounts_raw") <- log2(counts(sce.orig) + 1)
# Remove genes with no data
sce.orig <- sce.orig[rowSums(counts(sce.orig)) > 0,]
# Calculate QC metrics
is.mito <- rowData(sce.orig)$chromosome_name == "MT"
sce.orig <- calculateQCMetrics(sce.orig, exprs_values="counts",
                               feature_controls=list(MT=is.mito))
```

The `SingleCellExperiment` class, which we are using to hold the data, allows many versions (assays) of the expression matrix to be stored, and we create a new assay to store log counts (*see* **Note 13**).

Here we remove all genes that have no reads aligned to them and then designate a set of genes to use as control genes. We use the mitochondrial genes as controls. If ERCC spike-ins are used, these should also be used as control genes (*see* **Note 14**).

3. It is helpful to visualize the data at several points during quality control, filtering, and normalization to see how the data has changed. In our instructions we will use principal component analysis (PCA) plots for this purpose, but there are several other plot types that are also useful (*see* **Note 15**). Use the following code to create a pre-quality control PCA plot:

```
plotPCA(sce.orig, exprs_values="logcounts_raw", colour_by="treatment",
        size_by="total_features", shape_by="bulk")
```

The PCA plot for the example data (Fig. 2) shows the PBMC samples clustered apart from the other two treatments. There appears to be an association between principle component 1 and the number of features detected. In concordance with this, samples from bulk cells appear further along PC1 than the rest of the treatment they belong to.

4. In the QC metrics produced by scater there are three metrics that we will use for quality control of cells: (1) total read count in the cell, (2) total features (no. of genes with reads aligned), and (3) percentage of counts attributable to mitochondrial

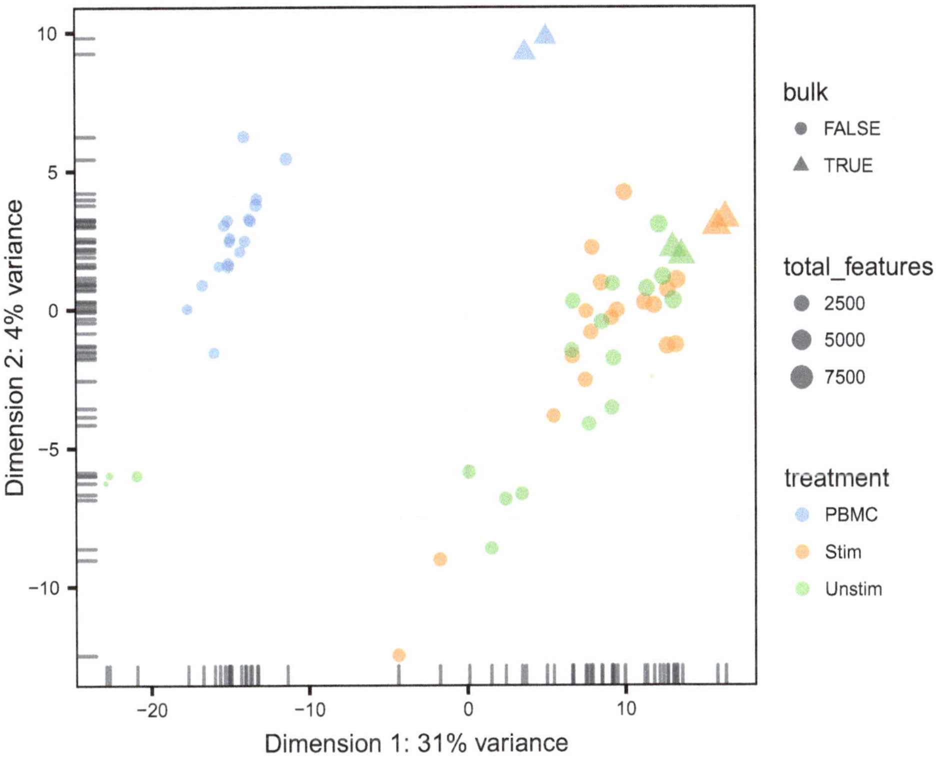

Fig. 2 PCA plot on the raw data prior to quality control

genes. If ERCC spike-ins are available, these should also be used (*see* **Note 14**). We will display these metrics in two plots.

The first plot shows total features and percentage of counts from mitochondrial genes. Use this code to produce the plot:

```
plotColData(sce.orig, x="total_features", y="pct_counts_MT",
        colour_by="treatment", shape_by="bulk")
```

This plot should be examined to determine which cells should be filtered out. Figure 3 shows this plot for our example data, with lines added to show our filtering cutoff points. Cells with high expression of mitochondrial genes can indicate poor-quality libraries [54]. For example, our data contained three cells with extremely high mitochondrial expression, and we chose to filter out all cells with more than 20% of reads aligned to mitochondrial genes. A low total feature count for a cell can indicate technical issues during the processing of the cell. We removed two cells with less than 1300 features, as these were clearly outliers when compared to the feature counts of the other cells (Fig. 3). For comparison, two studies comparing scRNA-seq datasets found an average of between 2,354 and 9,138 genes per cell (the lowest for Smart-Seq2 was in fact our

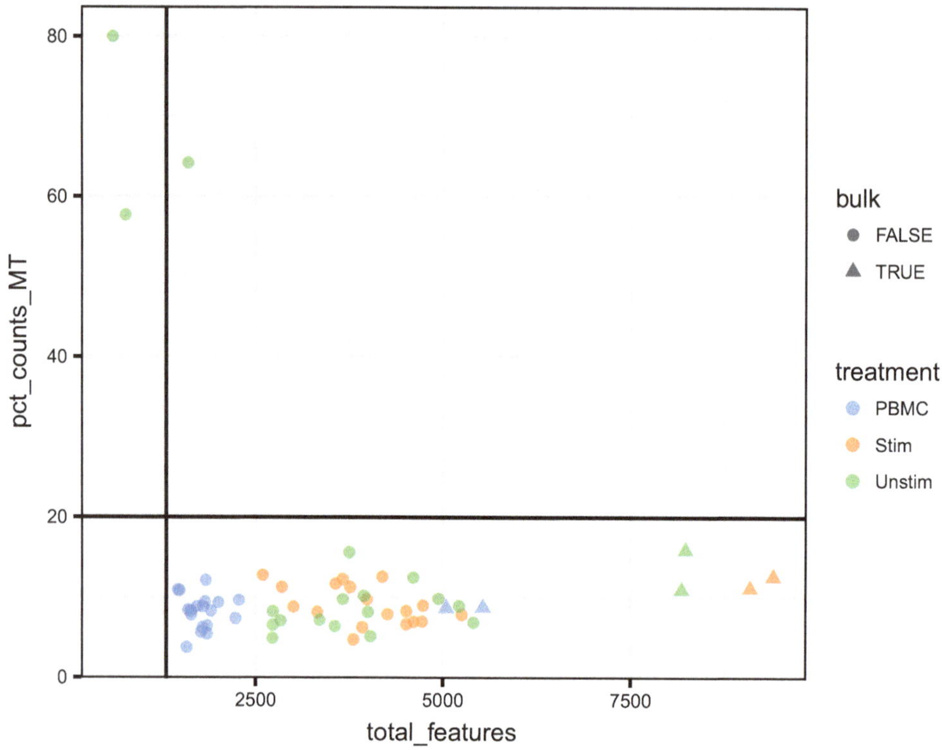

Fig. 3 Number of detected features (*x*-axis) and percentage of reads mapped to mitochondrial genes (*y*-axis) for each cell. Horizontal and vertical lines indicate the chosen filtering cutoff points for mitochondrial percentage and feature count, respectively

sample dataset with 2,563 genes per cell) [55, 56]. If a sample has an unusually high of feature count, this could indicate that there were multiple cells in the sample. The only samples in this data with high feature count were the bulk samples, which we know contain multiple cells. For reference, here is the code to produce the plot and add the filtering cutoff lines:

```
plotColData(sce.orig, x="total_features", y="pct_counts_MT",
            colour_by="treatment", shape_by="bulk") +
  geom_hline(yintercept=20) +
  geom_vline(xintercept=1300)
```

The next plot shows the total read count rather than total features and the percentage of reads from mitochondrial genes (again). A low total read count can also indicate technical issues. Use the following code to render the plot. We have included lines to indicate our chosen cutoff points (Fig. 4).

```
plotColData(sce.orig, x="total_counts", y="pct_counts_MT",
            colour_by="treatment", shape_by="bulk") +
  geom_hline(yintercept=20) +
  geom_vline(xintercept=3000000)
```

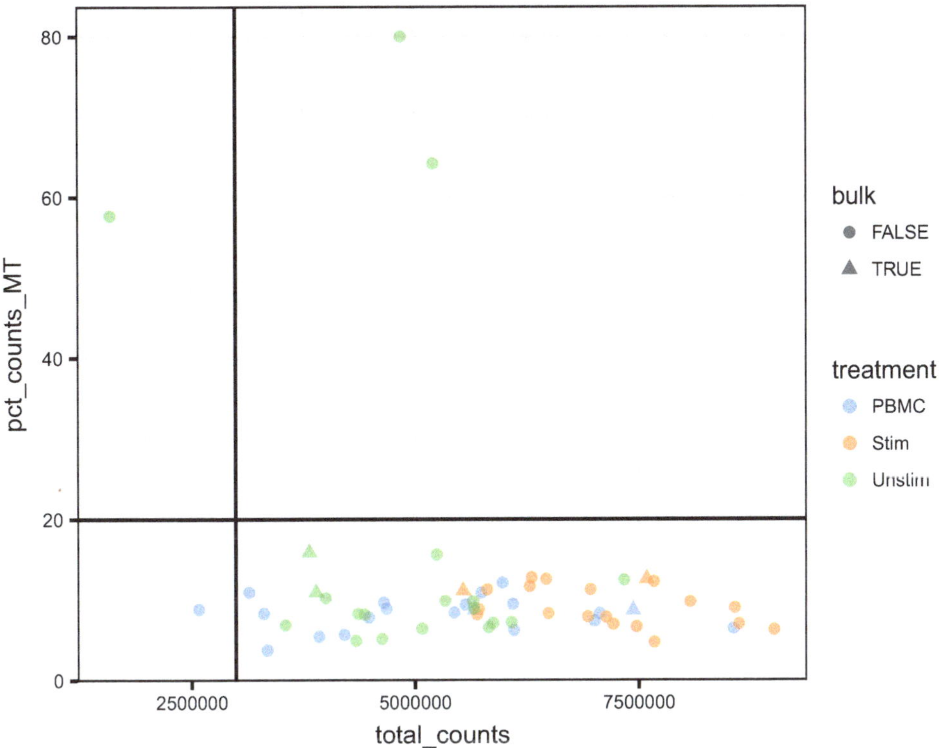

Fig. 4 Total count of reads mapped to genes (*x*-axis) and percentage of reads mapped to mitochondrial genes (*y*-axis) for each cell. Horizontal and vertical lines indicate the chosen filtering cutoff points for mitochondrial percentage and total count, respectively

Use these plots to decide which cells should be filtered out of the dataset. In our example, in addition to the filtering cutoffs we chose, we filter out the bulk samples as the rest of the analysis is designed to work with single cells.

5. Using the filtering cutoff points chosen in the previous step, create a variable to indicate which cells pass filtering. The following code sets up a column attribute to indicate which samples in the example data pass quality control and are not bulk samples:

```
colData(sce.orig)$use <- colData(sce.orig)$pct_counts_MT < 20 &
                         colData(sce.orig)$total_features > 1300 &
                         colData(sce.orig)$total_counts > 3000000 &
                         !(colData(sce.orig)$bulk)
table(colData(sce.orig)$use)
```

In our example four single-cell samples were marked for removal plus the six bulk samples.

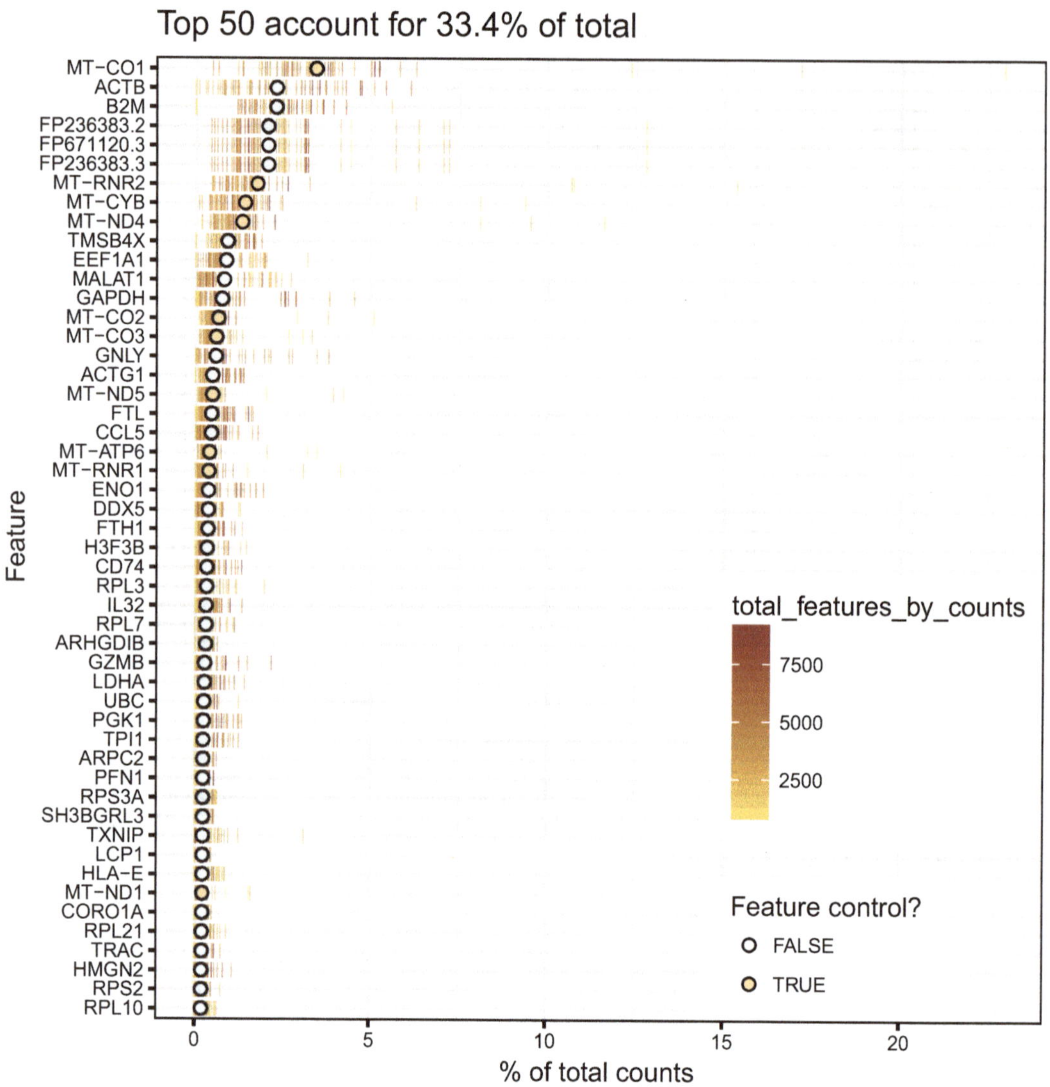

Fig. 5 Expression of the top 50 most highly expressed genes. The color of each bar represents the feature count for each cell (based on the "counts" assay)

6. To begin exploring, create a plot of the most highly expressed genes:

```
plotQC(sce.orig, type="highest-expression", exprs_values="counts")
```

For the example data, nothing anomalous is seen in the plot (Fig. 5). If this plot showed a few genes accounting for the majority of reads, those genes should be filtered out of the dataset.

7. We will filter out genes that were detected in less than four of the cells that passed quality control. This number can be

changed to a higher value for more conservative analyses. Here we define "detected" as having a read count greater than 5. This is arbitrary, but the idea here is to remove from the analysis genes that are lowly expressed and hard to detect and do not provide much information to distinguish cell types. Use the following code to filter the genes:

```
detected_counts <- counts(sce.orig[ , colData(sce.orig)$use]) > 5
cells_per_gene <- rowSums(detected_counts)
rowData(sce.orig)$use <- cells_per_gene >= 4
table(rowData(sce.orig)$use)
```

In the example dataset, the gene filtering resulted in the removal of bit more than half of the genes, leaving 8590 genes in the dataset.

8. Cell and gene filtering are now complete. Next, create a new dataset containing only the cells and genes to keep, and recalculate the quality control metrics:

```
sce.qc <- sce.orig[rowData(sce.orig)$use, colData(sce.orig)$use]
is.mito.qc <- rowData(sce.qc)$chromosome_name == "MT"
sce.qc <- calculateQCMetrics(sce.qc, exprs_values="counts",
                             feature_controls=list(MT=is.mito.qc))
```

9. Now create another PCA to look at the data after quality control.

```
plotPCA(sce.qc, exprs_values="logcounts_raw", colour_by="treatment",
        size_by="total_features", shape_by="bulk")
```

After filtering the PCA plot looks similar to before filtering, but the outlying samples have been removed (Fig. 6 compared with Fig. 2).

3.7 Normalization

Normalization is an important component for the analysis of scRNA-seq data as it can control some of the variation caused by factors that are not of interest. For example, systematic differences in gene counts between cells are caused by differences in reverse transcription efficiency, amplification rate, and read depth. Normalization is a way to remove these systematic differences prior to analyzing the data.

In the instructions above we used transcripts per million (TPM) to measure gene expression. TPM is actually a form of normalization that controls for sequencing depth and gene length. However, TPM is not useful for comparing between cells as it is a within-sample normalization method. The TPM value gives an estimate of the fraction of transcripts in a cell coming from a particular gene, but the total number of transcripts varies between

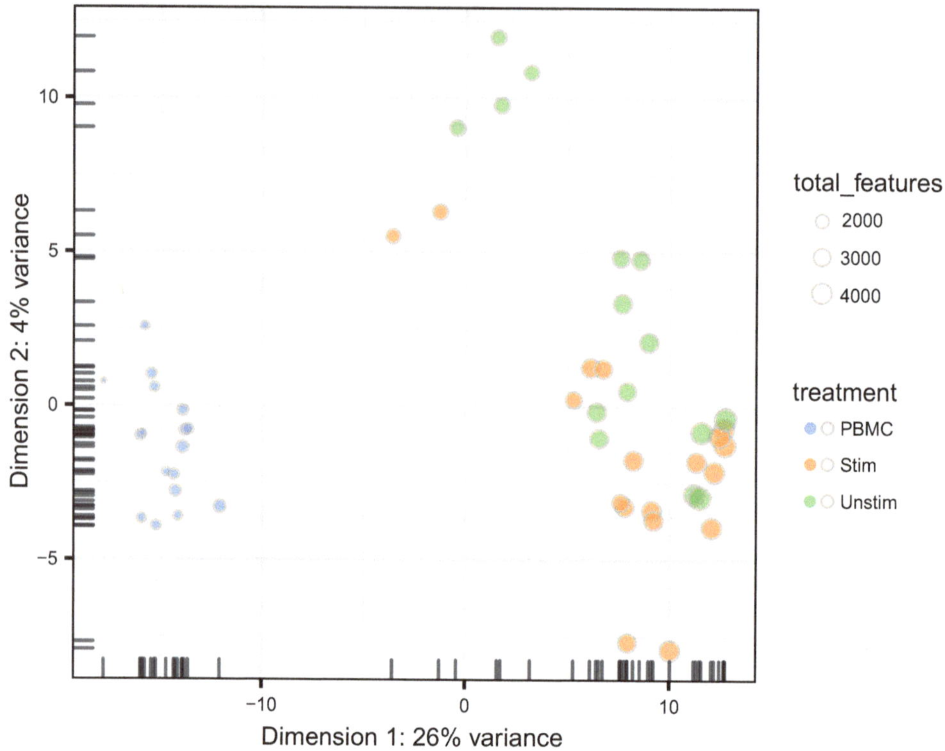

Fig. 6 PCA plot of cells passing quality control

cells. To compare the expression of a gene between cells, a between-sample normalization method is required. We use the deconvolution normalization method in the *scran* R package, which is designed to handle the high number of zeroes in the single-cell data and can deal with heterogeneity in the data by clustering the data and normalizing each cluster separately [57]. We pass the TPM values in to *scran* so that we retain the gene length normalization.

1. Normalize the data using *scran* with this code:

```
library(scran)
qclust <- quickCluster(sce.qc, min.size=15, assay.type="tpm")
table(qclust)
sce.norm <- computeSumFactors(sce.qc, sizes=seq(8, 36, 4),
                              clusters=qclust, assay.type="tpm")
sce.norm <- normalize(sce.norm, exprs_values="tpm")
```

The size parameters may need to be adjusted for the `quickCluster` and `computeSumFactors` functions to obtain good results. See the *scran* documentation for more details.

2. The normalized values are now in the "logcounts" assay in the sce.norm object. Create a new PCA plot to see how normalization has changed the data.

```
plotPCA(sce.norm, exprs_values="logcounts",
        colour_by="treatment", size_by="total_features")
```

The post-normalization plot for the example data is in Fig. 7. There is still an association between principal component 1 and feature count, but the amount of variance explained by principal component 1 has dropped. It would not be possible to completely remove this association as the PBMC samples had lower feature counts than the other two treatment conditions, so feature count is confounded with treatment.

3.8 Batch Effects

An important step in the analysis of scRNA-seq data from T cells is to account for intrinsic biases arising from the experimental design. Some of the factors are uncontrollable, such as performing the experiments across different days, users, time points, or samples. All these factors can bias the outcome of the analysis as several unwanted factors may explain a significant component of the observed variation in the gene expression profiles. It is nontrivial to control for all these factors. Several recommendations can however be made. Here are the most notable that arose from our experience with dealing with scRNA-seq data arising from human cryopreserved samples where we sorted rare T cell types at different time points:

- Controlling for library quality: Using primary cells, often cryopreserved, may result in poor sorting quality. It is advisable to consider control cells, for instance, cell lines that are sorted and prepared in parallel to the sample of interest. Usually cells derived from cell lines are larger in size and provide greater library size. Nevertheless, this forms a positive control for the experiment.
- Longitudinal analysis: In the analysis of the data used here, there were no longitudinal samples collected and sequenced. However, typically for T cell studies, multiple time points can be used for scRNA-seq. It is important to spend time in the design of experiment, for instance, ensuring that cells across different time points are prepared on the same day and on the same batch of reagents.

The example we are using in this chapter was processed in a single sequencing run, so there should be no batch effects arising from sequencing batch. In larger studies batch effects can be controlled by including multiple sequencing treatment conditions in each sequencing run or having multiple batches per treatment condition. In these cases, batch correction software can be applied to remove variation due to batch effect. *See* Table 1 for a list of batch correction software. However, if there is a 1-to-1 relationship between treatment condition and sequencing batch, batch effects are confounded with treatment condition and cannot be controlled for. Also, researchers need to be aware that there are other sources of variation besides sequencing batch that cannot always be avoided. In our example, we cannot be certain that all the differences between the PBMCs and the cultured cells are due to biological differences and not due to technical differences occurring, for example, during flow cytometric sorting.

3.9 Dimensionality Reduction

Given the large size of scRNA-seq expression matrices, often the data are transformed before clustering using a dimensionality reduction technique, such as principal component analysis (PCA) (which we introduced in Subheading 3.6; *see* Figs. 2, 6, and 7) or t-distributed stochastic neighbor embedding (tSNE) (*see* **Note 15** and Fig. 16). This analysis has been the most prolific in terms of software development (*see* online repository links in **Note 1**). Classical approaches include density-based clustering, hierarchical clustering, and *k*-means.

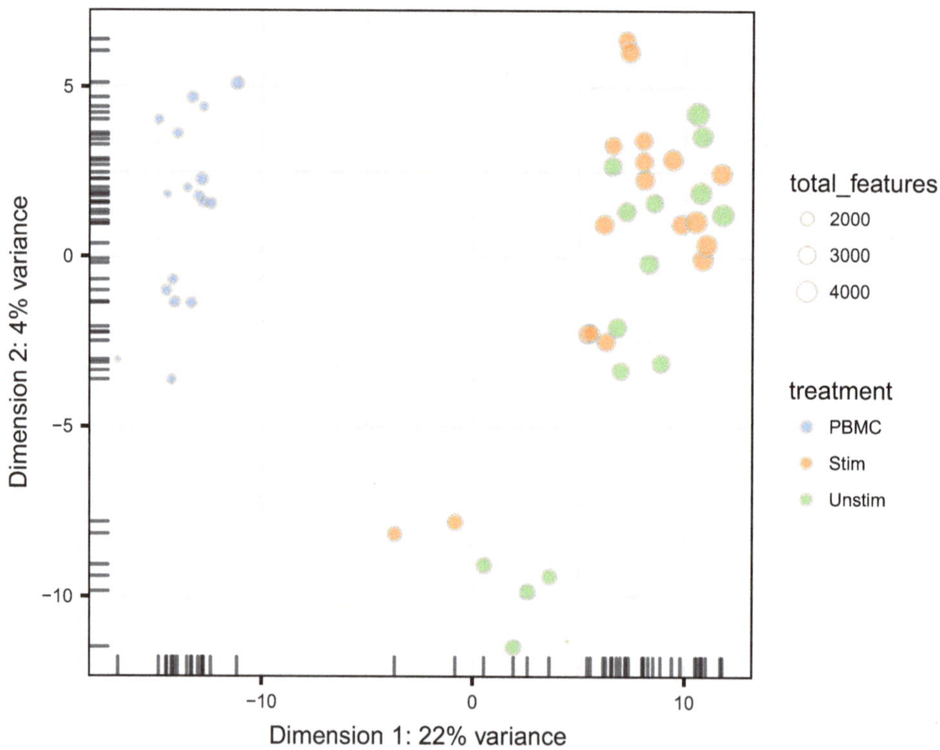

Fig. 7 PCA plot of cells after normalization

To perform PCA dimensionality reduction the following functions from the *scater* package can be used:

```
sce.norm <- runPCA(sce.norm)
```

To access the results, use this:

```
reducedDim(sce.norm, "PCA")
```

To plot the results, use this:

```
plotPCA(sce.norm)
```

Similarly, for tSNE:

```
sce.norm <- runTSNE(sce.norm, perplexity=1.5)
```

To access the results, use this:

```
reducedDim(sce.norm, "TSNE")
```

To plot the results, use this:

```
plotTSNE(sce.norm)
```

See the documentation for these functions for details on parameters that can be used to tune the results. In particular, changing the perplexity parameter for tSNE can have a big effect on the results.

3.10 Clustering

Clustering is a common technique utilized to identify cell subsets from scRNA-seq data. Clustering is usually obtained by measuring similarity between expression patterns across cells. This approach can be used to identify new cell types or different statuses within the same cell type, such as effector versus memory T cells. Similarity is obtained by measuring distances between gene expression profiles using correlation coefficients or distance functions, e.g., Euclidean distances.

The following steps show how clustering can be performed using the *SC3* package in R. *SC3* uses two dimensionality reduction methods (PCA and graph Laplacian-based method), as well as three different distance metrics (Euclidean, Pearson, and Spearman). *SC3* clusters cells using the consensus of the clusters obtained using each combination of methods [37].

1. Run *SC3* using the following command:

```
library(SC3)
sce.norm <- sc3(sce.norm, ks=2:5, biology=T)
```

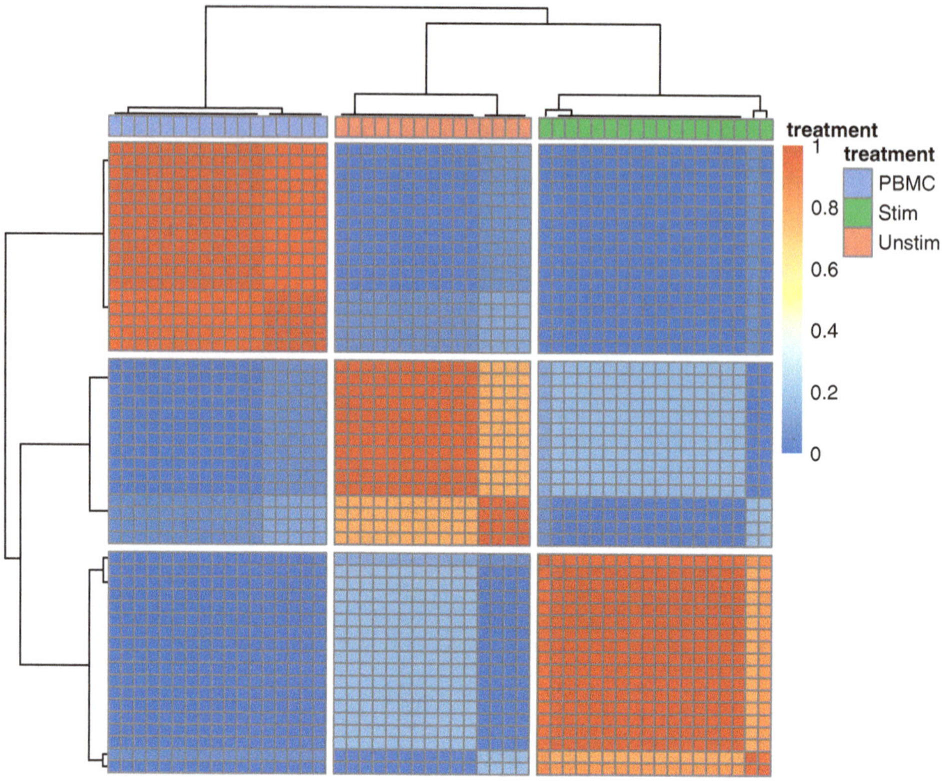

Fig. 8 Consensus clustering produced by the *SC3* package with three clusters ($k = 3$)

SC3 can be executed on a range of *k* values (=number of clusters). Here we cluster for *k* values between 2 and 5.

2. The results of *SC3* clustering can be viewed using the following command:

```
sc3_interactive(sce.norm)
```

This command will open a browser window with an app displaying the results. Use this app to compare the clustering at different values of *k*, look at differentially expressed genes and marker genes, and read short descriptions of the plots and how to interpret them.

3. Choose an optimal value of *k* based on the results shown in the interactive app and experimental considerations. Here we chose a *k* value of three clusters in our example. This *k* value can be used to generate various plots. For example, the following code plots the clustering consensus matrix (Fig. 8):

```
sc3_plot_consensus(sce.norm, k=3, show_pdata=c("treatment"))
```

3.11 Differential Expression

Differential expression (DE) analysis allows the performance of specific hypothesis testing on cell status, such as treated versus untreated, or early versus late, or effector versus memory cell types. For DE analysis we use the *MAST* package [58]. *MAST* is a model-based workflow that accounts for the bimodality of single-cell data, in that it accounts for a bimodal distribution of gene expression with a first peak around zero (hence estimating the proportion of zero values per gene). *MAST* performs modeling-based estimates of the total rate of gene expression per cell with respect to a background value (discrete) and of the actual mean expression level (continuous) values. Information from the discrete and continuous parts is combined to infer changes in expression between groups of cells. The following instructions describe the workflow to perform DE with *MAST*.

1. *MAST* uses a nonstandard class to store experiment data, so we need to convert the data into the required format.

```
library(MAST)
mast.norm <- FromMatrix(logcounts(sce.norm), colData(sce.norm),
                        rowData(sce.norm))
colData(mast.norm)$wellKey <- colData(mast.norm)$name
rowData(mast.norm)$primerid <- rowData(mast.norm)$feature_symbol
```

2. Genes that are not being expressed may still have a small nonzero expression value if, for example, the gene recently ceased expression and not all transcripts have been degraded yet. *MAST* includes a thresholding algorithm that removes this background noise by zeroing out values below a threshold and allows this threshold to vary across genes. Execute the following code to perform the thresholding:

```
thres <- thresholdSCRNACountMatrix(assay(mast.norm))
par(mfrow=c(5,2))
plot(thres)
assays(mast.norm) <- list(thresh=thres$counts_threshold,
                          logcounts=assay(mast.norm))
```

   This code also creates a plot that illustrates the results of this command for few genes with high and low expression levels compared to threshold (Fig. 9). This plot should be examined to ensure appropriate choice of the threshold.

3. The next step is to separate the data into groups across which gene expression will be compared, e.g., a background set and a test set. Groups can be chosen from the clusters identified via *SC3* (*see* **Note 16**) or a known experimental variable (treated vs. untreated). In our example we will compare gene expression between PBMCs and cultured cells (the stimulated and unstimulated treatments). The following example code

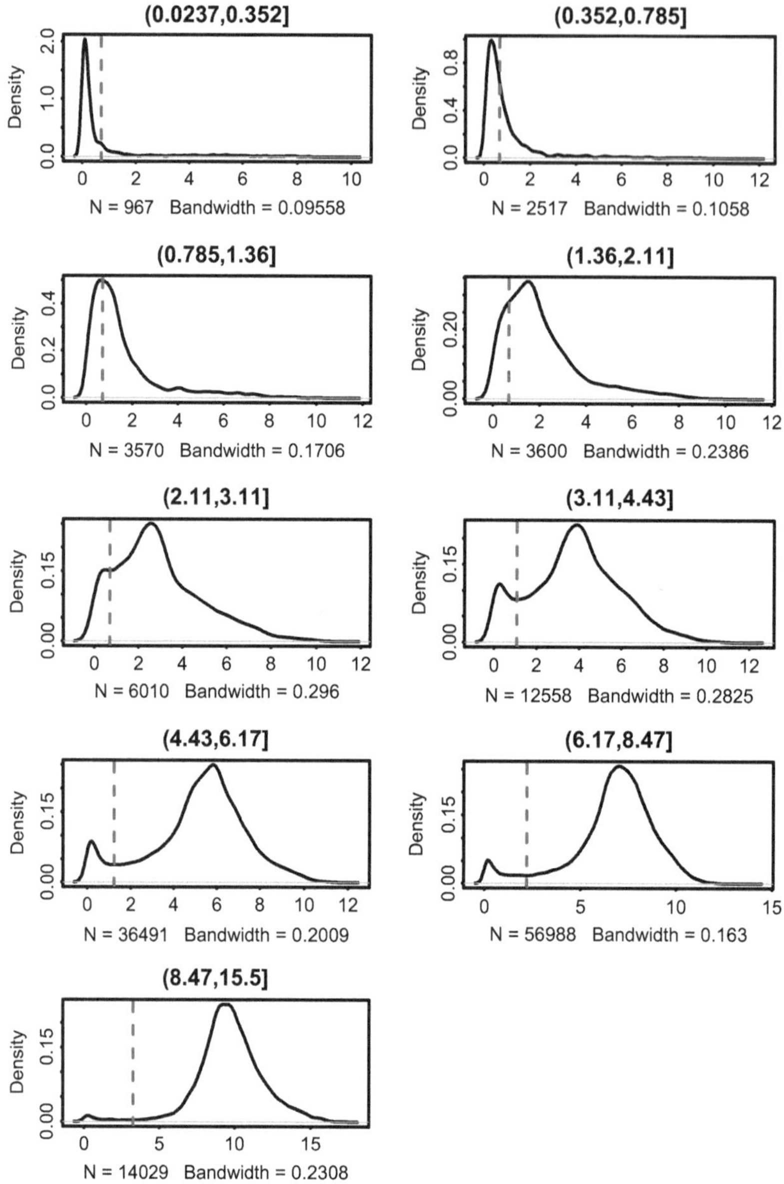

Fig. 9 Thresholds used by *MAST* to distinguish expressed from non-expressed genes. Here the genes are binned by median expression and the expression value density is plotted for each bin with the relavent threshold (dashed red lines)

creates a new factor to separate PBMCs from cultured cells, sets PBMC as the base factor level, and creates the *MAST* model:

```
colData(mast.norm)$cell_source <- factor(
 ifelse(colData(mast.norm)$treatment=="PBMC", "PBMC", "Culture"))
colData(mast.norm)$cell_source <-
  relevel(colData(mast.norm)$cell_source, "PBMC")
zlm_cell_source <- zlm(~cell_source, mast.norm)
```

4. Next use the model to perform differential expression tests using this code:

```
summary_cell_source <- summary(zlm_cell_source, logFC=T, doLRT=T)
```

5. The following code provides a p-value adjusted for multiple testing (using the Benjamini and Hochberg method [59] to generate a false discovery rate (FDR) value) and a log fold change. Only genes with $p < 0.05$ and log fold change >1 are returned. We have provided a function for extracting results (based on code in the *MAST* documentation [40]):

```
library(data.table)

# Function to get MAST results.
#  Based on code in MAST documentation.
getMASTResults <- function(sca, testSummary, contrastName,
                           pValue, fcThreshold) {
  # Extract p-values and log fold changes
  summary_dt <- testSummary$datatable
  p_values <- summary_dt[contrast==contrastName & component=='H',
                        .(primerid, 'Pr(>Chisq)')]
 log_fcs <- summary_dt[contrast==contrastName & component=='logFC',
                        .(primerid, coef, ci.hi, ci.lo)]
  fc_hurdle <- merge(p_values, log_fcs, by='primerid')

  # Adjust p-values for multiple testing
  fc_hurdle[,fdr:=p.adjust('Pr(>Chisq)', 'fdr')]

  # Get significant results and merge in gene details
  fc_hurdle_sig <- fc_hurdle[fdr<pValue & abs(coef)>fcThreshold]
  fc_hurdle_sig <- merge(fc_hurdle_sig, as.data.table(mcols(sca)),
                        by='primerid')

  # Sort by p-value
  setorder(fc_hurdle_sig, fdr)
}

results <- getMASTResults(mast.norm, summary_cell_source,
          "cell_sourceCulture", pValue=0.05, fcThreshold=1)
```

6. The results of differential expression can be visualized using violin plots. Use the following code to display violin plots for the top 16 differentially expressed genes:

```
library(ggplot2)

plot_genes <- results[1:min(30,nrow(results)), feature_symbol]
plot_data <- mast.norm[plot_genes,]
plot_data <- plot_data[,order(colData(plot_data)$cell_source)]
```

```
flat_data <- as(plot_data[1:16,], 'data.table')
ggplot(flat_data, aes(x=cell_source, y=thresh, color=cell_source)) +
  ylab("log expression") +
  geom_jitter() +
  facet_wrap(~feature_symbol, scale='free_y') +
  geom_violin()
```

Figure 10 show the violin plot for the example data.

7. An alternative way to visualize differentially expressed genes is to use a heatmap:

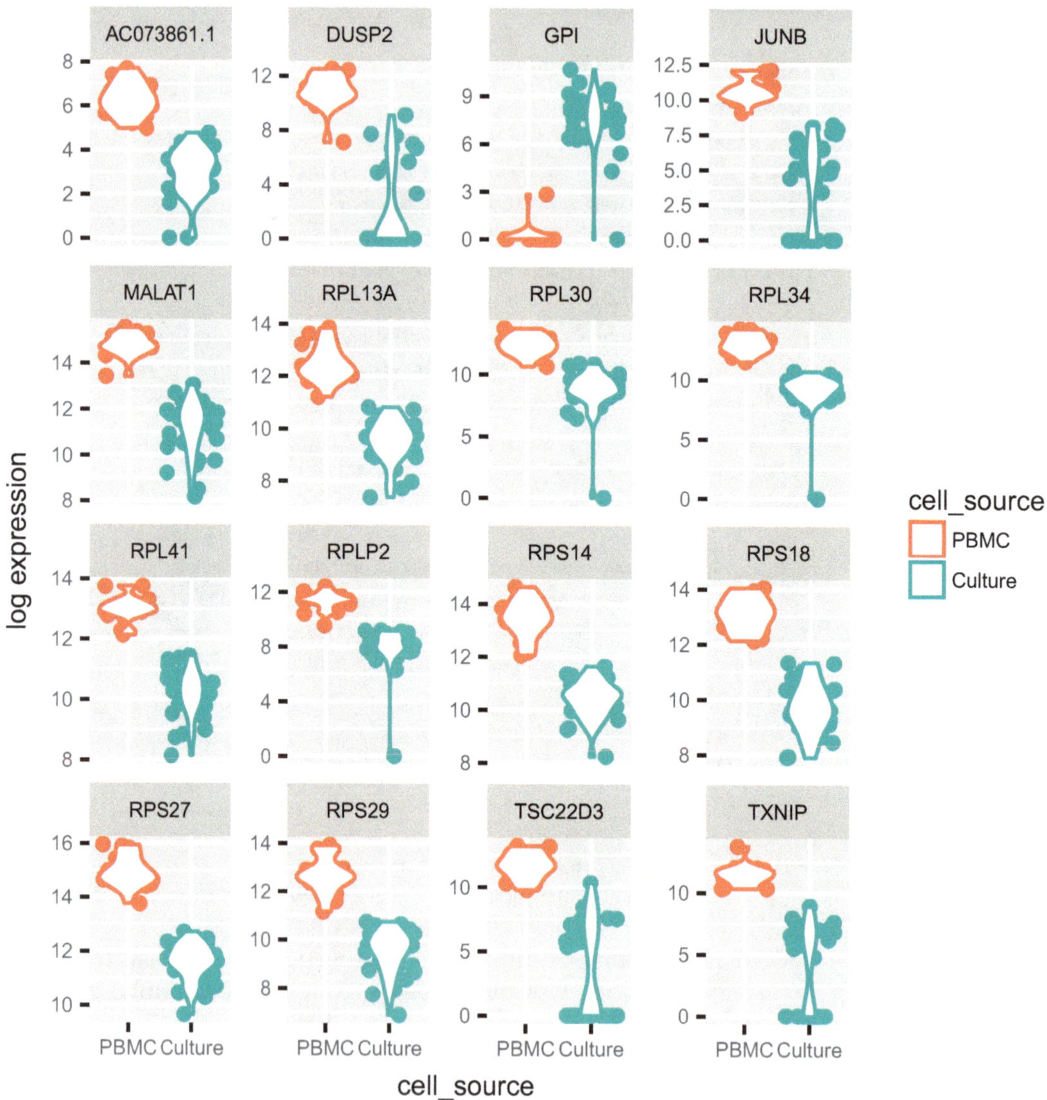

Fig. 10 Expression of top 16 differentially expressed genes, sorted by *p*-value, between PBMC and cultured samples, represented as violin plots

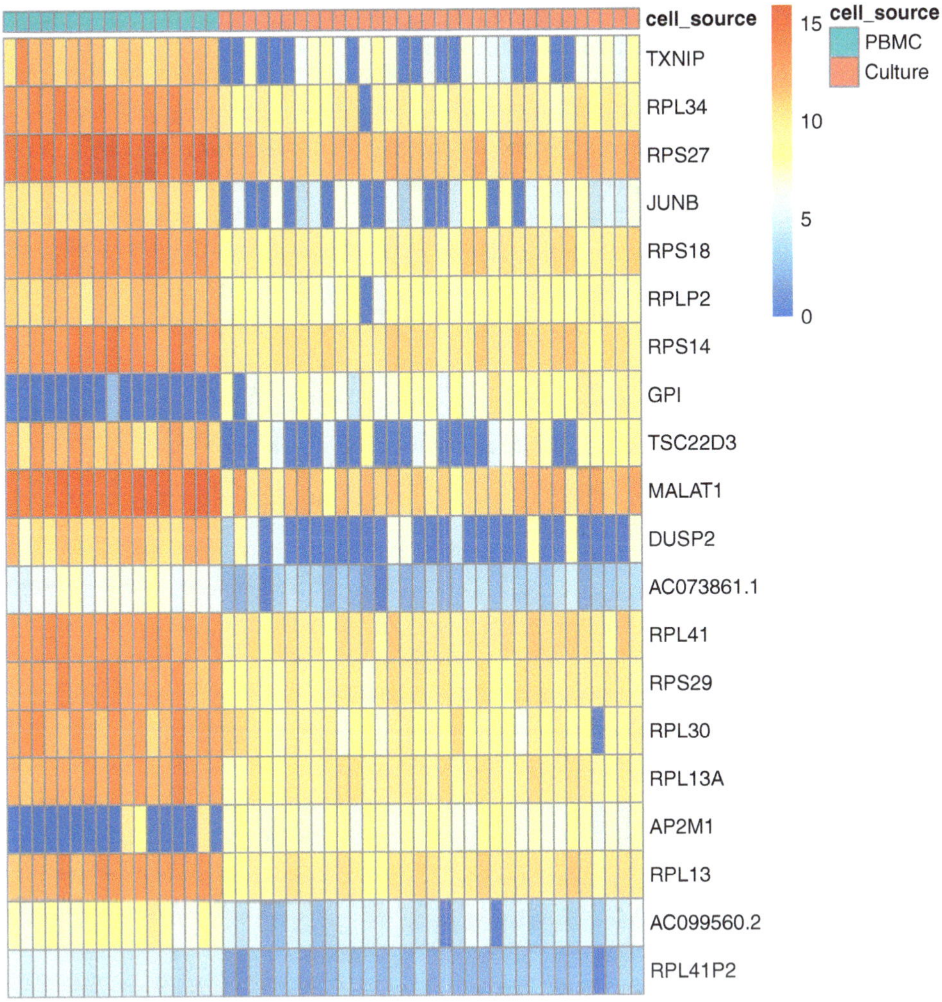

Fig. 11 Expression of top 20 differentially expressed genes, sorted by *p*-value, between PBMC and cultured samples, represented as a heatmap

```
library(pheatmap)
heatmap_ann <- as.data.frame(colData(plot_data)[,"cell_source",drop=F])
pheatmap(assay(plot_data), annotation=heatmap_ann,
         cluster_rows=F, cluster_cols=F, show_colnames=F)
```

Figure 11 shows the heatmap for the example data.

3.12 Gene Set Enrichment Analysis (GSEA)

A useful output in DE analysis is to perform GSEA to find categories of genes, associated with particular functions, phenotypes, or diseases, that are overrepresented in the DE genes. This step can be easily performed by exporting the set of DE genes identified from scRNA-seq into any of several available web-based programs that perform this analysis, e.g., https://david.ncifcrf.gov/ [45]. Here we perform this step using *MAST*. The underlining idea of this approach is to test average differences in gene expression between background and test sets. This is performed by calculating the *z*-scores (measuring how many standard deviations an element is from the mean) and performing a bootstrapping analysis to control for gene-gene correlation, which is known to affect the false significance (type I error) [58]. This analysis identifies the gene sets that differ from the average coefficient outside the set. As *MAST* creates a discrete, continuous, and also combined model, all these can be used for GSEA. This approach also allows the use of custom gene sets to test for enrichment in specific subsets or functions.

The following commands are an adaptation of code provided in the MAST vignettes [40] and perform GSEA analysis based on the differentially expressed genes identified via MAST:

1. Load the gene sets to test using the following code:

```
module <- "BTM"
min_gene_in_module <- 5
packageExt <- system.file("extdata", package='MAST')
module_file <- list.files(packageExt, pattern=module, full.names=T)
gene_set <- getGmt(module_file)
gene_ids <- geneIds(gene_set)
gene_ids <- gene_ids[!names(gene_ids)%like%"TBA" &
                     !names(gene_ids)%like%"B cell"]
sets_indices <- limma::ids2indices(gene_ids,
                        mcols(sce.norm)$external_gene_name)
# Only keep modules with at least min_gene_in_module
sets_indices <- sets_indices[sapply(sets_indices, length)
                             >= min_gene_in_module]
```

2. The bootstrapping is performed as follows:

```
boots <- bootVcov1(zlm_cell_source, R=50)
```

The parameter *R* identifies the number of bootstrap replicates, and it should be above 50 to have reasonable results.

3. Finally, we can test gee set enrichment in the three models generated above (zlm_cell_source, *see* Subheading 3.11)

```
gsea <- gseaAfterBoot(zlm_cell_source, boots, sets_indices,
                      CoefficientHypothesis("cell_sourceCulture"))
z_stat_comb <- summary(gsea, testType='normal')
print(z_stat_comb$set[1:5])
```

4. The results of this analysis can be interpreted by visualizing as a heatmap with *z*-scores for each gene set and each model:

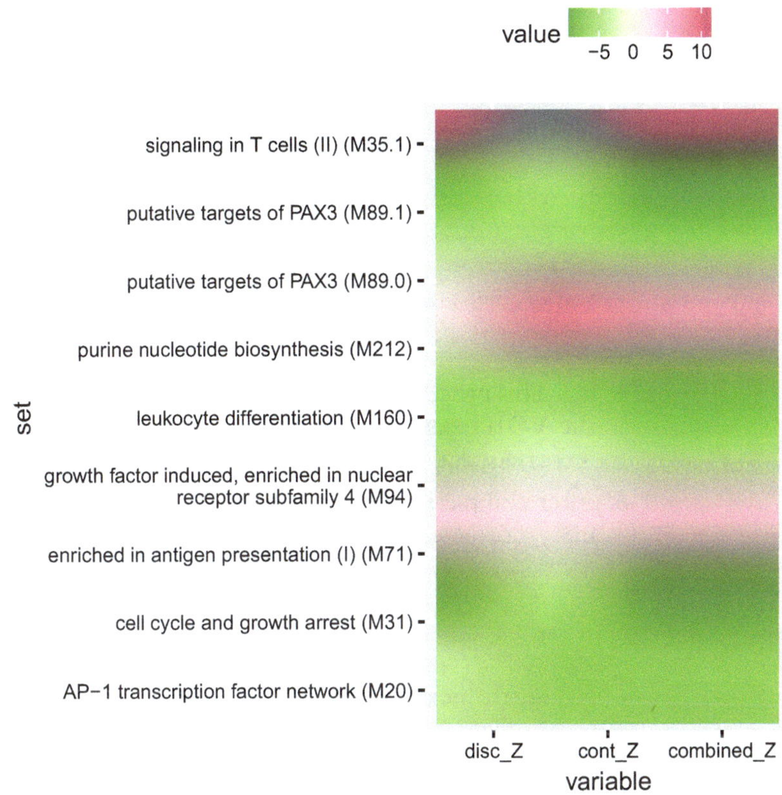

Fig. 12 GSEA using MAST pipeline. Heatmap showing z-scores for discrete, continuous, and combined models

```
sigModules <- z_stat_comb[combined_adj<.01]
sigModules <- sigModules[1:10,]
gseaTable <- melt(sigModules[, .(set, disc_Z, cont_Z, combined_Z)],
                  id.vars='set')
ggplot(gseaTable, aes(y=set, x=variable, fill=value)) +
  geom_raster() +
  scale_fill_distiller(palette="PiYG")
```

This heatmap (*see* Fig. 12) identifies several signatures of T cell activations, e.g., signaling in T cells, leukocyte differentiation, cell cycle, and growth arrest, which are consistent with the nature of the comparison between resting and ex vivo activated T cells.

3.13 Reconstruction of Full-Length TCR

We developed and recommend VDJPuzzle for TCR reconstruction [7, 15]. Our benchmark analysis showed comparable, and for some datasets better, success rate compared to other software, at least for long reads [55]. VDJPuzzle documentation can be found at https://bitbucket.org/kirbyvisp/vdjpuzzle2. Here we summarize the major steps:

1. Activate the vdjpuzzle Conda environment:

```
source activate vdjpuzzle
```

2. Change into the "vdjworking" folder and run vdjpuzzle:

```
cd vdjworking
${VDJPUZZLE_DIR}/bin/vdjpuzzle ../samples --THR=${THREADS} \
 --species=human --type=t --gtf=${GENOME_DIR}/annotation.gtf \
  --bowtie-index=${GENOME_DIR}/bt_index 2>&1 | tee log.txt
cd ..
```

This process will take some time (a few hours per cell). See the VDJPuzzle documentation for command-line options to increase speed by running processes in parallel or on a cluster.

3. After completion the results will be in the "vdjworking/summary_corrected" directory. Copy this into the "results" folder for easy access.

```
cp vdjworking/summary_corrected/* results/vdj
```

See the next section for instructions on how to visualize the results from TCR reconstruction.

3.14 TCR Visualization

In this section we provide instructions for a few different ways to visualize the TCR reconstruction results. There are many other possible ways to visualize this data and we introduce a few R packages that are useful on this purpose. Before executing these instructions, copy the "results" directory from the server to the workstations and set the working directory in R to the directory into which you copied the "results" directory.

1. Load the TCR data into R using this code:

```
TRA_data =
  read.table("results/vdj/TRA.csv", sep="\t",
           header=T)[c("CellID", "v", "d", "j", "cdr3aa")]
TRB_data =
  read.table("results/vdj/TRB.csv", sep="\t",
           header=T)[c("CellID", "v", "d", "j", "cdr3aa")]
```

2. Remove data for the bulk samples from the TCR results so we can focus on the single cells. In our example the bulk samples have the string "-ten" in the sample name:

```
TRA_data = TRA_data[! grepl("-ten", TRA_data$CellID),]
TRB_data = TRB_data[! grepl("-ten", TRB_data$CellID),]
```

3. In the VDJPuzzle results the sample name has a chain type prefix added and the gene names have a frequency appended. Remove these from the data:

```
# Remove chain prefix from cell ID
# i.e. Convert TRA_PBMC-single10_S50 into PBMC-single10_S50
TRA_data$CellID = gsub("^TRA_", "", TRA_data$CellID)
TRB_data$CellID = gsub("^TRB_", "", TRB_data$CellID)

# Remove frequencies from genes.
# i.e. Convert TRAV10*01 into TRAV10
TRA_data$v = gsub("[*][0-9]+$", "", TRA_data$v)
TRB_data$v = gsub("[*][0-9]+$", "", TRB_data$v)
TRA_data$d = gsub("[*][0-9]+$", "", TRA_data$d)
TRB_data$d = gsub("[*][0-9]+$", "", TRB_data$d)
TRA_data$j = gsub("[*][0-9]+$", "", TRA_data$j)
TRB_data$j = gsub("[*][0-9]+$", "", TRB_data$j)
```

4. Create a data table containing the TRA and TRB combination for each sample:

```
TRab_data = merge(TRA_data, TRB_data, by="CellID")
TRab_data$cdr3aa = paste(TRab_data$cdr3aa.x,
                         TRab_data$cdr3aa.y, sep = "_")
TRab_data = TRab_data[c("CellID", "v.x", "j.x",
                        "v.y", "d.y", "j.y", "cdr3aa")]
```

5. The column names need to be changed to the names expected by the visualization functions we use below:

```
colnames(TRA_data) = c("CellID","V.gene","D.gene","J.gene",
                       "CDR3.amino.acid.sequence", "Read.count")
colnames(TRB_data) = c("CellID","V.gene","D.gene","J.gene",
                       "CDR3.amino.acid.sequence", "Read.count")
colnames(TRab_data) = c("CellID","V.gene (A)","J.gene (A)",
                        "V.gene (B)","D.gene (B)","J.gene (B)",
                        "CDR3.amino.acid.sequence")
```

6. The *tcR* package [60] in R provides a set of functions for the analysis of TCR repertoires. Here we demonstrate using *tcR* to produce two histograms. The first is a histogram of J-gene usage in the beta chain (Fig. 13). The second is a histogram of the CDR3 amino acid sequence lengths in the alpha chain (Fig. 14). The function we use for the second plot was originally intended for nucleotide sequences but may be adapted for amino acids by temporarily renaming a column. Use the following code to create these plots:

```
library(tcR)
vis.gene.usage(TRB_data, c(HUMAN_TRBJ),
 .main="TRB J gene usage")
```

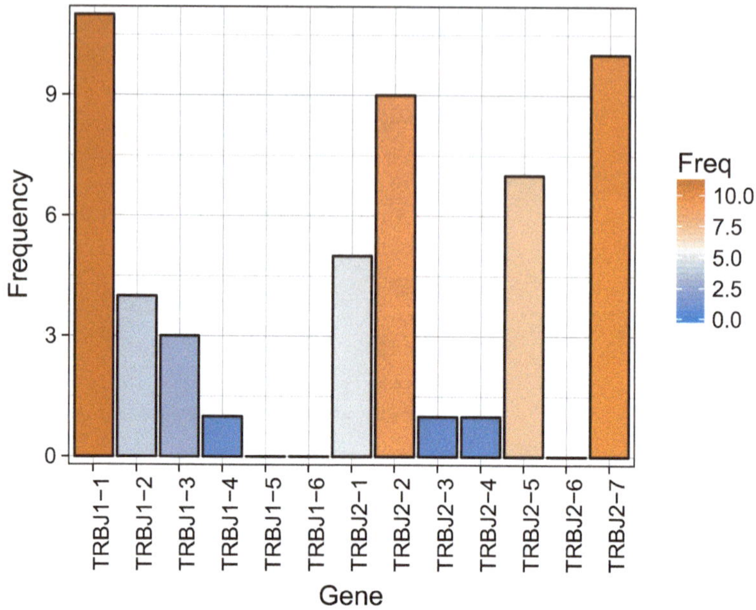

Fig. 13 Histogram of TRB J gene usage

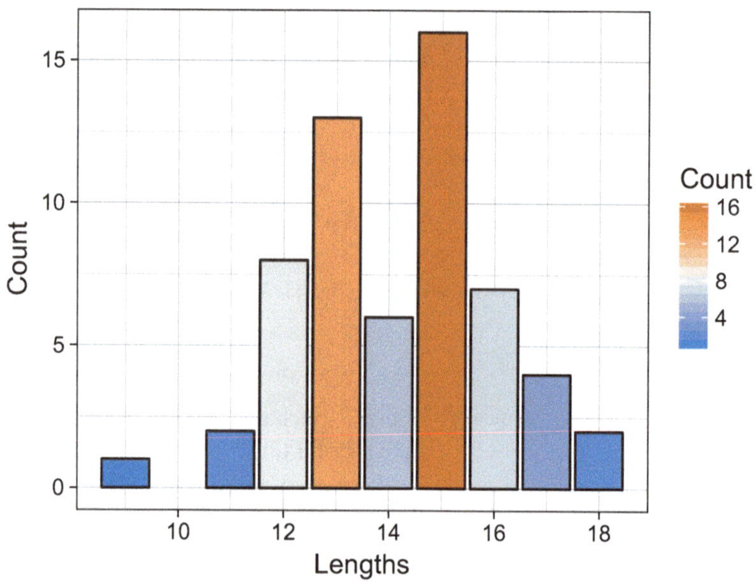

Fig. 14 Histogram of the length in amino acids of TRA CDR3 sequences

```
colnames(TRA_data)[5] = "CDR3.nucleotide.sequence"
vis.count.len(TRA_data, .name="TRA CDR3 amino acid lengths")
colnames(TRA_data)[5] = "CDR3.amino.acid.sequence"
```

7. Next, we use the *plotly* package in R to create a pie chart of the top ten CDR3 sequence combinations from TCR alpha

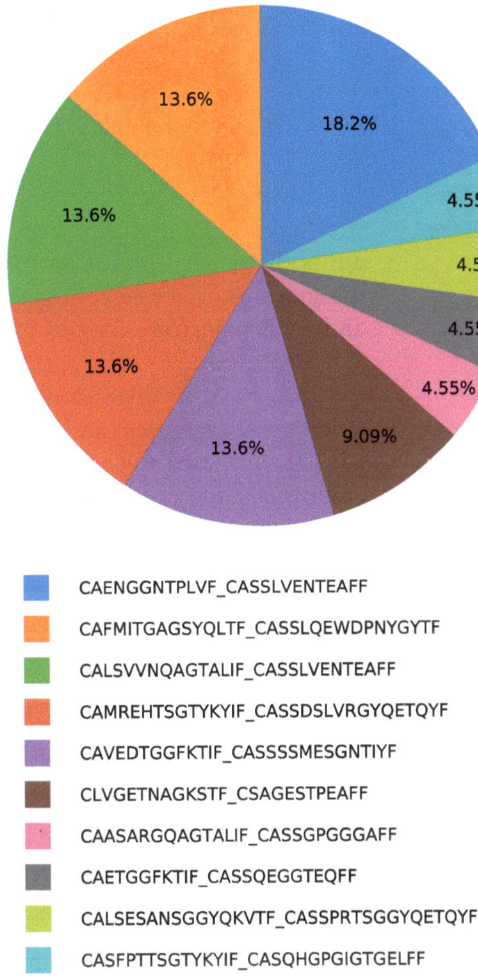

Fig. 15 Usage of the top 10 CDR3 clonotypes identified in the data from paired TCR alpha and beta chains

and beta pairs (Fig. 15). Use the following code to create this plot:

```
library(plotly)

plotly_pie = function(df, chainName, numClonotypes) {
  freq_data = count(df, df$CDR3.amino.acid.sequence)
  colnames(freq_data) = c("cdr3aa", "freq")
  freq_data = freq_data[order(-freq_data$freq),]
  plot_ly(head(freq_data, numClonotypes),
          labels = ~cdr3aa, values = ~freq, type = 'pie',
          textinfo = 'percent',
          textposition = 'inside', hoverinfo = 'text',
          text = ~paste(cdr3aa, ' (', freq, ')', sep = "")
  ) %>% config(displayModeBar = F) %>%
    layout(title = paste(chainName,
                  'Usage (top', numClonotypes, 'clonotypes)'),
           legend=list(orientation='h', font=list(size = 9)),
```

```
            showlegend=T)
}

        plotly_pie(TRab_data, 'TRab', 10)
```

3.15 Linking Flow Cytometry with Gene Expression

Usually scRNA-seq using Smart-Seq2 or similar allows for storing the fluorescent intensity values at the single-cell level from the antibody panel used for sorting. This information allows also for integrating surface cell phenotype data with gene expression profile and TCR repertoire data.

The bioinformatics analysis of this integrated dataset is still the subject of investigation and only bespoke solutions are available. A few examples can be found in Eltahla et al. [7] as well as in [61]. In our example, flow cytometric data could be included as columns in the sample annotation file (Subheading 3.5, **step 4**). These annotations could then be used during clustering and differential expression analysis for defining subpopulations of interest.

4 Conclusions

The proposed workflow summarizes the state-of-the-art methods and solutions for the analysis of scRNA-seq data from T cells. We have described a detailed pipeline assuming full-length T cell scRNA-seq data, although several steps of this pipeline can be also utilized for 3′-based methods, for instance, clustering and differential expression. The R packages *Seurat* [27] and *scran* [26] provide comprehensive sets of data analysis tools that cover much of the scRNA-seq workflow and support UMI-based data.

In this workflow both gene expression analysis and TCR reconstruction have been described. Current methods to integrate these two analyses remain largely manual. However, with the increase of the data generated and reduction of cost, it is likely that statistical and computational methods to integrate clonality and gene expression will be soon developed. For instance, to study immune responses during a pathogen infection, or following patients over the course of immunotherapies against cancer, it will be important to identify potential molecular signatures associated to specific T cell clones which may determine important clinical outcomes or successful immune responses.

Research on the experimental component of scRNA-seq has concentrated on achieving higher throughout, with an exponential increase in the number of cells generated per experiment [12]. This however has not been paralleled by alternative methods that improve quality. Reverse transcription efficiency is a major limit in scRNA-seq, as a single cell has only a very small amount of mRNA (max 50 pg per cell [49]). This causes the loss of many low expression genes and downstream complications at the bioinformatics level (e.g., inference of dropout genes). An advantage of the full-length approach is higher sensitivity compared to high-throughput 3′-based methods [15, 56].

For immune cells such as T cells, sometimes high-throughput is not helpful. For instance, the identification of rare cells, such as Ag-specific CD8 or CD4, remains more an issue in terms of accurate identification as antigenic epitopes are largely unknown or have a very low precursor frequency that makes detection very difficult.

At the time of writing, Smart-Seq2 methods and alike remain the only solution to study full-length transcriptome. Future directions are likely to involve solutions that combine the high-throughput of the microfluidics methods with the sensitivity of plate-based methods. High-throughput methods have already been modified to specifically amplify and sequence the TCR transcript. For instance, solutions are available to detect full-length TCR using 10× Genomics and Drop-seq methods [9] from Dolomite Bio. It is not yet possible to identify more complex transcripts using high-throughput methods, for instance, splice variants that result from multiple exon rearrangements, or the parallel identification of isoforms present in a given cell. The identification of complex splice variants requires full-length transcriptome. TCR is the not the only protein that results from complex splice rearrangements. T cell memory subsets, for instance, are often identified via staining of proteins such as CD45RA and CD45R0. These are two isoforms derived via alternative splicing from the same gene [62]. The detection of these splice variants from scRNA-seq may be helpful in identifying T cell subsets without flow staining. This and many other splice variants are important to understand immunological functions.

Future directions in development and application of scRNA-seq methods for T cell biology are likely to involve more affordable but sensitive high-throughput full-length methods, which allow accurate detection of complex gene transcripts, such as TCR and alternative splice variants.

5 Notes

1. Several repositories for scRNA-seq tools are available online. These include https://www.scrna-tools.org and https://omictools.com/rna-seq-category.
2. A standard office PC should be sufficient to meet the requirements for the workstation. However, an office PC is unlikely to meet the requirements for the server. A Linux cluster or other high-performance computing environment would be ideal to run the server steps.

3. This workflow uses several file formats to store sequences, alignments, and gene locations. *See* Table 2 for a list of these file types. In addition, some files may have an extra ".gz" extension, which indicates that the file has been compressed using the Gzip format. For example, a file ending in ".fastq.gz" is a compressed FASTQ file. Gzip files can be uncompressed using the gunzip command in Linux.
4. Downloading a reference genome from Ensembl can be confusing to those who have not done it before as there are many files to choose from. To download the files, go to the Ensembl FTP Download page located at https://www.ensembl.org/info/data/ftp/index.html. This page should contain a large table of links, with each row containing links for a species. Each link leads to an FTP site with several files available for download. Two files are required:
   (a) The genome sequence: Click on the "FASTA" link in the "DNA (FASTA)" column and download the file ending in "dna.primary_assembly.fa.gz".
   (b) The gene set: Click on the "GTF" link in the "Gene sets" column and download the basic ".gtf.gz" file (not the "abinitio" or "chr" files).

   Note: the Ensembl release number is included in the GTF filename. For example, the file "Homo_sapiens.GRCh38.92.gtf.gz" is from Ensembl release 92. The Ensembl release number is distinct from the genome assembly version (GRCh38 in this case).
5. A useful way to set up the environment variables is using a shell script. Be aware though that environment variables created while running a script are lost when the script exits. To get around this the source command can be used. For example, if the export commands to create the environment variables are in a shell script called environ.sh, the following command will set up the environment variables in the current terminal session:

```
source environ.sh
```

6. If ERCC spike-ins were added to the samples before sequencing, the ERCC sequences and gene set need to be appended to the reference genome before creating the STAR index. To obtain the ERCC sequences and gene set download them from https://tools.thermofisher.com/content/sfs/manuals/ERCC92.zip. Uncompress the files into the genome directory and append the ERCC genome data to the human genome data:

```
cat genome.fa ERCC92.fa > genome_appended.fa
```

```
cat annotation.gtf ERCC92.gtf > annotation_appended.gtf
```

The RSEM reference and STAR index should be created using the appended reference genome files ("genome_appended.fa" and "annotation_appended.gtf") rather than the original files on the command line.

7. Here is an example of how shell scripts can be used to automate the alignment and quantification steps. First create a script that will execute the process for a single sample. For example:

```
#!/bin/bash
SAMPLE=$(basename $1)
echo Processing sample: ${SAMPLE}
ADAPTORS=$CONDA_PREFIX/share/trimmomatic/adapters
trimmomatic PE -threads ${THREADS} -phred33 \
  samples/${SAMPLE}/reads_R1.fastq.gz \
  samples/${SAMPLE}/reads_R2.fastq.gz \
  samples/${SAMPLE}/R1_PAIRED.fastq.gz \
  samples/${SAMPLE}/R1_UNPAIRED.fastq.gz \
  samples/${SAMPLE}/R2_PAIRED.fastq.gz \
  samples/${SAMPLE}/R2_UNPAIRED.fastq.gz \
  ILLUMINACLIP:${ADAPTORS}/NexteraPE-PE.fa:2:30:10 \
  LEADING:3 TRAILING:3 SLIDINGWINDOW:4:15 MINLEN:36
rsem-calculate-expression -p ${THREADS} --paired-end --seed 12345 \
  --star --star-gzipped-read-file --no-bam-output \
  samples/${SAMPLE}/R1_PAIRED.fastq.gz \
  samples/${SAMPLE}/R2_PAIRED.fastq.gz \
  ${GENOME_DIR}/rsem_ref samples/${SAMPLE}/rsem
```

Note that this script takes the sample name as a parameter. Save this script as "align.sh" in the working directory. Use the following command to get the sample names from the directory structure and execute the script once for each sample:

```
find samples -mindepth 1 -maxdepth 1 -type d -exec ${PWD}/align.sh {} \;
```

8. Snakemake is a tool for executing data analysis workflows. It can be installed in Conda by installing the "snakemake" package. Snakemake requires a rule file containing information about the tasks to be executed and the dependencies between them. It can take advantage of multiple CPU cores or a clustering environment to execute the tasks in parallel [63]. For example, the following rule file could be used to execute the alignment and quantification steps for all the samples:

```
# Configuration - modify for local setup
THREADS=8
GENOME_DIR="/srv/genomes/human"
```

```
rule trim:
    input:
        r1='samples/{sample}/reads_R1.fastq.gz',
        r2='samples/{sample}/reads_R2.fastq.gz'
    output:
        r1='samples/{sample}/R1_PAIRED.fastq.gz',
        r2='samples/{sample}/R2_PAIRED.fastq.gz',
        r1_unpaired='samples/{sample}/R1_UNPAIRED.fastq.gz',
        r2_unpaired='samples/{sample}/R2_UNPAIRED.fastq.gz'
    threads: THREADS
    shell:
        'trimmomatic PE '
        '-threads {threads} -phred33 {input} '
        '{output.r1} {output.r1_unpaired} '
        '{output.r2} {output.r2_unpaired} '
        'ILLUMINACLIP:$CONDA_PREFIX/share/trimmomatic/adapters'
            '/NexteraPE-PE.fa:2:30:10 '
        'LEADING:3 TRAILING:3 SLIDINGWINDOW:4:15 MINLEN:36'

rule rsem:
    input:
        r1=rules.trim.output.r1,
        r2=rules.trim.output.r2
    output:
        'results/expr/{sample}/rsem.genes.results'
    params:
        out_prefix='results/expr/{sample}/rsem'
    threads: THREADS
    shell:
        'rsem-calculate-expression -p {threads} '
        '--paired-end --seed 12345 '
        '--star --star-gzipped-read-file --no-bam-output '
        '{input.r1} {input.r2} '
        '{GENOME_DIR}/rsem_ref {params.out_prefix}'

(samples,) = glob_wildcards('samples/{sample}/reads_R1.fastq.gz')

rule all:
    input: expand(rules.rsem.output, sample=samples)
```

To use this workflow, save it to a file called "Snakefile" in the working directory and execute the following command:

```
snakemake -j ${THREADS} all
```

9. The alignment tool used in these instructions, STAR, is not suitable for use on a desktop machine as the memory requirements are high (~35Gb) and, while it is fast compared to other

aligners [64], it will still take a long time for many samples. For very large datasets or for execution on a desktop machine, alternative alignment-free or pseudo-aligner algorithms are available (Table 1). These algorithms reduce processing time by avoiding a base by base alignment of every transcript to the reference genome.

The following things should be considered if using one of these tools:

- These tools are an alternative for the alignment and quantification steps. So, they replace two tools (STAR and RSEM).
- Read quality control should still be performed using Trimmomatic beforehand.
- These tools produce a file for each sample containing expression values. They need to be converted into expression matrices using code similar to that in Subheading 3.5.
- The quantifications are provided on the transcript level rather than the gene level, but our pipeline is based on genes. The *tximport* R package [52] contains tools for converting transcript counts to gene counts. *See* the *tximport* documentation for details.

10. There are several ways to measure the expression of transcripts from the read count. Generally, for full-length transcripts it is useful to consider TPM [65], RPKM [66], and FPKM [67], which all account for gene length in counting for the number of reads covering a certain transcript as well as normalizing for sequencing depth. (Note: these units should not be used for UMI-based protocols as they only sequence a short section of the transcript at either the 3′ or 5′ end.) Here we use TPM, namely, the expected number of transcripts of gene i in cell j per million transcripts, given the relative length of gene i and the abundances of all the other genes in cell j. Compared to the other units mentioned, TPM is easier to interpret since TPMs for a sample always add to one million. RSEM, the software we use to measure expression, outputs counts, TPMs, and FPKMs.
11. It may be preferable to execute the code to create the expression matrices on the server rather than the workstation, to provide opportunity to automate, or to reduce the amount of data that needs to be copied to the workstation. Be aware that the RDS data format, which we use to save the dataset to disk after constructing the matrices, may not work if loading the data using different versions of the R packages involved. Consider saving the data in a more generic format, such as tab-separated files.

12. If the samples contain ERCC spike-ins, use the following code to add some entries to the gene annotation for the ERCC genes:

```
ercc_genes <- grep("ERCC-.*", rownames(raw_counts), value=T)
gene_annotation <- rbind(gene_annotation,
    data.frame(ensembl_gene_id=ercc_genes,
               external_gene_name=ercc_genes,
               description=ercc_genes,
               chromosome_name=ercc_genes,
               start_position=0, end_position=0,
               gene_biotype='ERCC'))
```

13. Log counts are a useful way to look at gene quantification values in many circumstances. Many statistical techniques assume equal variance (homoscedasticity) for all data points, but for gene expressions the variance is greater when the gene expression is greater [68]. A log transform makes the data more homoscedastic. An offset of 1 is usually added to the expression values before log transformation to avoid taking a log of zero and to ensure there are no negative log counts.
14. The SingleCellExperiment class has a set of methods for dealing with spike-ins that should be used if ERCC spike-ins are in the data. The following is replacement code for calculating the QC metrics that includes the spike-ins:

```
isSpike(sce.orig, "ERCC") <- grepl("^ERCC-",
                                   rowData(sce.orig)$chromosome_name)

# Calculate QC metrics
is.mito <- rowData(sce.orig)$chromosome_name == "MT"
sce.orig <- calculateQCMetrics(sce.orig, exprs_values="counts",
      feature_controls=list(MT=is.mito,
                            ERCC=isSpike(sce.orig, "ERCC")))

sce.orig.ercc <- sce.orig
sce.orig <- sce.orig[!isSpike(sce.orig, "ERCC"),]
```

After using mitochondrial genes and ERCC spike-ins as feature controls for calculating QC metrics, the ERCC spike-ins are removed from the main dataset as it normally doesn't make sense to include the spike-ins in diagnostic plots or later analysis. However a copy of the dataset including the spike-ins is retained ("sce.orig.ercc") for use in plots where it does make sense, for example, the highly expressed gene plot.

ERCC spike-ins are useful for detecting low-quality samples. Samples with high ERCC counts indicate that not much

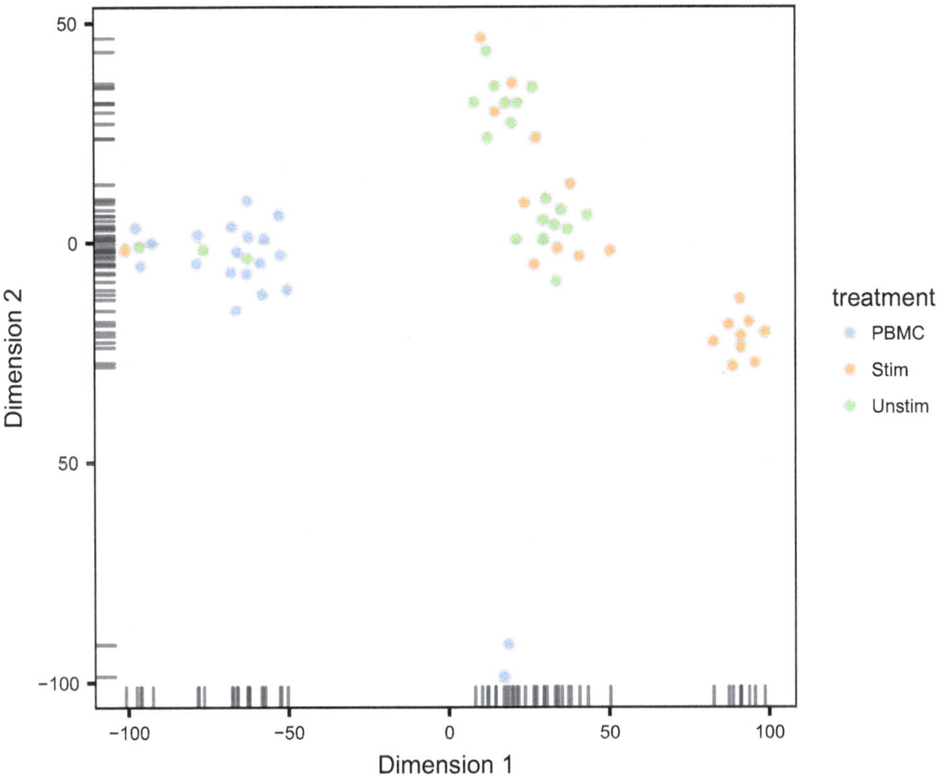

Fig. 16 T-distributed stochastic neighbor embedding (tSNE) plot on the raw data prior to quality control

mRNA was sequenced from the cell, indicating technical issues with the sample. ERCC spike-ins should be plotted for the samples and a spike-in percentage cutoff chosen for filtering. To do this, pass the "pct_counts_ERCC" metric in to the plotColData function for use as the x or y variable in the plot. For example:

```
plotColData(sce.orig, x="total_counts", y="pct_counts_ERCC",
            colour_by="treatment", shape_by="bulk")
```

15. The *scater* package provides several functions for plotting scRNA-seq data. Here we describe a few plots that can be useful to for visualization and quality control.

    Scater provides a few other forms of dimensionality reduction besides PCA, such as tSNE. A tSNE plot can be generated using the following code:

```
plotTSNE(sce.orig, exprs_values="logcounts_raw", colour_by="treatment",
         perplexity=1.5, rand_seed=37)
```

The tSNE plot on the example data (Fig. 16) shows that the major source of variation in the data is PBMCs versus the

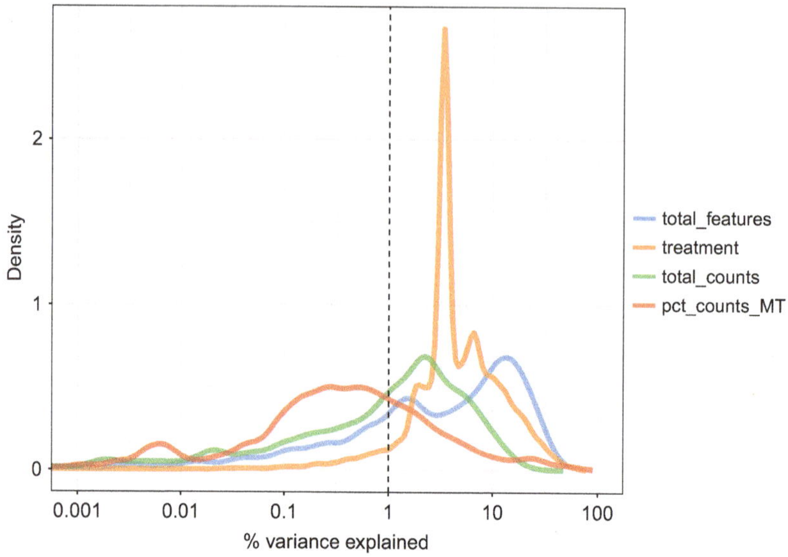

Fig. 17 Percentage of variance explained across genes by four sample metrics: total number of expressed genes (total_features), treatment, total counts, and percentage of counts from mitochondrial genes (pct_counts_MT)

stimulated and unstimulated treatments. Note that the `perplexity` and `rand_seed` parameters can be adjusted to produce different plots.

The explanatory variable plot shows how much of the variability in the data can be explained by the given variables. The plot for the example data (Fig. 17) shows that total features explains the most variance, while treatment is also an important source of variance.

```
plotQC(sce.orig, type="expl", exprs_values="logcounts_raw",
       variables=c("treatment", "total_features",
                   "total_counts", "pct_counts_MT"))
```

The last plot type displays the principle components that can be explained by a given variable. For example, the first principle component is proportional to the number of genes expressed per sample (Figs. 2 and 18). This type of PCA plot can be created using the following code (Fig. 18):

```
plotQC(sce.orig, exprs_values = "logcounts_raw", type = "find-pcs",
       variable = "total_features")
```

16. The clustering results from SC3 can be used as a factor for classifying samples for MAST. For example:

```
zlm_cell_type <- zlm(~sc3_3_clusters, mast.norm)
```

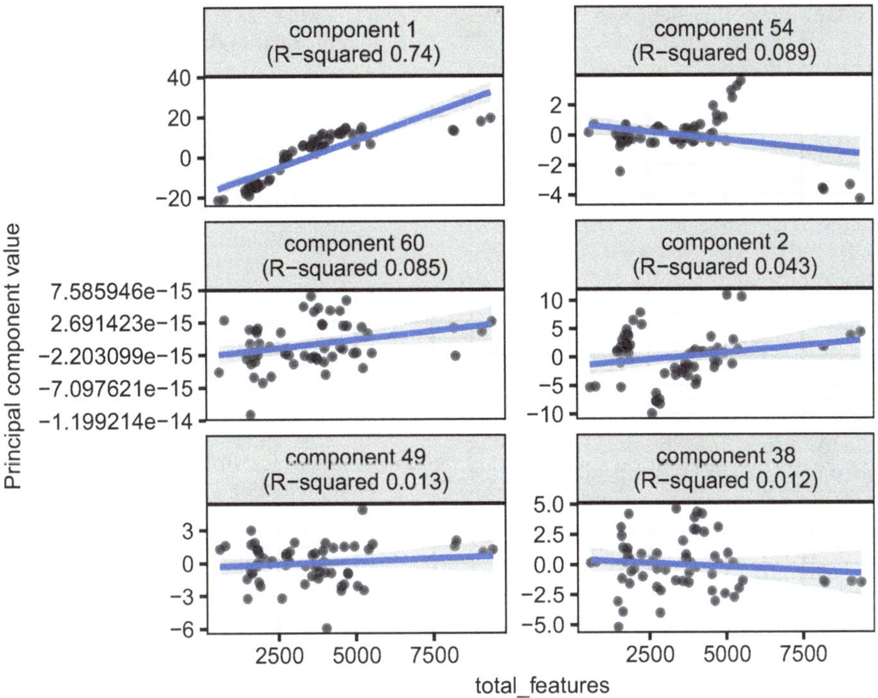

Fig. 18 Top six principle components that are explained by the total number of expressed genes in a sample, sorted by percentage of variance explained (R^2)

This command will create a model based on the clusters identified by *SC3* with a *k* value of 3.

References

1. Papalexi E, Satija R (2018) Single-cell RNA sequencing to explore immune cell heterogeneity. Nat Rev Immunol 18(1):35–45. https://doi.org/10.1038/nri.2017.76
2. Kolodziejczyk AA, Kim JK, Svensson V, Marioni JC, Teichmann SA (2015) The technology and biology of single-cell RNA sequencing. Mol Cell 58(4):610–620. https://doi.org/10.1016/j.molcel.2015.04.005
3. Vegh P, Haniffa M (2018) The impact of single-cell RNA sequencing on understanding the functional organization of the immune system. Brief Funct Genomics. https://doi.org/10.1093/bfgp/ely003
4. Haque A, Engel J, Teichmann SA, Lonnberg T (2017) A practical guide to single-cell RNA-sequencing for biomedical research and clinical applications. Genome Med 9(1):75. https://doi.org/10.1186/s13073-017-0467-4
5. Picelli S, Bjorklund AK, Faridani OR, Sagasser S, Winberg G, Sandberg R (2013) Smart-seq2 for sensitive full-length transcriptome profiling in single cells. Nat Methods 10 (11):1096–1098. https://doi.org/10.1038/nmeth.2639
6. Picelli S, Faridani OR, Bjorklund AK, Winberg G, Sagasser S, Sandberg R (2014) Full-length RNA-seq from single cells using Smart-seq2. Nat Protoc 9(1):171–181. https://doi.org/10.1038/nprot.2014.006
7. Eltahla AA, Rizzetto S, Pirozyan MR, Betz-Stablein BD, Venturi V, Kedzierska K, Lloyd AR, Bull RA, Luciani F (2016) Linking the T cell receptor to the single cell transcriptome in antigen-specific human T cells. Immunol Cell Biol 94(6):604–611. https://doi.org/10.1038/icb.2016.16

8. Zheng GX, Terry JM, Belgrader P, Ryvkin P, Bent ZW, Wilson R, Ziraldo SB, Wheeler TD, McDermott GP, Zhu J, Gregory MT, Shuga J, Montesclaros L, Underwood JG, Masquelier DA, Nishimura SY, Schnall-Levin M, Wyatt PW, Hindson CM, Bharadwaj R, Wong A, Ness KD, Beppu LW, Deeg HJ, McFarland C, Loeb KR, Valente WJ, Ericson NG, Stevens EA, Radich JP, Mikkelsen TS, Hindson BJ, Bielas JH (2017) Massively parallel digital transcriptional profiling of single cells. Nat Commun 8:14049. https://doi.org/10.1038/ncomms14049
9. Macosko EZ, Basu A, Satija R, Nemesh J, Shekhar K, Goldman M, Tirosh I, Bialas AR, Kamitaki N, Martersteck EM, Trombetta JJ, Weitz DA, Sanes JR, Shalek AK, Regev A, McCarroll SA (2015) Highly parallel genome-wide expression profiling of individual cells using nanoliter droplets. Cell 161 (5):1202–1214. https://doi.org/10.1016/j.cell.2015.05.002
10. Jaitin DA, Kenigsberg E, Keren-Shaul H, Elefant N, Paul F, Zaretsky I, Mildner A, Cohen N, Jung S, Tanay A, Amit I (2014) Massively parallel single-cell RNA-seq for marker-free decomposition of tissues into cell types. Science 343(6172):776–779. https://doi.org/10.1126/science.1247651
11. Hashimshony T, Wagner F, Sher N, Yanai I (2012) CEL-Seq: single-cell RNA-Seq by multiplexed linear amplification. Cell Rep 2 (3):666–673. https://doi.org/10.1016/j.celrep.2012.08.003
12. Svensson V, Vento-Tormo R, Teichmann SA (2018) Exponential scaling of single-cell RNA-seq in the past decade. Nat Protoc 13 (4):599–604. https://doi.org/10.1038/nprot.2017.149
13. Villani AC, Shekhar K (2017) Single-cell RNA sequencing of human T cells. Methods Mol Biol 1514:203–239. https://doi.org/10.1007/978-1-4939-6548-9_16
14. Stubbington MJT, Rozenblatt-Rosen O, Regev A, Teichmann SA (2017) Single-cell transcriptomics to explore the immune system in health and disease. Science 358 (6359):58–63. https://doi.org/10.1126/science.aan6828
15. Rizzetto S, Koppstein DNP, Samir J, Singh M, Reed JH, Cai CH, Lloyd AR, Eltahla AA, Goodnow CC, Luciani F (2018) B-cell receptor reconstruction from single-cell RNA-seq with VDJPuzzle. Bioinformatics. https://doi.org/10.1093/bioinformatics/bty203
16. Bolger AM, Lohse M, Usadel B (2014) Trimmomatic: a flexible trimmer for Illumina sequence data. Bioinformatics 30 (15):2114–2120. https://doi.org/10.1093/bioinformatics/btu170
17. FastQC (2018). https://www.bioinformatics.babraham.ac.uk/projects/fastqc/. Accessed 14 June 2018
18. Dobin A, Davis CA, Schlesinger F, Drenkow J, Zaleski C, Jha S, Batut P, Chaisson M, Gingeras TR (2013) STAR: ultrafast universal RNA-seq aligner. Bioinformatics 29(1):15–21. https://doi.org/10.1093/bioinformatics/bts635
19. Kim D, Pertea G, Trapnell C, Pimentel H, Kelley R, Salzberg SL (2013) TopHat2: accurate alignment of transcriptomes in the presence of insertions, deletions and gene fusions. Genome Biol 14(4):R36. https://doi.org/10.1186/gb-2013-14-4-r36
20. Trapnell C, Roberts A, Goff L, Pertea G, Kim D, Kelley DR, Pimentel H, Salzberg SL, Rinn JL, Pachter L (2012) Differential gene and transcript expression analysis of RNA-seq experiments with TopHat and cufflinks. Nat Protoc 7(3):562–578. https://doi.org/10.1038/nprot.2012.016
21. Li B, Dewey CN (2011) RSEM: accurate transcript quantification from RNA-Seq data with or without a reference genome. BMC Bioinformatics 12:323. https://doi.org/10.1186/1471-2105-12-323
22. Patro R, Duggal G, Love MI, Irizarry RA, Kingsford C (2017) Salmon provides fast and bias-aware quantification of transcript expression. Nat Methods 14(4):417–419. https://doi.org/10.1038/nmeth.4197
23. Pertea M, Pertea GM, Antonescu CM, Chang TC, Mendell JT, Salzberg SL (2015) StringTie enables improved reconstruction of a transcriptome from RNA-seq reads. Nat Biotechnol 33 (3):290–295. https://doi.org/10.1038/nbt.3122
24. Bray NL, Pimentel H, Melsted P, Pachter L (2016) Near-optimal probabilistic RNA-seq quantification. Nat Biotechnol 34(5):525–527. https://doi.org/10.1038/nbt.3519
25. Patro R, Mount SM, Kingsford C (2014) Sailfish enables alignment-free isoform quantification from RNA-seq reads using lightweight algorithms. Nat Biotechnol 32(5):462–464. https://doi.org/10.1038/nbt.2862
26. Lun AT, McCarthy DJ, Marioni JC (2016) A step-by-step workflow for low-level analysis of single-cell RNA-seq data with bioconductor. F1000Res 5:2122. https://doi.org/10.12688/f1000research.9501.2
27. Butler A, Hoffman P, Smibert P, Papalexi E, Satija R (2018) Integrating single-cell transcriptomic data across different conditions, technologies, and species. Nat Biotechnol 36

(5):411–420. https://doi.org/10.1038/nbt.4096

28. McCarthy DJ, Campbell KR, Lun AT, Wills QF (2017) Scater: pre-processing, quality control, normalization and visualization of single-cell RNA-seq data in R. Bioinformatics 33 (8):1179–1186. https://doi.org/10.1093/bioinformatics/btw777
29. Bacher R, Chu LF, Leng N, Gasch AP, Thomson JA, Stewart RM, Newton M, Kendziorski C (2017) SCnorm: robust normalization of single-cell RNA-seq data. Nat Methods 14 (6):584–586. https://doi.org/10.1038/nmeth.4263
30. Khalfaoui B, Vert J-P (2018) DropLasso: a robust variant of Lasso for single cell RNA-seq data. arXiv
31. Chen L, Zheng S (2018) BCseq: accurate single cell RNA-seq quantification with bias correction. Nucleic Acids Res. https://doi.org/10.1093/nar/gky308
32. Cole M, Risso D (2017) Scone: single cell overview of normalized expression data. R package version 1.2.0
33. Ritchie ME, Phipson B, Wu D, Hu Y, Law CW, Shi W, Smyth GK (2015) Limma powers differential expression analyses for RNA-sequencing and microarray studies. Nucleic Acids Res 43(7):e47. https://doi.org/10.1093/nar/gkv007
34. Johnson WE, Li C, Rabinovic A (2007) Adjusting batch effects in microarray expression data using empirical Bayes methods. Biostatistics 8 (1):118–127. https://doi.org/10.1093/biostatistics/kxj037
35. Angerer P, Haghverdi L, Buttner M, Theis FJ, Marr C, Buettner F (2016) Destiny: diffusion maps for large-scale single-cell data in R. Bioinformatics 32(8):1241–1243. https://doi.org/10.1093/bioinformatics/btv715
36. Qiu X, Mao Q, Tang Y, Wang L, Chawla R, Pliner HA, Trapnell C (2017) Reversed graph embedding resolves complex single-cell trajectories. Nat Methods 14(10):979–982. https://doi.org/10.1038/nmeth.4402
37. Kiselev VY, Kirschner K, Schaub MT, Andrews T, Yiu A, Chandra T, Natarajan KN, Reik W, Barahona M, Green AR, Hemberg M (2017) SC3: consensus clustering of single-cell RNA-seq data. Nat Methods 14(5):483–486. https://doi.org/10.1038/nmeth.4236
38. Lin P, Troup M, Ho JW (2017) CIDR: ultrafast and accurate clustering through imputation for single-cell RNA-seq data. Genome Biol 18(1):59. https://doi.org/10.1186/s13059-017-1188-0
39. Ji Z, Ji H (2016) TSCAN: pseudo-time reconstruction and evaluation in single-cell RNA-seq analysis. Nucleic Acids Res 44(13):e117. https://doi.org/10.1093/nar/gkw430
40. Finak G, McDavid A, Yajima M, Deng J, Gersuk V, Shalek A, Schlicter CK, Miller HW, McElrath MJ, Prlic M, Linsley P, Gottardo R (2018) Using MAST with RNASeq: MAIT analysis. https://bioconductor.org/packages/release/bioc/vignettes/MAST/inst/doc/MAITAnalysis.html. Accessed 25 May 2018
41. Kharchenko PV, Silberstein L, Scadden DT (2014) Bayesian approach to single-cell differential expression analysis. Nat Methods 11 (7):740–742. https://doi.org/10.1038/nmeth.2967
42. Robinson MD, McCarthy DJ, Smyth GK (2010) edgeR: a bioconductor package for differential expression analysis of digital gene expression data. Bioinformatics 26 (1):139–140. https://doi.org/10.1093/bioinformatics/btp616
43. Haghverdi L, Buttner M, Wolf FA, Buettner F, Theis FJ (2016) Diffusion pseudotime robustly reconstructs lineage branching. Nat Methods 13(10):845–848. https://doi.org/10.1038/nmeth.3971
44. duVerle DA, Yotsukura S, Nomura S, Aburatani H, Tsuda K (2016) CellTree: an R/bioconductor package to infer the hierarchical structure of cell populations from single-cell RNA-seq data. BMC Bioinformatics 17 (1):363. https://doi.org/10.1186/s12859-016-1175-6
45. Huang d W, Sherman BT, Lempicki RA (2009) Systematic and integrative analysis of large gene lists using DAVID bioinformatics resources. Nat Protoc 4(1):44–57. https://doi.org/10.1038/nprot.2008.211
46. Subramanian A, Tamayo P, Mootha VK, Mukherjee S, Ebert BL, Gillette MA, Paulovich A, Pomeroy SL, Golub TR, Lander ES, Mesirov JP (2005) Gene set enrichment analysis: a knowledge-based approach for interpreting genome-wide expression profiles. Proc Natl Acad Sci U S A 102(43):15545–15550. https://doi.org/10.1073/pnas.0506580102
47. Stubbington MJT, Lonnberg T, Proserpio V, Clare S, Speak AO, Dougan G, Teichmann SA (2016) T cell fate and clonality inference from single-cell transcriptomes. Nat Methods 13 (4):329–332. https://doi.org/10.1038/nmeth.3800
48. Afik S, Yates KB, Bi K, Darko S, Godec J, Gerdemann U, Swadling L, Douek DC, Klenerman P, Barnes EJ, Sharpe AH, Haining WN, Yosef N (2017) Targeted reconstruction of T cell receptor sequence from single cell

RNA-seq links CDR3 length to T cell differentiation state. Nucleic Acids Res 45(16):e148. https://doi.org/10.1093/nar/gkx615

49. Zhu YY, Machleder EM, Chenchik A, Li R, Siebert PD (2001) Reverse transcriptase template switching: a SMART approach for full-length cDNA library construction. BioTechniques 30(4):892–897

50. Lun ATL, Calero-Nieto FJ, Haim-Vilmovsky L, Gottgens B, Marioni JC (2017) Assessing the reliability of spike-in normalization for analyses of single-cell RNA sequencing data. Genome Res 27(11):1795–1806. https://doi.org/10.1101/gr.222877.117

51. Langmead B, Trapnell C, Pop M, Salzberg SL (2009) Ultrafast and memory-efficient alignment of short DNA sequences to the human genome. Genome Biol 10(3):R25. https://doi.org/10.1186/gb-2009-10-3-r25

52. Soneson C, Love MI, Robinson MD (2015) Differential analyses for RNA-seq: transcript-level estimates improve gene-level inferences. F1000Res 4:1521. https://doi.org/10.12688/f1000research.7563.2

53. Lun A, Risso D (2017) SingleCellExperiment: S4 classes for single cell data. R package version 1.0.0. https://doi.org/10.18129/B9.bioc.SingleCellExperiment

54. Ilicic T, Kim JK, Kolodziejczyk AA, Bagger FO, McCarthy DJ, Marioni JC, Teichmann SA (2016) Classification of low quality cells from single-cell RNA-seq data. Genome Biol 17:29. https://doi.org/10.1186/s13059-016-0888-1

55. Rizzetto S, Eltahla AA, Lin P, Bull R, Lloyd AR, Ho JWK, Venturi V, Luciani F (2017) Impact of sequencing depth and read length on single cell RNA sequencing data of T cells. Sci Rep 7(1):12781. https://doi.org/10.1038/s41598-017-12989-x

56. Ziegenhain C, Vieth B, Parekh S, Reinius B, Guillaumet-Adkins A, Smets M, Leonhardt H, Heyn H, Hellmann I, Enard W (2017) Comparative analysis of single-cell RNA sequencing methods. Mol Cell 65(4):631–643 e634. https://doi.org/10.1016/j.molcel.2017.01.023

57. Lun AT, Bach K, Marioni JC (2016) Pooling across cells to normalize single-cell RNA sequencing data with many zero counts. Genome Biol 17:75. https://doi.org/10.1186/s13059-016-0947-7

58. Finak G, McDavid A, Yajima M, Deng J, Gersuk V, Shalek AK, Slichter CK, Miller HW, McElrath MJ, Prlic M, Linsley PS, Gottardo R (2015) MAST: a flexible statistical framework for assessing transcriptional changes and characterizing heterogeneity in single-cell RNA sequencing data. Genome Biol 16:278. https://doi.org/10.1186/s13059-015-0844-5

59. Benjamini Y, Hochberg Y (1995) Controlling the false discovery rate: a practical and powerful approach to multiple testing. J R Stat Soc Ser B Methodol 57(1):289–300

60. Nazarov VI, Pogorelyy MV, Komech EA, Zvyagin IV, Bolotin DA, Shugay M, Chudakov DM, Lebedev YB, Mamedov IZ (2015) tcR: an R package for T cell receptor repertoire advanced data analysis. BMC Bioinformatics 16:175. https://doi.org/10.1186/s12859-015-0613-1

61. Velten L, Haas SF, Raffel S, Blaszkiewicz S, Islam S, Hennig BP, Hirche C, Lutz C, Buss EC, Nowak D, Boch T, Hofmann WK, Ho AD, Huber W, Trumpp A, Essers MA, Steinmetz LM (2017) Human haematopoietic stem cell lineage commitment is a continuous process. Nat Cell Biol 19(4):271–281. https://doi.org/10.1038/ncb3493

62. Streuli M, Hall LR, Saga Y, Schlossman SF, Saito H (1987) Differential usage of three exons generates at least five different mRNAs encoding human leukocyte common antigens. J Exp Med 166(5):1548–1566

63. Koster J, Rahmann S (2012) Snakemake--a scalable bioinformatics workflow engine. Bioinformatics 28(19):2520–2522. https://doi.org/10.1093/bioinformatics/bts480

64. Baruzzo G, Hayer KE, Kim EJ, Di Camillo B, FitzGerald GA, Grant GR (2017) Simulation-based comprehensive benchmarking of RNA-seq aligners. Nat Methods 14(2):135–139. https://doi.org/10.1038/nmeth.4106

65. Li B, Ruotti V, Stewart RM, Thomson JA, Dewey CN (2010) RNA-Seq gene expression estimation with read mapping uncertainty. Bioinformatics 26(4):493–500. https://doi.org/10.1093/bioinformatics/btp692

66. Mortazavi A, Williams BA, McCue K, Schaeffer L, Wold B (2008) Mapping and quantifying mammalian transcriptomes by RNA-Seq. Nat Methods 5(7):621–628. https://doi.org/10.1038/nmeth.1226

67. Trapnell C, Williams BA, Pertea G, Mortazavi A, Kwan G, van Baren MJ, Salzberg SL, Wold BJ, Pachter L (2010) Transcript assembly and quantification by RNA-Seq reveals unannotated transcripts and isoform switching during cell differentiation. Nat Biotechnol 28(5):511–515. https://doi.org/10.1038/nbt.1621

68. Marioni JC, Mason CE, Mane SM, Stephens M, Gilad Y (2008) RNA-seq: an assessment of technical reproducibility and comparison with gene expression arrays. Genome Res 18(9):1509–1517. https://doi.org/10.1101/gr.079558.108

Chapter 17

Structural Modeling of Lymphocyte Receptors and Their Antigens

Songling Li, Jan Wilamowski, Shunsuke Teraguchi, Floris J. van Eerden, John Rozewicki, Ana Davila, Zichang Xu, Kazutaka Katoh, and Daron M. Standley

Abstract

Structural modeling plays a key role in protein function prediction on a genome-wide scale. For B and T lymphocyte receptors, the critical functional question is: which antigens and epitopes are targeted? With emerging B cell receptor (BCR) and T cell receptor (TCR) sequencing methods improving in both breadth and depth, there is a growing need for methods that can help answer this question. Since lymphocyte-antigen recognition depends on complementarity, structural modeling is likely to play an important role in understanding antigen specificity and affinity. In the case of BCRs, such modeling methods have a long history in the study and design of antibodies. However, for TCRs there are relatively few publicly available modeling tools, and, to our knowledge, none that incorporate interaction between TCRs and peptide-MHC (pMHC) complexes. Here, we provide a web-based tool, *ImmuneScape* (https://sysimm.org/immune-scape/), to carry out TCR-pMHC modeling as a first step toward structure-based function prediction.

Key words Immunological synapse, Structural modeling, Molecular recognition, TCR repertoire, Epitope prediction, Structural systems immunology

1 Introduction

BCRs and TCRs share a common immunoglobulin-like fold in their N-terminal variable (V) region that mediates antigen recognition, as shown in Fig. 1. Specific binding to an extremely wide variety of antigens is accomplished by generation of diverse complementarity determining regions (CDRs) that sit at the apex of the V region. In both BCRs and TCRs, this region is made up of two polypeptide chains, commonly referred to as light (L) and heavy (H) in BCRs and alpha (A) and beta (B) in TCRs, respectively. BCR modeling is a relatively mature field and typically the V region can be generated with root-mean square deviations (RMSDs) of under 2 Å from experimentally determined structures [1].

Shin Kaneko (ed.), *In Vitro Differentiation of T-Cells: Methods and Protocols*, Methods in Molecular Biology, vol. 2048, https://doi.org/10.1007/978-1-4939-9728-2_17, © Springer Science+Business Media, LLC, part of Springer Nature 2019

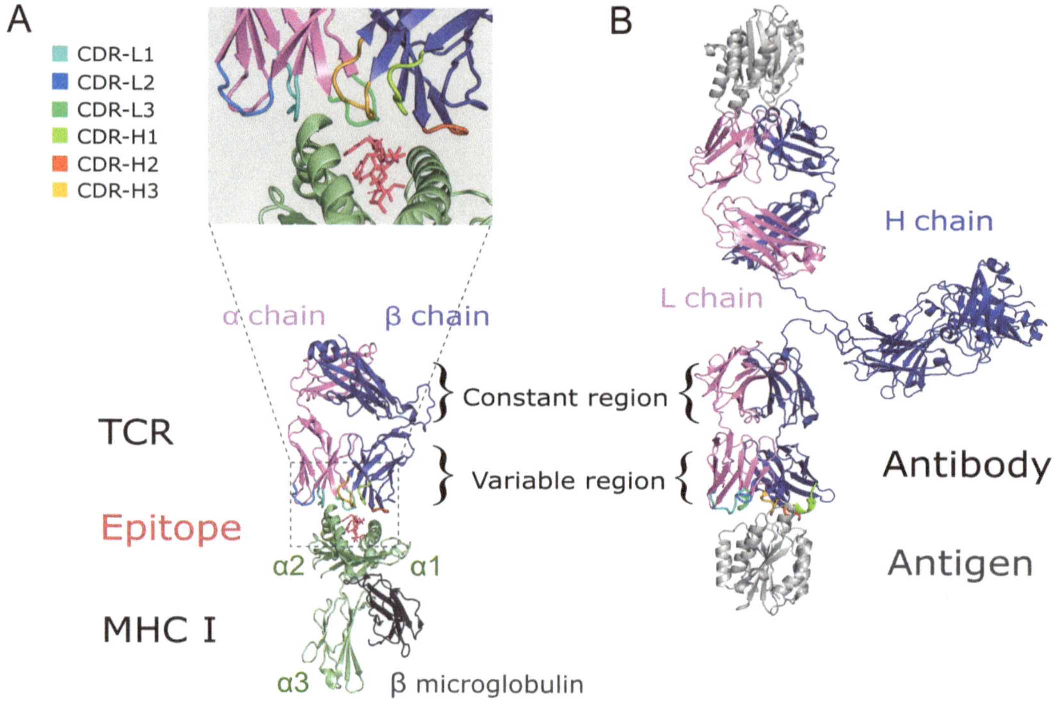

Fig. 1 Lymphocyte receptor structure. (**a**) Key structural elements of TCR-pMHC complexes. (**b**) Antibody-antigen interactions

Two key differences between BCRs and TCRs from a structural point of view are their stability and mode of antigen recognition. The extracellular portion of BCRs (antibodies) are stable in isolation, allowing them to function autonomously and making them relatively easy to crystalize; TCRs, in contrast, exist only as cell surface receptors and are notoriously difficult to crystallize. One consequence of this is that there are far fewer TCRs than BCRs in the Protein Data Bank (PDB) [2]. The greater degrees of freedom available to antibodies enables them to interact with antigens using a variety of binding modes, whereas TCRs recognize only short, linear peptides presented on major histocompatibility (MHC) molecules on the surfaces of cells. Data-driven prediction algorithms have proven to be highly accurate in reproducing experimental MHC epitope presentation data. The precise sequence (allele) of the MHC, called human leukocyte antigen (HLA) in humans, varies from donor to donor, but can be determined from a blood test.

Because TCR-pMHC binding has limited conformational degrees of freedom and the epitopes in known structures are almost completely surrounded by MHC and TCR, we can naively assume the TCR-pMHC modeling problem is easier than that of BCR-antigen modeling. In spite of this observation, there are very few published methods describing TCR modeling, and none that the authors are aware of that incorporate pMHC information. Nevertheless, the tools needed to carry out the individual steps required for such modeling already exist. We focus here on combining such established tools for the task of modeling TCR-pMHC complexes.

We take advantage of the fact that prediction of epitope-MHC interactions is a very mature field [3]. Generally speaking, the most successful prediction methods are sequence-based methods that have been trained on large amounts of experimental data [4, 5]. Integration of MHC-epitope sequence pairs into structural TCR-pMHC models is a logical next step, but one that remains an open problem due to the paucity of TCR modeling and TCR-MHC docking tools. Here, we approach the problem in a very rudimentary way in order to generate candidate models quickly that can be further refined or validated by experiment or by more elaborate modeling methods. At the end of this chapter we give a brief tutorial on using molecular dynamics to refine such candidate models. In principle, MHC-epitope prediction could be used in place of experimentally observed epitopes to generate trial epitopes given a target MHC and TCR pair; however, selecting among candidate TCR-epitope pairs is a research problem beyond the scope of this chapter. A future extension of the protocol presented here will be to use experimental data to train a classifier that can predict epitopes given both MHC and TCR.

2 Tools and Methods

Modeling TCR-pMHC complexes from sequence is potentially a very complicated process. The protocol outlined here should be considered a basic prototype on which more elaborate methods can be developed. There are some core tasks (e.g., mapping between MHC alleles and known structures) that are likely to be common to most approaches and can involve many tedious operations. For this reason, we have provided a single web-based tool, *ImmuneScape*, that automates all of the basic modeling steps. The *ImmuneScape* server, along with additional resources, can be accessed at https://sysimm.org/immune-scape/ and https://github.com/sysimm/immune-scape-data/, respectively.

2.1 Simple Instructions

Since many users will be familiar with predicting epitopes starting from MHC allele information, we have taken this as a starting point.

2.1.1 Input

To get started you will need to define parameters that can be used to extract the amino acid sequences to be modeled. The requirements are as follows:

1. Epitope search space: An amino acid sequence of the epitope or a longer sequence from which epitopes will be drawn.
2. The maximum length(s) of the epitope. The lengths here provide the window size of the epitope search space. Actual epitopes may be shorter than the length provided.
3. MHC allele(s). Both human and mouse MHC alleles are supported.
4. Amino acid sequence of the TCR alpha and beta chains.

The input form is shown in Fig. 2a. Sample input has been provided for both class I and II alleles. Users familiar with predicting epitopes from MHC allele information will note that, with exception to the TCR sequences, the inputs are the same as those for most MHC epitope prediction tools. Indeed, *ImmuneScape* uses MHC epitope prediction internally, as described below. Since TCR amino acid sequence data might not be familiar to all users, we briefly describe the assembly of TCR amino acid sequence data in Subheading 2.2. Sequence data is assumed to be correct (both at the amino acid level and at the TCR-pMHC level). Paired (alpha-beta chain) TCR sequence data can now be obtained routinely [6]. TCR-pMHC interactions can be observed directly by tetramer staining [7] or inferred by expansion of TCR clones upon stimulation by a pMHC of interest [8]. Techniques for identifying TCR-pMHC interactions are rapidly evolving and here we make use of a database of published TCR-pMHC interactions, VDJdb [9].

2.1.2 Output

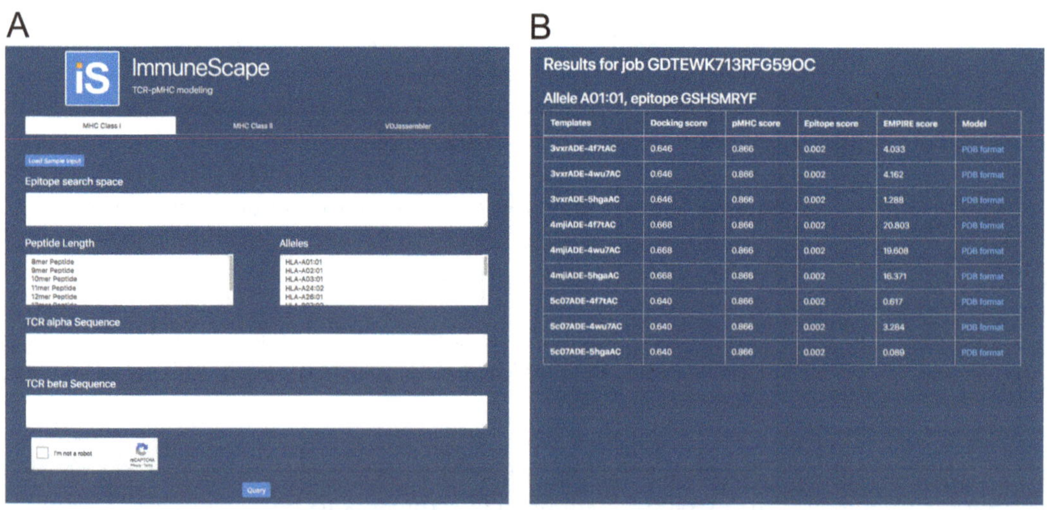

Fig. 2 *ImmuneScape* web interface. (**a**) The input form to *ImmuneScape* is similar to that for MHC-epitope prediction but includes an additional TCR sequence field. (**b**) the results are given as a table of scores with a link to the coordinates of the 3D model

The output consists of a table of results and a link to the associated structural model (Fig. 2b). The columns of the table are defined as follows:

1. Templates: The docking and pMHC template, respectively.
2. Docking score: The weighted BLOSUM62 sequence score for docking template.
3. pMHC score: The weighted BLOSUM62 sequence score for pMHC template.
4. Epitope score: NetMHCpan raw score (class I) or 1-log50k (aff) (class II).
5. EMPIRE score: TCR-pMHC binding energy (kcal/mol) [10].
6. Link to model (PDB format).

2.1.3 Methodology

The *ImmuneScape* calculation consists of six main steps, as illustrated in Fig. 3 and described in detail in Subheadings 2.3.1–2.3.5:

1. Model the TCR (Subheading 2.3.1).
2. Select pMHC templates (Subheading 2.3.2).
3. Select docking templates (Subheading 2.3.3).

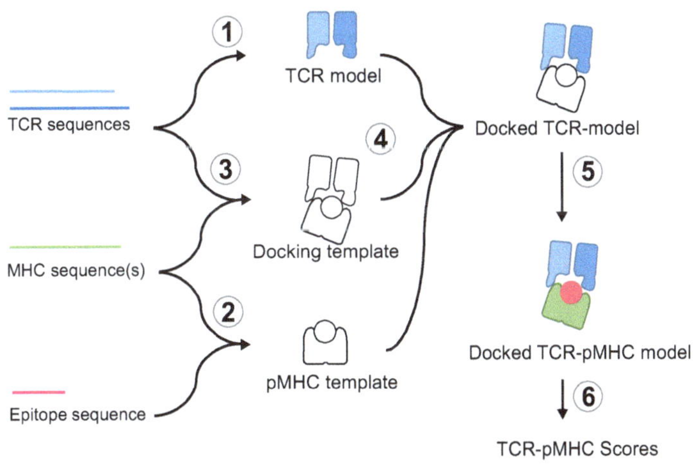

Fig. 3 Overview of *ImmuneScape* procedure. The six main steps used by *ImmuneScape* are illustrated. Given input sequences for TCR, MHC, and epitope, the first step is to build a TCR model using Repertoire Builder. The second step is to select a pMHC template based on sequence similarity to the MHC and epitope sequences. The third step is to select a docking template based on sequence similarity to the TCR and MHC sequences. The fourth step is to create a "docked TCR model" that combines the TCR model and the pMHC template using the docking template. The fifth step is to mutate and model the side chains in the docked TCR model to match those of the input MHC and epitope sequences, generating a "docked TCR-pMHC model." The sixth step is to compute score based on the TCR-pMHC model

4. Dock TCR model to pMHC templates (Subheading 2.3.4).
5. Remodel side chains of pMHC (Subheading 2.3.5).
6. Score model (Subheading 2.3.6).

2.2 Assembling TCR Amino Acid Sequences

It is common practice to describe TCR sequences by their variable (V) and joining (J) genes along with the CDR3 amino acid sequence. Assembling these three pieces of information into a coherent amino acid sequence that can be used for structural modeling thus requires some decision-making. These decisions are generally straightforward, but at the same time critical, for a mistake here could doom any downstream structural or functional analyses. Obviously, if the amino acid sequences of the whole rearranged TCR of interest are already known (e.g., as provided by the TCR sequencing platform), this step is not required. Ideally, the CDR3 sequence would be given in a canonical form where 5′ end is the second cysteine of the V gene and the 3′ end is the conserved phenylalanine or tryptophan of the J gene, and both ends align significantly and unambiguously with the V and J genes, respectively (Fig. 4). Under such conditions, we can extend the 5′ end and 3′ end of the CDR with the corresponding V and J genes in a straightforward manner. For a small number of sequences, one can perform such assembly manually; however, for high-throughput analysis, this is impractical. Accordingly, we have provided a tool, *VDJassembler*, to automate the process as part of the *ImmuneScape* server. *VDJassembler* should work for sequences even if the CDR3 sequence is not provided in its canonical form. If a tab-separated table with three columns (CDR3 sequence, V gene name, and J gene name) is given, *VDJassembler* will output a list of the corresponding amino acid sequences.

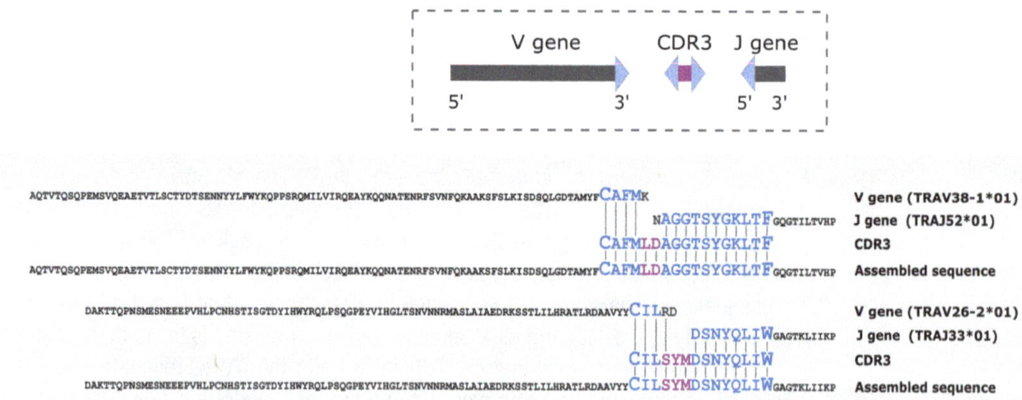

Fig. 4 TCR amino acid sequence assembly. A full TCR alpha or beta sequence is assembled from a part of a V sequence, a CDR3 sequence, and a part of a J sequence

2.3 Detailed Instructions

Here will more precisely define the steps that *ImmuneScape* uses to construct a TCR-pMHC structural model.

2.3.1 Building a Lymphocyte Receptor Structural Model

There are a number of methods for BCR/antibody modeling: PIGSPro [11], Rosetta Antibody [12], LYRA [13], Kotai Antibody Builder [14], and ABodyBuilder [15]. TCR modeling is fundamentally no different than BCR modeling; however, the number of methods available is far fewer. To our knowledge, there are only two such methods, Repertoire Builder [16] (https://sysimm.ifrec.osaka-u.ac.jp/rep_builder/) and LYRA. This is probably due to the relatively smaller number of experimentally determined TCR structures, as mentioned above. Here, as our focus is on TCR modeling, we will cover the general procedure for two methods: LYRA and Repertoire Builder.

Repertoire Builder

Repertoire Builder (Fig. 5a) is a template-based structure modeling method. TCR structural templates are collected from PDB. After finishing, the structures and a text file containing information

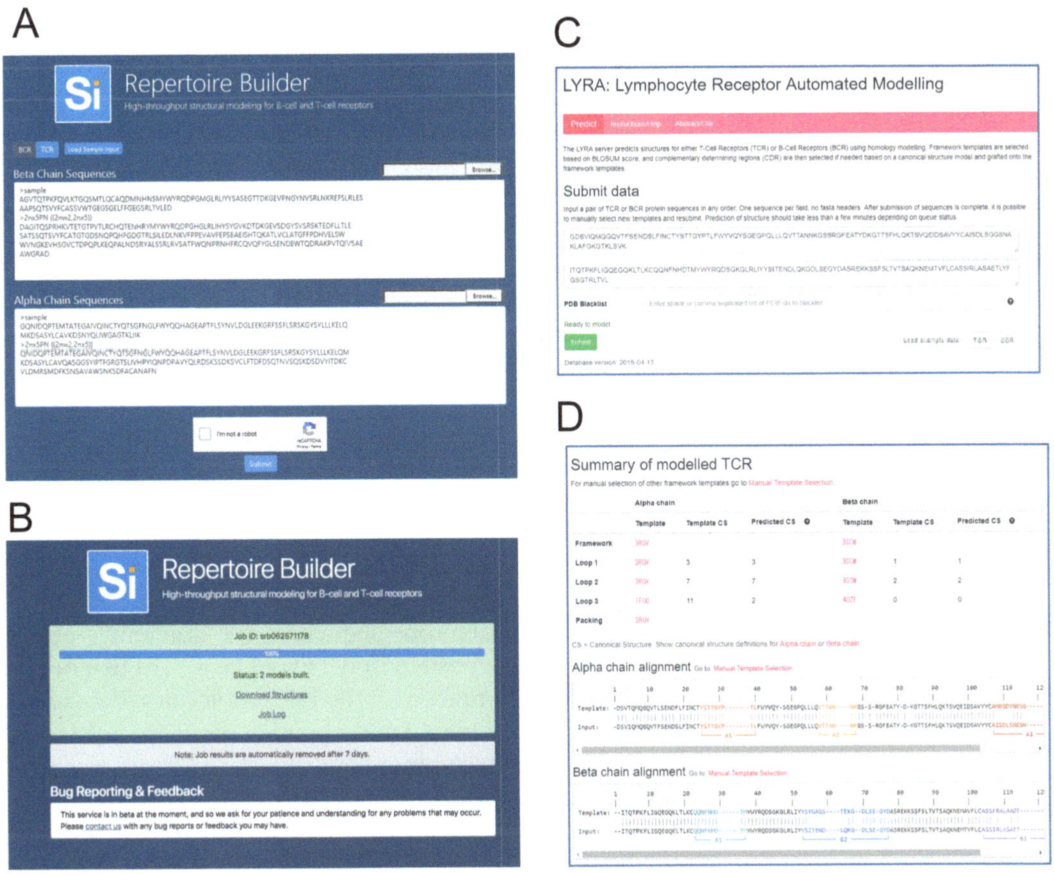

Fig. 5 TCR structural modeling. (**a**) The input page for Repertoire Builder. (**b**) The results page for Repertoire Builder. (**c**) The input page for LYRA. (**d**) The results page for LYRA

about the job can be downloaded (Fig. 5b). As an additional feature, Repertoire Builder will notify you if any of your beta or alpha chains are highly similar to an existing entry in the PDB. Cases where models are not built from a sequence pair are usually due to a lack of high-quality template structures in the PDB or possibly to improperly assembled sequences. Note that Repertoire Builder uses separate structural templates for each of the CDRs. If a template with the appropriate length is not available for any one of the six CDRs, the model will not be built. Repertoire Builder makes use of several features in the MAFFT multiple alignment software [17]. Because these features may be of use in other contexts, we describe them explicitly in Subheading 2.5.

LYRA

LYRA is available at http://www.cbs.dtu.dk/services/LYRA/. LYRA works similarly to Repertoire Builder, but currently does not allow batch sequence uploading (Fig. 5c). After inputting your beta and alpha chain sequences and clicking the "Submit" button, LYRA will redirect you to a status page showing the progress of your job as it completes. Details of the modeling calculation are displayed on the result page. Producing a single model with the LYRA web server typically takes a few minutes. However, the standalone executable (available from the authors) requires only seconds per model. Upon completion, LYRA will redirect to a results page (Fig. 5d) with links to download the structure along with information about loop templates and chain alignment.

2.3.2 Predicting Peptide-MHC Interactions

The prediction of MHC epitopes has been studied using various computational approaches. With regard to structural modeling, there are two aspects that concern us. First, there is the prediction of the epitopes themselves, given a set of potential antigens (from which the epitope sequences are drawn) and a set of MHC alleles. Second, there is the "register" or one-dimensional alignment of the epitope to the MHC binding cavity. There are myriad tools for predicting peptide-MHC interactions, so we will illustrate their use with one of the more popular ones: NetMHCpan [18–21] which performs both epitope prediction and register prediction for class I and II MHCs.

MHC Epitope Prediction

Even though numerous epitope sequences for different types of HLA alleles have been deposited in databases (e.g., IEDB), not all epitope residues are in contact with the HLA molecules. To address this problem, anchor residues within a peptide sequence need to be defined and placed in a common framework. The alignment of epitope anchor residues is of importance for structure modeling. NetMHCpan [18, 20, 21] and NetMHCIIpan [19] can accept a query peptide sequence and HLA allele name or sequence as input and predict binding epitopes for class I and class II MHCs, respectively. NetMHCpan and NetMHCIIpan web servers are available at http://www.cbs.dtu.dk/services/NetMHCpan/ and http://www.cbs.dtu.dk/services/NetMHCIIpan/, respectively. Command-line software packages are also available for both as well. NetMHC-pan and NetMHCIIpan both use a 9mer numbering scheme to register anchor positions within an epitope "core." Following this scheme, the register of anchor residues within a predicted epitope can be defined by aligning them to the corresponding 9-mer core epitope.

Prediction examples of NetMHCpan and NetMHCIIpan are given here using the stand-alone tools. The NetMHCpan command for class I MHCs is.

```
netMHCpan -f query.fa -inptype 0 -a HLA-A02:01 -l 8,9,10 -s
```

Options are as follows:

"`-f query.fa`": query.fa is a query peptide sequence file in FASTA format including the following sequence information:

```
>Query
AAFLPSDFFPSVAA
```

"`-inptype 0`": if it is set to 0, the peptide sequence is given in FASTA format.

"`-a HLA-A02:01`": input HLA information as an allele name. Alternatively, the allele sequence in FASTA format can also be used.

"`-l 8,9,10`": this defines epitope length(s) to be screened. Multiple lengths are concatenated by "`,`".

"`-s`": predicted epitopes are sorted by descending affinity.

Other possible input formats and options can be found using "`-h`" option.

The main output is shown in Fig. 6a.

To register anchor positions within a predicted peptide, information from these columns (`Peptide, Core, Of, Gp, Gl, Ip, Il`) is required. Sequences under the Peptide column are substrings extracted from the full-length query peptide sequence. Sequences under the Core column are always nine residues long (including gaps) and indexed sequentially. Non-gap characters correspond to anchor positions. For more detailed information about these columns, users can refer to NetMHCPan's output description at:

http://www.cbs.dtu.dk/services/NetMHCpan-4.0/output.php.

When aligning Peptide and Core sequences, three scenarios can be summarized as follows:

If the `Gl` value for a given hit is nonzero, the number of `Gl` gap(s) should be added after the `Gp-th` residue within the Core sequence. Using the first hit as an example, aligned sequences along with anchor indices are shown as follows:

A

Pos	HLA	Peptide	Core	Of	Gp	Gl	Ip	Il	Icore	Identity	Score	%Rank	BindLevel
3	HLA-A*02:01	FLPSDFFPSV	FLPSDFPSV	0	5	1	0	0	FLPSDFFPSV	Query	0.8715930	0.0842	<= SB
3	HLA-A*02:01	FLPSDFFPS	FLPSDFFPS	0	0	0	0	0	FLPSDFFPS	Query	0.3078000	0.9972	<= WB
4	HLA-A*02:01	LPSDFFPSV	LPSDFFPSV	0	0	0	0	0	LPSDFFPSV	Query	0.0175330	6.5911	
3	HLA-A*02:01	FLPSDFFP	FLPSDFFP-	0	0	0	8	1	FLPSDFFP	Query	0.0070570	10.1064	
6	HLA-A*02:01	SDFFPSVAA	SDFFPSVAA	0	0	0	0	0	SDFFPSVAA	Query	0.0056190	11.1073	
2	HLA-A*02:01	AFLPSDFFPS	ALPSDFFPS	0	1	1	0	0	AFLPSDFFPS	Query	0.0036900	13.2018	
1	HLA-A*02:01	AAFLPSDFF	AAFLPSDFF	0	0	0	0	0	AAFLPSDFF	Query	0.0024050	15.5835	
2	HLA-A*02:01	AFLPSDFFP	AFLPSDFFP	0	0	0	0	0	AFLPSDFFP	Query	0.0019030	17.0926	
1	HLA-A*02:01	AAFLPSDFFP	AAFPSDFFP	0	3	1	0	0	AAFLPSDFFP	Query	0.0005200	28.0435	
5	HLA-A*02:01	PSDFFPSV	-PSDFFPSV	0	0	0	0	1	PSDFFPSV	Query	0.0004620	29.3500	
4	HLA-A*02:01	LPSDFFPSVA	LPSDFFPSA	0	8	1	0	0	LPSDFFPSVA	Query	0.0004200	30.4444	
5	HLA-A*02:01	PSDFFPSVA	PSDFFPSVA	0	0	0	0	0	PSDFFPSVA	Query	0.0004160	30.5556	
5	HLA-A*02:01	PSDFFPSVAA	PSFFPSVAA	0	2	1	0	0	PSDFFPSVAA	Query	0.0003960	31.1176	
7	HLA-A*02:01	DFFPSVAA	DFFPS-VAA	0	0	0	5	1	DFFPSVAA	Query	0.0003640	32.0690	
2	HLA-A*02:01	AFLPSDFF	AFLPSDF-F	0	0	0	7	1	AFLPSDFF	Query	0.0003300	33.2593	
6	HLA-A*02:01	SDFFPSVA	SDFFPSV-A	0	0	0	7	1	SDFFPSVA	Query	0.0002110	38.8235	
1	HLA-A*02:01	AAFLPSDF	AAFLPSD-F	0	0	0	7	1	AAFLPSDF	Query	0.0001650	42.1667	
4	HLA-A*02:01	LPSDFFPS	-LPSDFFPS	0	0	0	0	1	LPSDFFPS	Query	0.0000910	50.9259	

B

Seq	Allele	Peptide	Identity	Pos	Core	Core_Rel	1-log50k(aff)	Affinity(nM)	%Rank	Exp_Bind	BindingLevel
3	DRB1_0301	PVSKMRMATPLLMQA	Query	4	MRMATPLLM	0.885	0.572	102.59	1.70	NA	<=SB
2	DRB1_0301	APVSKMRMATPLLMQ	Query	5	MRMATPLLM	0.890	0.549	132.12	2.50	NA	<=WB
1	DRB1_0301	AAPVSKMRMATPLLM	Query	6	MRMATPLLM	0.885	0.495	235.79	5.50	NA	<=WB

Fig. 6 NetMHCPan output. (**a**) NetMHCPan output. (**b**) NetMHCIIpan output

```
Peptide FLPSDFFPSV
 Core FLPSD-FPSV
 Anchor 12345-6789
```

If the `Il` value of a hit is nonzero, a number of `Il` gap(s) should be added after the `Il-th` residue within the Peptide sequence. For the fourth hit, aligned sequences along with anchor indices are shown as follows:

```
Peptide FLPSDFFP-
 Core FLPSDFFP-
 Anchor 12345678-
```

If the `Of` value is nonzero, `Of` gap(s) should be added before the Core sequence.

NetMHCIIPan, for class II MHCs, works in a similar way to NetMHCPan, except that users can define both MHC alpha and beta chains by allele names or sequences. Detailed explanations on the prediction output are available at http://www.cbs.dtu.dk/services/NetMHCIIpan-3.2/output.php.

The command-line version can be invoked by:

```
netMHCIIpan -f query.fa -inptype 0 -a DRB1_0301 -length 15 -s
```

Options are as follows:

"`-f query.fa`": defines a query peptide sequence file in FASTA format including the following sequence.

```
>Query
AAPVSKMRMATPLLMQA
```

"`-inptype 0`": query sequence input in FASTA format.

"`-a DRB1_0301`": MHC allele name to be screened.

"`-length 15`": by default, 15mer epitopes with a 9mer core will be predicted. If query peptide length is shorter than 15 residues, the user needs to set the length no longer than that of the query sequence.

"`-s`": output is sorted by descending affinity.

The main output is shown in Fig. 6b. Peptide alignment and anchor registration is relatively straightforward for class II MHCs, since 9mer core is always a substring within the predicted epitope. For the first hit, alignments are shown as follows:

```
Peptide PVSKMRMATPLLMQA
 Core ----MRMATPLLM--
 Anchor ----123456789--
```

Selecting Peptide-MHC Structural Templates

Given a predicted peptide-MHC interaction, it is reasonable to attempt to build a three-dimensional model using known pMHC structures as templates. However, the details of how potential templates should be scored and how the structural conformations should be sampled within a set of one or more templates are open questions. There are various factors one might consider in choosing a structural template: the sequence similarity between the "query" MHC and the template MHC, the sequence similarity between the query epitope and the template epitope, the length of the epitope, the sequence and structure of the TCR (if known), etc. Here, we will sidestep these thorny issues and focus on a quick and relatively naive approach. Namely, we align the query MHC sequence to a set of pMHC templates and rank them by their sequence similarity scores based on the alignments using the BLOSUM62 amino acid exchange matrix as implemented in MAFFT. The BLOSUM62 sequence scores are weighted, and weights were determined by a simple machine learning procedure [16]. No matter what approach is taken, however, it seems clear that a mapping between MHC alleles and PDB entries is valuable. While this mapping might sound trivial, there are a number of potential pitfalls, as described below.

Mapping Between MHC Alleles and PDB Entries

Here, we will assume that we have a reference set of MHC amino acid sequences of interest. We will use the HLA sequences from ftp://ftp.ebi.ac.uk/pub/databases/ipd/imgt/hla/hla_prot.fasta

as a starting point. We have separated the sequences into class I alpha, class II alpha, and class II beta. Given a particular reference HLA sequence, we can align the reference to all PDB entries to find close matches, but there is an issue of how to weight the aligned residues. For example, it seems reasonable to assume that, for TCR-pMHC modeling, the HLA N-terminal domain (NTD), which interacts with the TCR receptor and the epitope, is more important than the C-terminal domain, which interacts with the antigen-presenting cell. NetMHCpan goes a step further and identifies residues that make physical contact with the epitope, and this reduction of the critical set of residues allows the NetMHCPan algorithm to achieve more accurate predictions than if the MHC sequences are considered equally important. Here, we are interested in both interactions between MHC-epitope and MHC-TCR, so we consider all NTD residues. In other words, the residues in the NTD have a weight of 1 and other residues have a weight of 0.

We begin with a set of MHC NTD structures extracted from the PDB. The NTD boundaries were determined iteratively: first, an automated structural domain parser [22] was used to make a rough cut through each PDB entry. Next, we built a multiple sequence alignment (MSA) of the putative domains using MAFFT [17] with default options. Then we impose the consensus domain boundaries on the original PDB entries. We do this for class I alpha, class II alpha, and class II beta. With these domains in hand, we again prepare a multiple sequence alignment including the PDB NTD sequences and the full-length HLA reference sequences of a given class and chain type (class I alpha, class II alpha, and class II beta), determine the consensus domain boundaries in the context of this MSA, and impose the boundaries on the full set of HLA reference sequences of a given class and chain type. From these steps we can obtain a complete and consistent set of HLA reference N-terminal domains. The complete set of HLA reference N-terminal domains can then be reduced to a representative set by running a sequence clustering tool such as cd-hit [23]. The next step is to align the representative domains to the entire PDB and output the significant matches. This procedure is heuristic and we have implemented what appears to be a reasonable choice of parameters. Since there are various use-case scenarios where one might need a mapping between MHC alleles and PDB entries (e.g., to generate benchmarks for structural modeling or to model alleles with more coverage in the PDB), we maintain a list of all pairs with significant sequence similarity. The following files are available for download at https://github.com/sysimm/immune-scape-data/ and will be updated regularly:

1. `hla_prot.fasta` (the HLA reference sequences).
2. HLA I:

- HLA I alpha NTD sequences.
- HLA I alpha NTD cd-hit cluster file at 100% sequence identity.
- PDB HLA I alpha NTD sequences.
- HLA I alpha NTD to PDB map for cluster representatives.

3. HLA II:
   - HLA II alpha and beta NTD sequences.
   - HLA II alpha and beta NTD cd-hit cluster files at 100% sequence identity.
   - PDB HLA II alpha and beta NTD sequences.
   - PDB HLA II alpha-beta chain indices.
   - HLA II alpha NTD to PDB map for cluster representatives.
   - HLA II beta NTD to PDB map for cluster representatives.

In all the steps discussed below, we will assume that HLAs and TCRs have been partitioned into consistent NTD and V domains, respectively.

Clustering MHC Templates by the Length of the Bound Peptide

If the HLA allele has more than one representative in the PDB (e.g., for well-studied alleles), it may be beneficial to consider the sequence and length of the bound peptide when selecting a template. To do this, it is necessary to adopt a consistent epitope numbering scheme. Here we use the NetMHCpan scheme, as described above. In this way, it is possible to define the core length of a pMHC query or template as the number of residues contained between position 1 and position 9. In order to construct a realistic model, and for reasons of simplicity, we use only pMHC templates with epitope core lengths that match those of the query.

There are many ways one can conceive of scoring a query-template pMHC pair. Since such a discussion is beyond the scope of this protocol, we will present only a simplistic approach: MHC NTD BLAST score. That is, we ignore the sequence of the epitope and rank all pMHC templates with matching epitope core lengths by the blast score of the MHC NTD. For class II, this involves treating the alpha and beta chains as a single "virtual" chain.

2.3.3 Selecting Docking Templates

Simulating TCR-pMHC binding is, in principle, a computationally demanding task. To save time, *ImmuneScape* uses a set of representative TCR-pMHC class I and class II complexes from the PDB as "docking templates." We align the query TCR sequences and MHC sequence(s) to a set of TCR-MHC templates and rank them by their weighted overall sequence similarity scores based on the alignments using the BLOSUM62 amino acid exchange matrix as implemented in MAFFT.

2.3.4 Template-Based TCR-pMHC Docking

The TCR model and pMHC template are superimposed onto the docking template using conserved residues in the TCR and MHC as anchors. The identification of conserved residues involves the use of multiple alignment and structural superposition and is beyond the scope of this chapter. Here it is sufficient to note that the conformational sampling performed is much less than that which would be carried out by a general protein-protein docking server, such as ClusPro [24].

2.3.5 Side-Chain Remodeling

At this stage *ImmuneScape* has built a model of the query TCR, but not yet the pMHC. The pMHC is still represented by a set of templates that have been docked to the TCR model using a set of TCR-pMHC docking templates. In this section we will describe the steps needed to mutate the residues of the docked pMHC template to those of the query. For each pMHC template-docking template pair, we must perform the following operations:

1. Mutate any residues in the template MHC that do not match those of the query. Note that in this step the positions of the pMHC template and TCR model are fixed in the bound conformation.
2. Mutate any residues in the template epitope that do not match those of the query.
3. Identify any residues at the TCR-MHC-epitope interface that clash; for these residues the side-chain conformations are rebuilt even if they do match.

For mutation and side-chain modeling, we can use SCWRL4 [25]. Note that, with the exception of the TCR modeling step, the entire TCR-pMHC modeling routine consists only of structural superposition (onto the docking template) and side-chain modeling. For this reason, the modeling protocol described above building is quick and relatively easy.

2.3.6 TCR-pMHC Scoring

In addition to the BLAST (docking and pMHC), NetMHCpan, and Repertoire Builder scores (above), the TCR-pMHC binding energy is computed using a protein-protein docking energy function [10]. Generally speaking, we can assume that any model with a large positive energy has potentially serious problems. Unfortunately, when VDJdb data (or PDB data with >90% sequence identity templates blacklisted) are run, typical *ImmuneScape* models have positive energies suggesting clashes even after side-chain remodeling. However, preliminary work suggests that, for true binders, these clashes can be removed by a short molecular dynamics simulation.

2.4 TCR-pMHC Molecular Dynamics

The protocol described above is straightforward and relatively fast because PDB entries are used as templates for modeling. If the query sequence is close enough to known structures, it might work fine. However, as there are still few TCR structures in the PDB while there are effectively an infinite number of TCR-pMHC combinations possible, we cannot expect high-quality models in most cases. To account for such cases, we can use molecular dynamics (MD) to sample TCR-pMHC structures that do not exist in the PDB.

2.4.1 Example of Refinement by MD

Using all atom MD simulations we were able to dramatically reduce the energy of the modeled complexes. PDB entry 1ao7 was modeled using *ImmuneScape* using a 90% sequence identity cutoff to blacklist highly similar templates, which resulted in a model with an EMPIRE energy of 53 kcal mol^{-1}. During the MD simulation the energy dropped below zero within the first picosecond, with only minimal structural changes, as is evident from the RMSD in Fig. 7a. After 0.5 ns of further MD simulation, the energy dropped to below −16 kcal mol^{-1}, while the RMSD of the TCR from the native structure (with MHC superimposed) improved modestly (Fig. 7b). This dramatic reduction in the binding energy is encouraging; however, upon further simulation we observed that the TCR tended to drift away from the pMHC, as indicated by the steady rise in both EMPIRE binding energy and RMSD. Thus, for longer MD simulations, it is expected that some form of restraints will be needed to compensate for the missing transmembrane regions, membrane and accessory proteins which presumably stabilize the TCR-pMHC immune synapse.

2.4.2 Detailed MD Instructions

In the explanation below, we assume the input structure contains no heteroatoms, which is valid if you follow the protocol above for TCR-pMHC modeling, but may not be valid for PDB entries. The simulations performed in this chapter were done with the GROMOS96 54a7 force field [26]. The energies mentioned in this chapter were calculated using EMPIRE [10]. This tutorial is based on the Gromacs tutorial by Justin Lemkul: http://www.bevanlab.biochem.vt.edu/Pages/Personal/justin/gmx-tutorials/lysozyme/.

2.4.3 Preparatory Work

1. Determine at which resolution you want to run. Coarse grained (CG) and fine grained (FG) require different protocols and force fields.
2. Choose a force field.
   (a) Common force fields with atomistic resolution are Gromos, Charmm, and OPLS.
   (b) Martini is a common CG force field.
3. Obtain your structure.
4. Check for missing residues in your structure, and fix this if needed. This should not be necessary if running with output from ImmuneScape/Repertoire Builder.

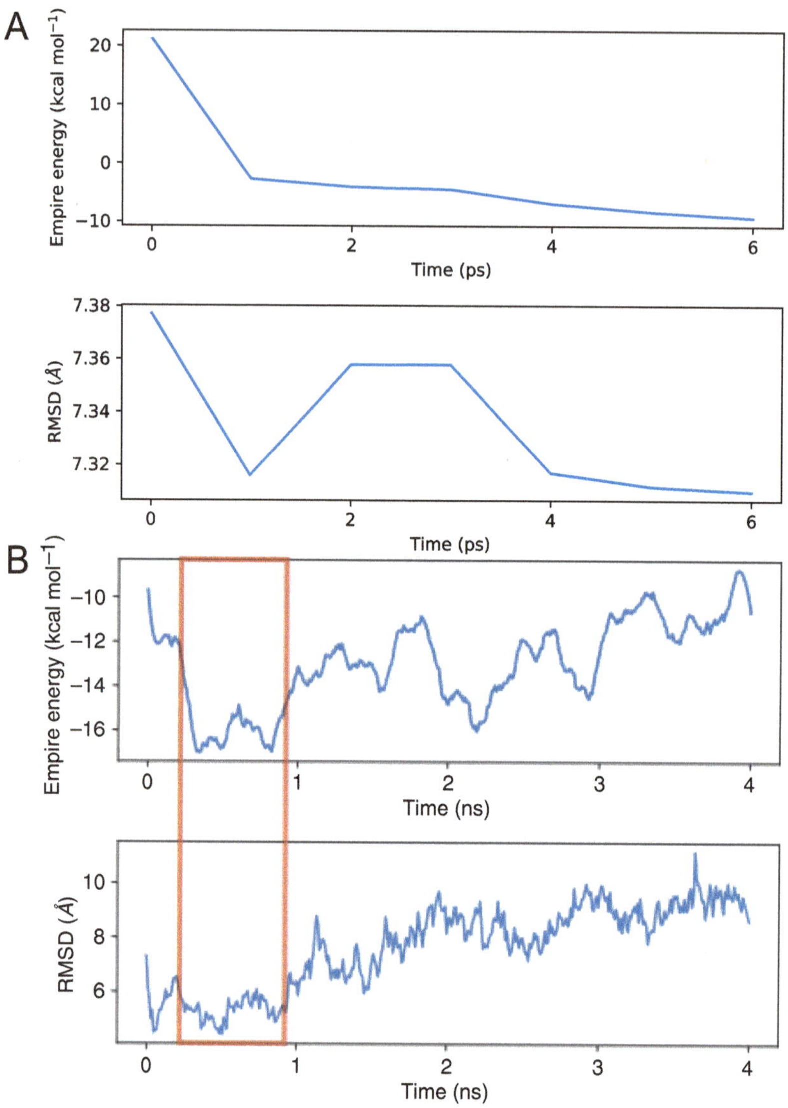

Fig. 7 Molecular dynamics-based refinement. (**a**) EMPIRE energy and TCR RMSD as a function of time for the short equilibration run. (**b**) EMPIRE energy and TCR RMSD as a function of time for a 4 ns production run. The most EMPIRE energy-favorable region is highlighted by a red rectangle

5. Visually inspect your input model. The protocol above can generate severe clashes between the TCR and the pMHC. Such clashes should be reflected in a high binding energy.
6. Source the Gromacs GMXRC executable.

The following protocol is to perform a fine-grained simulation.

2.4.4 Generating the Input Files

A Gromacs simulations requires at least three files:

1. Coordinate file (*.gro file).

2. Topology file (∗.top file).
3. Parameter file (∗.mdp file).

The coordinate file contains the location for every atom/particle/bead in your simulation. It is a big list of atoms, with for every atom an X, Y, and Z coordinate. The topology file describes your system. It does not contain any coordinates, but it describes all the atoms in your system and how they are related to each other. For example, for each atom in your system, it describes its type and associated parameters used by the force field. In case of molecules, it lists which atoms are connected and by which potential. For purposes of readability, the topology files often refer to ∗.itp files, which can contain the description of the molecules. The parameter file describes the framework under which the simulation is run. Here, for example, the algorithm to control the temperature and how often your coordinates are written out are defined.

The first steps are now to convert your structure toward a force field-compatible structure and to generate a topology file.

1. You are asked to select a force field.
2. Depending on the force field you get questions on how to treat water molecules and your protein termini.

```
gmx pdb2gmx -f <structure.pdb>
```

Running pdb2gmx will give you the following files:

1. Coordinate file (∗gro).
2. Topology file (∗top).
3. Position restraint file(s).
4. Itp file(s).

In some cases while running pdb2gmx you will get an error message about unrecognized hydrogen options. This can be easily fixed with PyMOL (Schrodinger Inc.) after you load the structure with the following commands:

```
remove hydrogens
save <stucture_without_hydrogens>.pdb,<structure>
```

2.4.5 Putting It in a Box and Solvating the System

In order to run a simulation you have to define the size and shape of the simulation box. Virtually all simulations are done under periodic boundary conditions (PBC). This creates virtual copies of your simulation box in all directions, such that if a molecule diffuses through the left side of the box, it will reenter at the right side. The box can be defined with the program editconf:

```
gmx editconf -f conf.gro -o conf_box.gro -c -d 1.0 -bt cubic
```

This defines a cubic box (-bt cubic), centers the protein in the box (-c), and places it at least 1 nm from the box edges (-d). The last part is important; the protein should never see its own periodic image. Setting -d to 1 means that there is at least 2 nm between any two periodic images of the protein.

The next step is to solvate the system in SPC water, which is a common water model. Note that also the topology is given as an input, which is automatically updated to reflect the added water to the system.

```
gmx solvate -cp conf_box.gro -cs spc216.gro -o conf_solv.gro -p topol.top
```

Usually ions are added to neutralize the system and to replicate more physiological conditions. The genion program replaces water molecules with ions; apart from the topology and the coordinate file, it also requires the parameter file. In order to use genion, first, a *.tpr file has to be generated using grompp, discussed below:

```
gmx grompp -f gromos.mdp -c conf_solv.gro -p topol.top -o ion -maxwarn 1

gmx genion -s ion.tpr -o conf_ions.gro -p topol.top -nname CL -pname NA
-conc 0.154 -neutral
```

The genion with -neutral option neutralizes the system, and the -conc 0.154 option will bring the concentration to a physiological salt concentration of 0.154 molar. With -pname and -nname the names of the positive and negative ions are specified. Genion will ask you from which molecule group it should replace ions; you should select the solvent/water group for this.

2.4.6 Energy Minimization

Input often contains some unfavorable configurations; starting a simulation straight from the initial structure often results in the system to become unstable. Also the process of solvation and the addition of ions can result in overlapping atoms, which results in very high energies. An energy minimization step resolves this issue. Like the addition of ions, energy minimization requires the generation of a *.tpr file.

```
gmx grompp -f em.mdp -c conf_ions.gro -p topol.top -o em
gmx mdrun -v -deffnm em
```

During the energy minimization the energy should go down. If the final energy is still high, there is a problem; if you see a huge amount of LINCS warnings, there is also likely to be a problem with your input structure.

2.4.7 Equilibration

After the energy minimization, all unrealistic and unfavorable configurations are relaxed. However the system is still not ready for a production run, because the solvent still has to be reoriented around the protein. During the equilibration runs, the protein is kept restrained, so that the solvent molecules can slowly adjust themselves to the presence of the protein. The equilibration runs consist usually of two runs. In the equilibration, first, the system is slowly heated toward the desired temperature, but the pressure is not controlled. This is called an NVT simulation, because the number of particles, the volume, and the temperature are kept constant. This is followed by an NPT equilibration, in which not the volume but the pressure is kept constant.

Gromacs requires at least three files for a simulation: a topology (∗.top), a configuration (∗.gro), and a parameter file (∗.mdp). Rather than supplying these files directly to mdrun, which is the program that performs the simulation, these three files are combined and checked by grompp into a tpr file.

```
gmx grompp -f nvt.mdp -c em.gro -p topol.top -o nvt

gmx mdrun -v -deffnm nvt
```

Apart from the ∗.mdp, ∗.gro, and ∗.top file, we also supply ∗.cpt file for the NPT simulation. The ∗.cpt file is a checkpoint file that contains all the necessary information, like the particle velocities, to continue the simulation.

```
gmx grompp -f npt.mdp -c nvt.gro -p topol.top -t nvt.cpt -o npt

gmx mdrun -v -deffnm npt
```

2.4.8 Production Run

After the equilibration the production simulation can be performed. The simulation length is defined in the ∗.mdp file. Your simulation produces a number of files: ∗.xtc, ∗.trr, ∗.gro, and ∗.edr are the most important ones, next to the ∗.log files. The ∗.xtc files contain the trajectory in a compressed format, the ∗.trr file contains the trajectory in a non-compressed format, the ∗.gro file contains the final configuration, the ∗.edr file contains all the energy, and the ∗.log file contains all the logs.

```
gmx grompp -f md.mdp -c npt.gro -p topol.top -t npt.cpt -o md

gmx mdrun -v -deffnm md
```

Depending on the initial structure, it might be necessary to perform more extensive equilibration simulations.

2.4.9 Analysis

After the simulation, the most important part is to analyze your simulation. To look if your simulation has equilibrated you can look at the root mean square deviation (RMSD). Select the protein backbone for the fit and for the RMSD calculation.

```
gmx rms -s md.tpr -f md.xtc -o rmsd.xvg -tu ns
```

If the RMSD has not reached a plateau by the end of the simulation, it means that your simulations have not yet converged, and you could consider extending it.

The gmx energy program lets you extract a whole number of properties from your simulation, for example, the potential and kinetic energy, temperature, pressure, box size, etc. It requires the ∗.edr file as an input.

One of the better things to do in a simulation is to inspect the simulation yourself. Due to the periodic boundary conditions, this becomes however quite messy, as molecules become split over the pbc. Luckily the tool gmx trjconv can center your system back in the box.

```
gmx trjconv -s md.tpr -f md.xtc -o md_pbc.xtc -pbc mol -ur compact
```

Note that in more complex systems, you might require several different passes of your trajectory by gmx trjconv with different options.

Subsequently you can look at your trajectory using software like VMD or PyMOL. Please make sure that apart from the trajectory file, you also supply a coordinates file (∗.gro).

2.5 MAFFT Usage in Repertoire Builder

Multiple alignment plays a key role in the preparation of TCR and MHC sequences and structures. By aligning sequences that share a strong evolutionary relationship, we are able to assign an absolute position to every residue. This indexing is essential for routine tasks such as domain definition as well as for more elaborate tasks such as scoring a TCR-pMHC complex. Because of the importance of maintaining high-quality multiple sequence alignments, we briefly describe the usage of MAFFT [17] and ASH [27], which handle sequence and structural data, respectively. MAFFT and ASH can be combined to convert templates into an overall MSA [28] by using the following command:

```
mafft --localpair --maxiterate 16 --seedtable ashresult input >
output
```

This is an iterative refinement option, L-INS-i, incorporating structural alignment information provided by ASH. The `--seedtable` flag works like the `--seed` flag explained previously [17] but the input format has been customized for this purpose. For

CDRs, non-CDR "framework" regions, and for the orientation between beta and alpha chains, different template MSAs are maintained.

When building a TCR model from a query sequence, MAFFT is used again to add the query sequence into pre-aligned MSAs of different TCR regions, where a treelike relationship is assumed for each template and the query sequence is aligned to the template based on the location in the tree. The command-line arguments for this calculation are as follows:

```
mafft --addfragments query --keeplength template > output
```

See https://mafft.cbrc.jp/alignment/software/addsequences.html#fragments for detailed information. MSA-specific scoring functions will then be applied to the extended MSAs to select the best matched templates for homology modeling.

Repertoire Builder allows building TCR models from FASTA-formatted sequence information for beta and alpha chains. It is tuned for high-throughput use and currently requires ~30 min to build 10,000 models. The interface for building TCR models, available by clicking the TCR tab, is shown below along with sample input.

Sequences can be supplied as files or directly pasted into the text boxes provided. FASTA headers for beta and alpha chains must match exactly in order to pair beta and alpha chains unambiguously when building the model (but such headers are not required at all for a single sequence pair). There is currently a limit of 10,000 models per job.

Repertoire Builder supports blacklisting by PDB ID with a specially formatted FASTA header line:

```
>2nx5PN ((2nw2,2nx5))
```

The above syntax means that PDB entries 2nw2 and 2n × 5 will be excluded from all modeling steps. After specifying the input and clicking the "Submit" button, you will be redirected to a status page showing the progress of the job as it completes. Jobs with only a handful of sequences typically complete in a minute or two.

References

1. Almagro JC, Teplyakov A, Luo J, Sweet RW, Kodangattil S, Hernandez-Guzman F, Gilliland GL (2014) Second antibody modeling assessment (AMA-II). Proteins 82(8):1553–1562. https://doi.org/10.1002/prot.24567
2. Berman H, Henrick K, Nakamura H (2003) Announcing the worldwide protein data bank. Nat Struct Biol 10(12):980. https://doi.org/10.1038/nsb1203-980
3. Larsen MV, Lundegaard C, Lamberth K, Buus S, Lund O, Nielsen M (2007) Large-scale validation of methods for cytotoxic T-lymphocyte epitope prediction. BMC Bioinformatics 8:424. https://doi.org/10.1186/1471-2105-8-424

4. Roomp K, Antes I, Lengauer T (2010) Predicting MHC class I epitopes in large datasets. BMC Bioinformatics 11:90. https://doi.org/10.1186/1471-2105-11-90
5. Wang P, Sidney J, Dow C, Mothe B, Sette A, Peters B (2008) A systematic assessment of MHC class II peptide binding predictions and evaluation of a consensus approach. PLoS Comput Biol 4(4):e1000048. https://doi.org/10.1371/journal.pcbi.1000048
6. Rosati E, Dowds CM, Liaskou E, Henriksen EKK, Karlsen TH, Franke A (2017) Overview of methodologies for T-cell receptor repertoire analysis. BMC Biotechnol 17(1):61. https://doi.org/10.1186/s12896-017-0379-9
7. Altman JD, Moss PA, Goulder PJ, Barouch DH, McHeyzer-Williams MG, Bell JI, McMichael AJ, Davis MM (1996) Phenotypic analysis of antigen-specific T lymphocytes. Science 274(5284):94–96
8. Maryanski JL, Jongeneel CV, Bucher P, Casanova JL, Walker PR (1996) Single-cell PCR analysis of TCR repertoires selected by antigen in vivo: a high magnitude CD8 response is comprised of very few clones. Immunity 4 (1):47–55
9. Shugay M, Bagaev DV, Zvyagin IV, Vroomans RM, Crawford JC, Dolton G, Komech EA, Sycheva AL, Koneva AE, Egorov ES, Eliseev AV, Van Dyk E, Dash P, Attaf M, Rius C, Ladell K, McLaren JE, Matthews KK, Clemens EB, Douek DC, Luciani F, van Baarle D, Kedzierska K, Kesmir C, Thomas PG, Price DA, Sewell AK, Chudakov DM (2018) VDJdb: a curated database of T-cell receptor sequences with known antigen specificity. Nucleic Acids Res 46(D1):D419–D427. https://doi.org/10.1093/nar/gkx760
10. Liang S, Liu S, Zhang C, Zhou Y (2007) A simple reference state makes a significant improvement in near-native selections from structurally refined docking decoys. Proteins 69(2):244–253. https://doi.org/10.1002/prot.21498
11. Lepore R, Olimpieri PP, Messih MA, Tramontano A (2017) PIGSPro: prediction of immunoGlobulin structures v2. Nucleic Acids Res 45 (W1):W17–W23. https://doi.org/10.1093/nar/gkx334
12. Weitzner BD, Jeliazkov JR, Lyskov S, Marze N, Kuroda D, Frick R, Adolf-Bryfogle J, Biswas N, Dunbrack RL Jr, Gray JJ (2017) Modeling and docking of antibody structures with Rosetta. Nat Protoc 12(2):401–416. https://doi.org/10.1038/nprot.2016.180
13. Klausen MS, Anderson MV, Jespersen MC, Nielsen M, Marcatili P (2015) LYRA, a webserver for lymphocyte receptor structural modeling. Nucleic Acids Res 43(W1):W349–W355. https://doi.org/10.1093/nar/gkv535
14. Yamashita K, Ikeda K, Amada K, Liang S, Tsuchiya Y, Nakamura H, Shirai H, Standley DM (2014) Kotai antibody builder: automated high-resolution structural modeling of antibodies. Bioinformatics 30(22):3279–3280. https://doi.org/10.1093/bioinformatics/btu510
15. Leem J, Dunbar J, Georges G, Shi J, Deane CM (2016) ABodyBuilder: automated antibody structure prediction with data-driven accuracy estimation. MAbs 8(7):1259–1268. https://doi.org/10.1080/19420862.2016.1205773
16. Schritt D, Li S, Rozewicki J, Katoh K, Yamashita K, Volkmuth W, Cavetc G, Standley DM (2019) Repertoire builder: high-throughput structural modeling of B and T cell receptors. Mol Syst Des Eng. https://doi.org/10.1039/C9ME00020H
17. Katoh K, Standley DM (2013) MAFFT multiple sequence alignment software version 7: improvements in performance and usability. Mol Biol Evol 30(4):772–780. https://doi.org/10.1093/molbev/mst010
18. Hoof I, Peters B, Sidney J, Pedersen LE, Sette A, Lund O, Buus S, Nielsen M (2009) NetMHCpan, a method for MHC class I binding prediction beyond humans. Immunogenetics 61(1):1–13. https://doi.org/10.1007/s00251-008-0341-z
19. Jensen KK, Andreatta M, Marcatili P, Buus S, Greenbaum JA, Yan Z, Sette A, Peters B, Nielsen M (2018) Improved methods for predicting peptide binding affinity to MHC class II molecules. Immunology. https://doi.org/10.1111/imm.12889
20. Jurtz V, Paul S, Andreatta M, Marcatili P, Peters B, Nielsen M (2017) NetMHCpan-4.0: improved peptide-MHC class I interaction predictions integrating eluted ligand and peptide binding affinity data. J Immunol 199 (9):3360–3368. https://doi.org/10.4049/jimmunol.1700893
21. Nielsen M, Andreatta M (2016) NetMHCpan-3.0; improved prediction of binding to MHC class I molecules integrating information from multiple receptor and peptide length datasets. Genome Med 8(1):33. https://doi.org/10.1186/s13073-016-0288-x

22. Alexandrov N, Shindyalov I (2003) PDP: protein domain parser. Bioinformatics 19 (3):429–430
23. Fu L, Niu B, Zhu Z, Wu S, Li W (2012) CD-HIT: accelerated for clustering the next-generation sequencing data. Bioinformatics 28 (23):3150–3152. https://doi.org/10.1093/bioinformatics/bts565
24. Kozakov D, Hall DR, Xia B, Porter KA, Padhorny D, Yueh C, Beglov D, Vajda S (2017) The ClusPro web server for protein-protein docking. Nat Protoc 12(2):255–278. https://doi.org/10.1038/nprot.2016.169
25. Krivov GG, Shapovalov MV, Dunbrack RL Jr (2009) Improved prediction of protein side-chain conformations with SCWRL4. Proteins 77(4):778–795. https://doi.org/10.1002/prot.22488
26. Schmid N, Eichenberger AP, Choutko A, Riniker S, Winger M, Mark AE, van Gunsteren WF (2011) Definition and testing of the GROMOS force-field versions 54A7 and 54B7. Eur Biophys J 40(7):843–856. https://doi.org/10.1007/s00249-011-0700-9
27. Standley DM, Toh H, Nakamura H (2007) ASH structure alignment package: sensitivity and selectivity in domain classification. BMC Bioinformatics 8:116. https://doi.org/10.1186/1471-2105-8-116
28. Rozewicki J, Li S, Amada KM, Standley DM, Katoh K (2019) MAFFT-DASH: integrated protein sequence and structural alignment. Nucleic Acids Res 47(W1):W5–W10. https://doi.org/10.1093/nar/gkz342

Chapter 18

Assessing T Lymphocyte Aging Using Telomere Length and Telomerase Activity Measurements in Low Cell Numbers

Tinhinane Fali, Camille K'Ros, Victor Appay, and Delphine Sauce

Abstract

As T lymphocytes proliferate and differentiate in vivo or in vitro, their functional capacity can change dramatically. In particular, extensive cell division is often associated with telomere shortening and the onset of cellular senescence, thus impacting the proliferative potential of the cells. Telomere length and integrity represent therefore key molecular markers of the status and aging of the cells. To assess these markers, we established qPCR-based methods to measure telomere length as well as telomerase activity, applied to low cell numbers, which is necessary when working with rare or small subsets of T lymphocytes.

Key words T lymphocytes, Telomere length, Telomerase activity, qPCR, Single cell

1 Introduction

Maintenance of chromosome integrity and control of cellular replication are crucially dependent on telomeres and telomerase activity [1]. Telomeres are hexanucleotide (TTAGGG) tandem repeats of DNA at chromosome termini. Every cell division is usually associated with a telomere attrition of 50 to 150 bp, until telomeres become too short and cells enter a process of growth arrest, referred to as cellular senescence. Telomerase, a ribonucleoprotein complex, can nonetheless maintain the length of telomeres, by catalyzing the addition of telomeric repeat sequences at the ends of the chromosomes. Assessing telomere length and telomerase activity levels is therefore key to evaluate the aging and proliferative potential of cells [2]. However, assessing these key attributes in rare T lymphocyte subsets or antigen-specific T cells may represent a particular difficulty due to the small numbers of cells available for analysis. Working with small cell numbers requires indeed to take

Victor Appay and Delphine Sauce contributed equally to this work.

Shin Kaneko (ed.), *In Vitro Differentiation of T-Cells: Methods and Protocols*, Methods in Molecular Biology, vol. 2048, https://doi.org/10.1007/978-1-4939-9728-2_18,

into consideration several factors, including precision, accuracy, and specificity of the result as well as cost, throughput, and requirement in time or highly skilled labor.

The first approach to measure telomere length was based on quantifying the length distribution of the terminal restriction fragments (TRFs) using Southern blots [3], which require high quantity of DNA, and is labor intensive and costly. Since then, several methods have been developed, including quantitative fluorescence in situ hybridization (Q-FISH) that quantifies telomere length signals in individual chromosomes [4, 5], flow-FISH that associates flow cytometry to measure telomere length in individual cells [6], quantitative PCR (qPCR) that measures mean of telomere length [7, 8], and PCR of single telomere length analysis STELA [9, 10] (Table 1). Likewise, our capacity to measure telomerase activity in cells has also been considerably improved by the development of the telomeric repeat amplification protocol (TRAP) assay [11, 12].

However, these approaches are not easily applicable to working with small size or rare cell populations. To address this issue, we have developed robust methods that enable measurements of telomerase activity and telomere length in very small number of cells. We describe here a real-time quantitative PCR-based TRAP (RTQ-TRAP) applied to low number of cells (as low as 200 cells), based on a modified TRAP assay described by Wege et al. [13]. This quick and sensitive RTQ-TRAP method is based on a two-step procedure: telomerase-mediated primer extension and amplification of the extended product (Fig. 1). Moreover, we present a qPCR method adapted from Cawthon et al. [7], which allows telomere length measurement from as few as 20 cells. This method is based on the calculation of the *T/S* ratio telomeric product (*T*) normalized for a single copy gene (*S*), which can be further converted into base pair using a reference cell line. Moreover, we provide a supplementary protocol adapted to telomere length measurement in single cells (Fig. 2). These methods are relatively inexpensive and high throughput and require very little quantity of cell extract or DNA. These approaches are perfectly adapted to the study of low number of human T lymphocytes sorted by FACS.

2 Materials

2.1 Assessment of Telomerase Activity

1. Cells.

   The telomerase-positive cell line (293T cells) cultured in MEM medium containing 10% FBS as positive control. At least 200T lymphocytes sorted from human PBMCs or in culture.

2. 1× DPBS (Ca2+ and Mg2+ free), pH 7.2 (Ref: 14200-075, Gibco by Life Technologies).
3. Nuclease-free or DEPC-treated water (*see* **Note 1**).

Table 1
Comparison of telomere length measurement methods

Method	Minimum sample required	Result produced	Additional information derived	Sequence measured	Time to complete	Labor intensity	Expertise level required
TRF analysis	>1 μg	Mean absolute telomere length of all chromosomes	Distribution of telomere lengths	Subtelomere and telomere	3 days	High	Low
Slot blot	10 ng	Ratio of telomere to centromere copy number of all chromosomes, relative to that of standard	None	Telomere	2 days	Medium	Medium
STELA (single telomere length analysis)	10–20 ng	Absolute telomere lengths of specific chromosome	Load of critically short telomeres and mean telomere length of specific chromosome	Subtelomere and telomere	1 day	High	High
U-STELA (universal single telomere length analysis)	10–20 ng	Absolute telomere lengths of all chromosomes	Load of critically short telomeres of all chromosomes	Subtelomere and telomere	1 day	High	High
qRT-PCR (quantitative real-time PCR)	5–20 ng	Ratio of telomere sequence copy number to single copy gene copy number of all chromosomes, relative to that of standard	None	Telomere	6H	Low	High
Q-FISH (quantitative FISH)	Ten metaphase cells or 100 interphase cells	Absolute telomere length of individual chromosomes	Mean telomere length of all chromosomes and distribution of telomere lengths	Telomere	1 day	Medium	Medium
Flow-FISH (flow cytometry FISH)	10^6 cells	Mean absolute telomere length of all chromosomes	Distribution of telomere length and telomere length of specific cell types in a population	Telomere	1 day	Medium	High

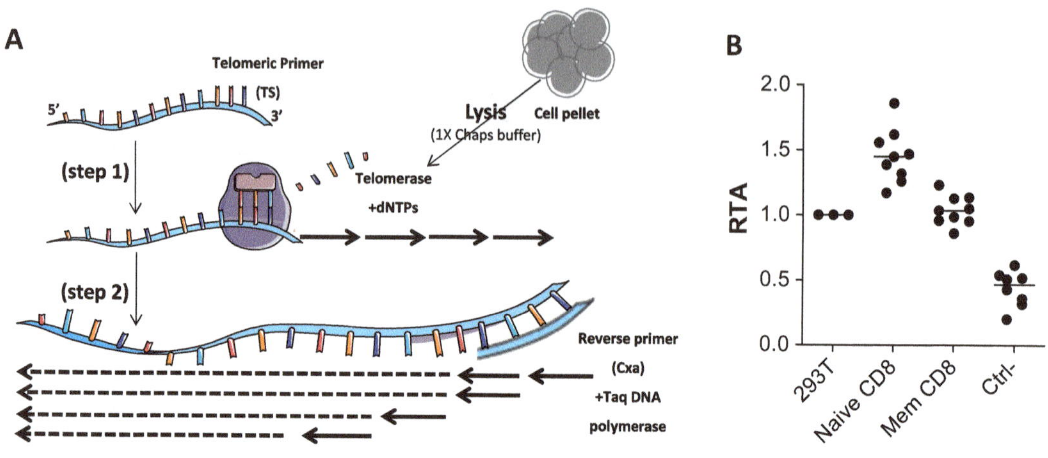

Fig. 1 Measurement of telomerase activity using RTQ-TRAP assay. (**a**) Schematic representation of the RTQ-TRAP assay. After cell lysis using 1× CHAPS buffer, the telomerase binds telomeric ends and elongated telomeric primers (TS) added in the reaction mixture (**step 1**). This step is followed by amplification of the elongated products in a qPCR reaction with reverse primer complementarity (Cxa) (**step 2**). (**b**) Examples of relative telomerase activity (RTA) from sorted naïve or memory CD8$^+$ T cells from a young adult compared to the 293T cell line and negative control (Ctrl−). Bars indicate median

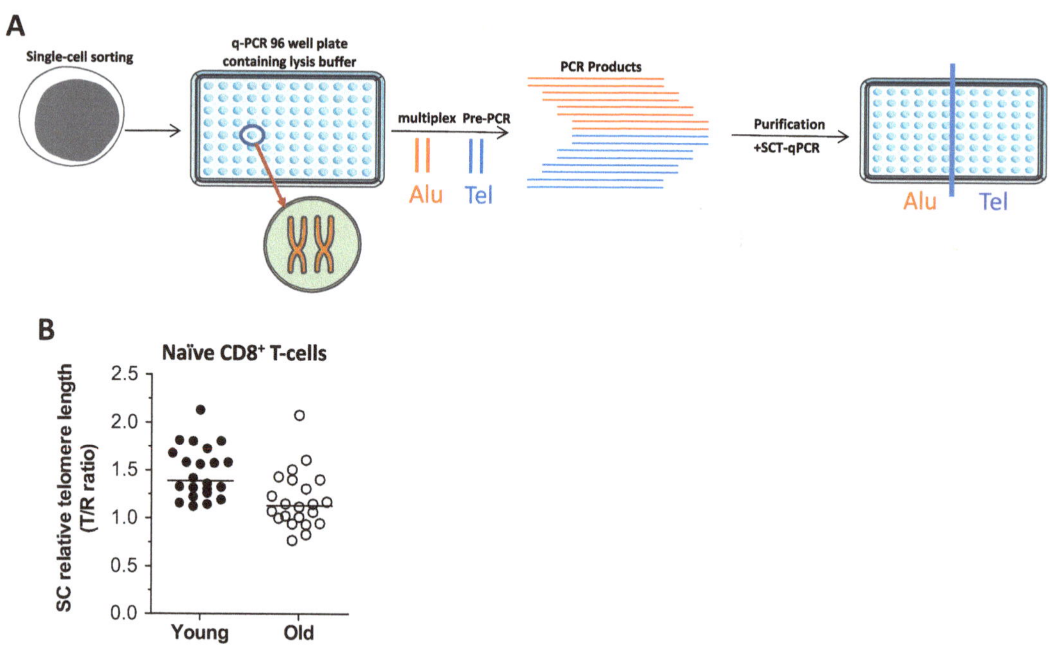

Fig. 2 Single-cell telomere length measurement protocol. (**a**) Schematic representation of the RTQ-TRAP assay. Single cells were first sorted into 96-well plates containing lysis buffer to extract gDNA, followed by multiplex pre-PCR containing reference gene primers (Alu) in red and telomere primers (Tel) blue. The red and blue lines represent PCR productions by *Alu* and *Tel* primers, respectively, after preamplification. Single-cell telomere products are then purified, followed by simplex qPCR (SCT-qPCR). (**b**) Telomere length data (*T/R* ratio $= C_{tTel}/C_{tAlu}$) generated on single naïve CD8$^+$ T cells FACS sorted from PBMCs of a young individual (25 years old) or an old (75 years old) individual are shown

4. Lysis buffer: 10× CHAPS cell extract buffer (Ref: 9852, Cell Signaling).

   1× concentration: 50 mM pipes/HCl (pH 6.5, 2 mM EDTA, 0.5–1% CHAPS, 20 μg/ml leupeptin, 10 μg/ml pepstatin A, 10 μg/ml aprotinin, 5 mM DTT), and 1 mM PMSF in DEPC water. Buffer should be prepared immediately before use and PMSF should be added freshly.
5. 2× Power SYBR Green PCR Master Mix (Ref 4367659 Life Technologies): SYBR Green 1 Dye, AmpliTaq Gold DNA Polymerase LD, dNTPs with dUTP/dTTP blend, Passive Reference 1, optimized buffer components (*see* **Note 2**).
6. Oligonucleotide primers used both at 400 nM.
   (a) TS primer (telomerase substrate for extension): 5′-AA TCCGTCGAGCAGAGTT-3′.
   (b) ACX primer (reverse primer for amplification): 5′-GCG CGG(CTTACC)$_3$CTAACC-3′.
7. A real-time qPCR machine, multichannel pipettors, 96-well PCR plate.

2.2 Telomere Length Measurement

1. Cells. 293T cell line used as standard and cultured in MEM medium containing 10% FBS. At least 20T lymphocytes sorted from human PBMCs or in culture.
2. 1′ DPBS (Ca^{2+}- and Mg^{2+}-free), pH 7.2. (Ref: 14200-075, Gibco by Life Technologies).
3. Nuclease-free or DEPC-treated water.
4. Genomic DNA extraction system, such as Macherey-Nagel; Genomic DNA from tissue, NucleoSpin Tissue XS or QIAGEN; QIAamp DNA Mini Kit.
5. Oligonucleotide primers all used at 400 nM.
   (a) Single copy gene primers (36B4):
   36B4 (F): 5′-CAGCAAGTGGGAAGGTGTAATCC-3′.
   36B4 (R): 5′-CCCATTCTATCATCAACGGGTAC AA-3′.
   (b) Telomere gene primers (Tel):
   Tel (F): 5′-CGGTTTGTTTGGGTTTGGGTTT GGGTTTGGGTTTGGGTT-3′.
   Tel (R): 5′-GGCTTGCCTTACCCTTACCCTTA CCCTTACCCTTACCCT-3′.
6. 2× Power SYBR Green PCR Master Mix (Ref 4367659 Life Technologies): SYBR Green 1 Dye, AmpliTaq Gold DNA Polymerase LD, dNTPs with dUTP/dTTP blend, Passive Reference 1, optimized buffer components.
7. A real-time qPCR machine, multichannel pipettors, 96-well PCR plate.

2.3 Single-Cell Telomere Length Measurement

1. Single T lymphocytes sorted from human PBMCs.
2. Lysis buffer: 2× lysis buffer (100 mM Tris–HCl (pH 7.4), 300 mM NaCl, 0.8 mM ADTA, 2% Nonidet P-40, 5 mM DTT).
3. Nuclease-free or DEPC-treated water.
4. TaKaRa PCR amplification kit (Ref R011): TaKaRa Taq DNA Polymerase, dNTPs, 10× optimized PCR buffer.
5. Oligonucleotides primers.
   (a) Reference gene primers (Alu) (*see* **Note 3**):
   Alu (F): 5′- GACCATCCCGGCTAAAACG-3′.
   Alu (R): 5′- CGGGTTCACGCCATTCTC-3′.
   (b) Telomere gene primers (Tel):
   Tel (F): 5′-CGGTTTGTTTGGGTTTGGGTTTGGGTTTGGGTTTGGGTT-3′.
   Tel (R): 5′-GGCTTGCCTTACCCTTACCCTTACCCTTACCCTTACCCT-3′.
6. DNA Clean and Concentrator™-5 kit (Ref D4013) (*see* **Note 4**).
7. 2× Power SYBR Green PCR Master Mix (Ref 4367659 Life Technologies): SYBR Green 1 Dye, AmpliTaq Gold DNA Polymerase LD, dNTPs with dUTP/dTTP blend, Passive Reference 1, optimized buffer components.
8. Thermocycler, a real-time qPCR machine, multichannel pipettors, 96-well PCR plate.

3 Methods

3.1 Assessment of Telomerase Activity

3.1.1 Cell Lysate Preparation

1. Harvest the cells (including the positive control 293T cells), and wash with ice-cold PBS. Pellet cells in microcentrifuge tubes at 500 × *g* for 5 min at room temperature. At this point, cell pellets may be frozen and stored at −80 °C (*see* **Note 5**).
2. Add 4-5× volume of lysis buffer relative to pellet size. Resuspend by pipetting up and down repeatedly plus vortexing (*see* **Note 6**).
3. Incubate the sample at 4 °C for 30 min with agitation to lyse the cells (*see* **Note 7**).
4. Centrifuge at maximum speed (>13,000 × *g*) for 30 min at 4 °C.
5. Carefully transfer the supernatant to a fresh collection tube, and be careful to not transfer any cell debris (keep everything on ice).

6. Measure the protein concentration of each sample (*see* **Note 8**).
7. Divide the lysate into 15 μl aliquots. Samples can be frozen at −80 °C for up to a month.

3.1.2 TRAP Reaction

1. Prepare control samples: telomerase-positive 293T cell line control; telomerase-negative control (85 °C for 10 min heat-inactivated sample) and lysis buffer-only control.
2. Serially dilute the telomerase-positive control (293T cell line lysate) to make a standard curve. First dilution should start at 2000 μg/ml.
3. Calculate the amount of proteins needed from each sample, and prepare triplicates of each sample in a final aliquot volume of 5 μl/replicate. For 200 cells, 0.5–1 μg of protein lysate is sufficient (500–1000 μg/ml).
4. Calculate the total number of reactions for the experiment and prepare a master mix. For each 25 μl reaction well (with 20 μl master mix), prepare:

2× Power SYBR Green PCR Master Mix	12.5 μl
0.4 μM TS primer	1 μl
0.4 μM CX primer	1 μl
Nuclease-free water	5.5 μl

5. Dispense 5 μl of cell lysate in triplicates for each sample, and add on top 20 μl of the master mix into each well of qPCR plate to make it 25 μl total qTRAP reaction mixture. Mix well and centrifuge the plate (everything should be done on ice and in the dark avoiding photobleaching of the SYBR Green dye and telomerase degradation).
6. Seal the qPCR plate and incubate for 20–30 min at 25 °C in the dark for primer extension step.
7. Place the plate in a real-time qPCR machine and run the following program for amplification step: 10 min at 95 °C; 35 cycles of 30 s at 95 °C, 30 s at 60 °C, and 30 s at 72 °C.

3.1.3 Detection and Analysis

1. Analyze the qPCR data from the dilution series of the telomerase-positive control (293T cell line) to plot a standard curve. On the *y* axis, plot the C_t values (standard curve's average of the triplicates). On the *x* axis, plot the $\log_{10}$ values of the protein concentration of the dilutions. Perform linear regression (coefficient of linearity (R^2) should be ≥ 0.95 to validate each PCR plate) (*see* **Note 9**).
2. Calculate relative telomerase activity (RTA) of the samples based on the C_t values of positive control at the same protein

concentration level (RTA = $C_{t293T\ (1\mu g)}/C_{tsample\ (1\mu g)}$) (RTA > 1 = telomerase activity higher than the positive-control cell line. RTA < 1 = lower level of telomerase activity) (*see* **Note 10**).

3.2 Telomere Length Measurement

3.2.1 Genomic DNA Preparation

1. Harvest the cells (including 293T cell control), and wash with PBS. Pellet cells in microcentrifuge tubes at 500 × *g* for 5 min at room temperature. At this point, dry cell pellets may be frozen and stored at −80 °C for several months.
2. Extract gDNA from cell pellets following manufacture instruction from Macherey-Nagel or QIAGEN kits (*see* **Note 11**).
3. Prepare reference DNA sample control: serially dilute the reference DNA from 293T cell line to make a standard curve. First dilution starts at 16 ng/μl (*see* **Note 12**).
4. Calculate the gDNA concentration needed from each sample, and prepare triplicates of each sample in a final aliquot volume of 5 μl/replicate. For 100 cells, 60–120 ng of gDNA is sufficient (2–4 ng/μl).

3.2.2 qPCR Reaction

1. Calculate the total number of reactions for the experiment and prepare a master mix PCR. For each 25 μl reaction well (with 20 μl master mix), prepare:

2× Power SYBR Green PCR Master Mix	12.5 μl
0.4 μM forward primer	1 μl
0.4 μM reverse primer	1 μl
Nuclease-free water	5.5 μl

2. Two master mix PCR should be prepared, one with the single copy gene 36B4 (*S*) used as a control and one for the telomere gene (*T*) for amplification of every sample performed (*see* **Notes 13** and **14**).
3. Dispense 5 μl of gDNA in triplicates for each sample and standard control, and add 20 μl of the master mix into each well of qPCR plate to make it to a 25 μl total qPCR reaction mixture. Mix well and centrifuge the plate (everything should be done on ice and in the dark avoiding photobleaching of the SYBR Green dye).
4. Seal the qPCR plates and run a suitable thermal cycle for each PCR. The programs for both qPCRs (Tel and 36B4) start with 95 °C for 10 min to activate the AmpliTaq Gold DNA Polymerase. For Tel PCR, it is followed with 30 cycles of denaturation at 95 °C for 15 s, extension at 54 °C for 2 min, and elongation at 72 °C for 1 min. For 36B4 PCR, it is followed with 95 °C for 15 s and 58 °C for 1 min.

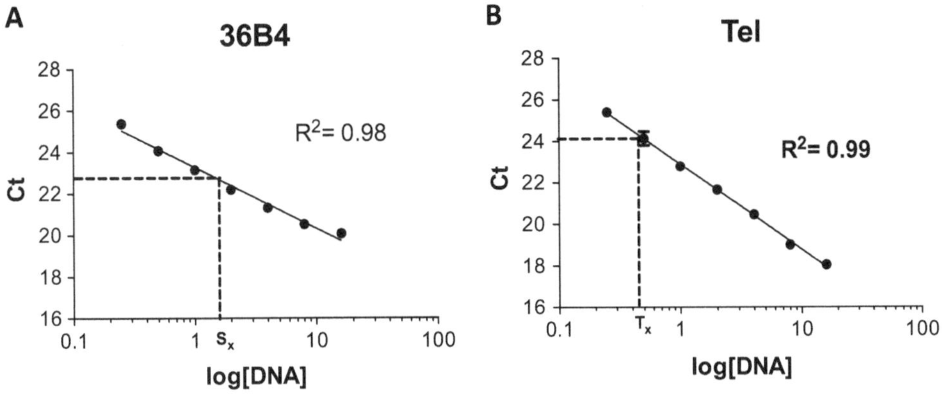

$$\text{TLx (ratio)} = T_x/S_x$$

Fig. 3 Standard curves used to determine relative *T*/*S* ratio. Seven concentrations of standard genomic DNA control (293T cell line) were prepared by twofold serial dilutions (16 ng/μl, 8 ng/μl, 4 ng/μl, 2 ng/μl, 1 ng/μl, 0.5 ng/μl, 0.25 ng/μl per well) and aliquoted in duplicate to a 96-well PCR plate. Both reference and target fluorescent signals were collected from each PCR reaction plate. (**a**) Data for single copy gene 36B4 (*S*) and (**b**) data for telomere target (*T*) are shown

3.2.3 Detection and Analysis

1. Analyze the data from the dilution series of the reference DNA control (293T cell line) to plot a standard curve. On the *y* axis, plot the C_t values (standard curve's average of the triplicates). On the *x* axis, plot the $\log_{10}$ values of the protein concentration of the dilutions. Perform linear regression (coefficient of linearity (R^2) should be ≥0.95 to validate each PCR plate) (*see* **Note 9**).
2. Calculate relative telomere length (RTL) of each sample based on determining the relative ratio (*T*/*S* ratio) of telomere repeat copy number amplification (*T*) to single copy gene 36B4 copy number amplification (*S*) in experimental samples using the standard curve from reference DNA (Fig. 3) ($T/S = 1$ means that the relative telomere length of the sample is comparable with the reference DNA).
3. To convert *T*/*S* ratio into kilobase pairs, mean TRF length of the reference DNA control (serial dilution of 293T cell line) was determined by Southern blot in parallel with qPCR as described by Kimura et al. [14] (Fig. 4).

3.3 Single-Cell Telomere Length Measurement

3.3.1 Single-Cell Genomic DNA Extraction

1. Sort single cell into 96-well plate containing 10 μl of 2× lysis buffer.
2. Heat the plate at 75 °C for 5 min (*see* **Note 15**).

A

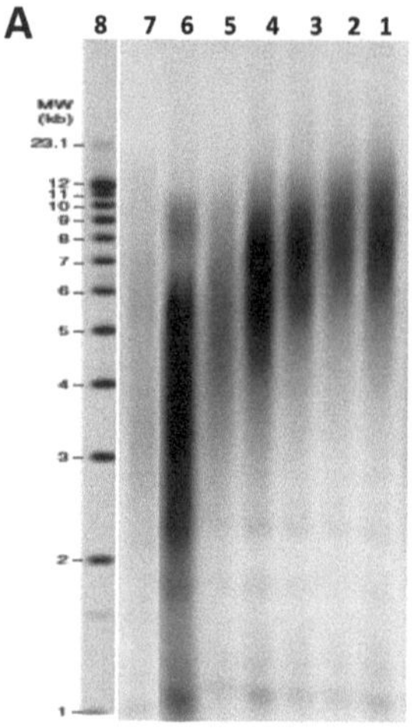

Well	1	2	3	4	5	6	7
gDNA	6µg	3µg	1.5µg	0.75µg	0.35µg	0.18µg	0.09µg
mean TRF (Kbp)	8	7.5	6.8	6.6	4.25	3.2	0.56

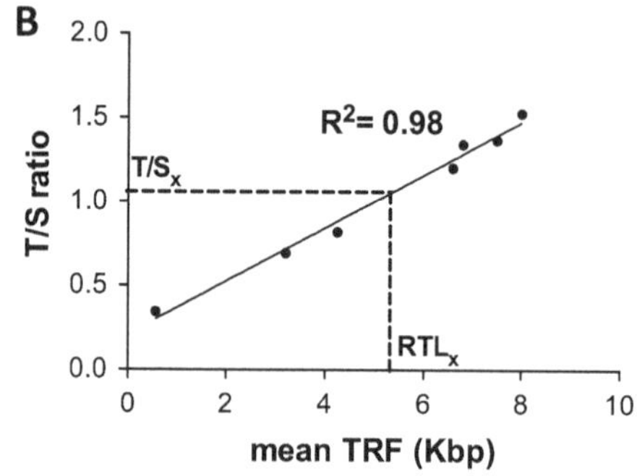

Fig. 4 Calculation of relative telomere length (kbp). Seven concentrations of standard genomic DNA control (293T cell line) were prepared by twofold serial dilutions (6 µg, 3 µg, 1.5 µg, 0.35 µg, 0.18 µg, 0.09 µg per well) and analyzed in parallel by TRF length analysis as well as by qPCR. (**a**) Determination of mean terminal restriction fragment length by Southern blot analysis. Genomic DNA was digested with *Hinfl*/*Rsa*I and separated on 0.5% agarose gel. DNA was transferred to a membrane and probed with telomeric probe that hybridized to TTAGGG repeats. Line 8 contains molecular weight ladders (values are indicated along the left in kb); lines 1 to 7 contain standard DNA from 293T cell line (6 µg to 0.09 µg). Mean TRF length was determined using ImageQuant software tool. (**b**) Standard curve obtained from *T*/*S* ratio determined by qPCR and mean TRF determined by Southern blot. *T*/*S* ratio from specific samples can then be converted into kbp length using this curve

3.3.2 Multiplex Pre-PCR Amplification and Purification of gDNA

1. Use TaKaRa PCR amplification kit and follow the instructions. The final PCR reaction volume is 40 µl containing in each PCR reaction well: 4 µl of 10× PCR buffer, 4 µl of dNTP at 2.5 mM, 0.25 µl of DNA polymerase, and 1 µl of each primer pair (specific primer for telomere sequence and primer for multicopy reference sequence) at 1 µM each, completed with RNase-free water.
2. Run a multiplex pre-PCR amplification according to the following thermal profile: melting at 94 ° C for 5 min, followed by 16–18 cycles of melting at 94 °C for 15 s, annealing at 60 °C for 30 s, extension at 72 °C for 30 s, ended by a final extension at 72 °C for 10 min.
3. Use DNA Clean and Concentrator™-5 kit to purify the pre-amplified gDNA (*see* **Note 16**). For 40 µl of pre-PCR product add 200 µl of DNA binding buffer, mix briefly by vortexing, and transfer the mixture into Zymo-Spin column in a collection tube. Centrifuge for 30 s at 13,000 rpm and discard the flow-through, and wash twice with 200 µl of DNA wash buffer (centrifuge for 30 s at 16,060 × *g* and discard the flow-through). Transfer the column to a new collection tube, and elute the DNA with 35 µl of DEPC water.

3.3.3 q-PCR Reaction

1. Calculate the total number of reactions for the experiment and prepare a master mix PCR. For each 20 μl reaction well (with 15 μl master mix), prepare:

2× Power SYBR Green PCR Master Mix	10 μl
0.4 μM forward primer	1 μl
0.4 μM reverse primer	1 μl
Nuclease-free water	3 μl

2. Two master mix PCR should be prepared and run in separate qPCR reaction, one with the multi-copy gene (Alu) used as a reference gene and one with the telomere gene (Tel).
3. Dispense 5 μl of gDNA in triplicates for each sample and standard control, and add 15 μl of the master mix into each well of qPCR plate to make it to a total volume of 20 μl for the qPCR reaction mixture. Mix well and centrifuge the plate (everything should be done on ice and in the dark avoiding photobleaching of the SYBR Green dye).
4. Seal the qPCR plates and run a suitable thermal cycle for each PCR. The programs for both qPCRs start with 95 °C for 10 min to activate the AmpliTaq Gold DNA Polymerase, followed with 40 cycles of melting at 95 °C for 15 s, annealing at 60 °C for 30 s, and extension at 72 °C for 30 s, ended with 80 cycles of melting curve from 60 °C to 95 °C.

3.3.4 Detection and Analysis

1. Determine a relative telomere length (RTL) for each sample based on calculating the relative ratio of the amount of qPCR product obtained in telomere reaction (T) to amount of qPCR product obtained in multiple-copy gene reference reaction (R) as T/R ratio $= C_{\mathrm{tTel}}/C_{\mathrm{tAlu}}$.

4 Notes

1. Telomerase is a reverse transcriptase with an RNA component; special precaution must be observed to ensure RNase-free manipulation. For example, reagents should be made in DEPC-treated water, and working surfaces and equipment including pipettes and lab ware should be decontaminated with a solution such as RNase AWAY.
2. Due to the sensitivity of the PCR-based TRAP assay, contamination should be avoided. Examples include working in a designated area and using barrier tips.
3. The use of Alu instead of 36B4 as reference gene is due to the poor output of single copy gene 36B4 in the case of single-cell amplification.

4. The DNA Clean and Concentrator™-5 provides an easy method for the rapid purification and concentration of high-quality DNA from PCR.
5. Quick frozen samples may be stored for over 12 months. While frozen lysates can be used, they should not be compared directly with freshly prepared lysates with regard to telomerase activity.
6. The volume of lysis buffer depends on cell number (use 100 μl of lysis buffer for 1×10^6 cells and use 20 μl for 200 cells).
7. Alternative process: cell extract can be immediately frozen at −80 °C or in liquid nitrogen and thawed before centrifugation.
8. Bradford assays may be performed to calculate the lysate protein concentration.
9. When working with big sample size or longitudinal study, the same standard DNA should be used to set up standard curve reaction in every plate during the study. Prepare also a bulk PBMC samples (internal control) to run it in all plates in order to validate absence of interpolate variability; T/S ratio of internal control should be stable with mean standard deviation error <5%.
10. RTA is calculated as the ratio of sample C_t value over the 293T cell line C_t value due to the relative expression in the qPCR; the lower is the C_t value, the greater is the telomerase activity in the sample.
11. gDNA can be stored at 4 °C or −20 °C for months (avoiding multiple sample freezing and thawing that can cause DNA breaks).
12. For each plate (Tel plate and S plate), 293T cell line DNA dilution references should be run in duplicate. The strategy to determine relative telomere length (RTL) by qPCR is to measure for each DNA sample the factor by which the sample value differs from a reference DNA control (293T cell line) value using a standard curve.
13. The choice of the single copy gene is critical for reliability of the result; any change in reference gene can impact the telomere length result. 36B4 encodes the acidic ribosomal phosphoprotein P0 located at chromosome 12.
14. The qPCR reaction for the telomere and 36B4 genes should be run in two separate 96 -well PCR plates and single standard curve should be plotted for each PCR.
15. Single-cell genomic DNAs with lysis buffer can be stored at −20 °C at this step.
16. This step of purification removes the excess of primers and oligonucleotides from the multiplex preamplified gDNA and allows us to perform a qPCR reaction on the purified product.

Acknowledgments

We thank Olivier Pelle from the flow cytometry platform of the Imagine Institute in Paris for assistance with single-cell sorting. This work was supported by the Fondation pour la Recherche Médicale (FRM) (Project DEQ20120323690), the French Agence Nationale de la Recherche sur le SIDA (ANRS, Project N14007DR), and the INSERM transversal research program on aging (Project AGEMED). Victor Appay and Delphine Sauce contributed equally to this work.

References

1. Hodes RJ, Hathcock KS, Weng NP (2002) Opinion: telomeres in T and B cells. Nat Rev Immunol 2(9):699–706
2. Fajkus J (2006) Detection of telomerase activity by the TRAP assay and its variants and alternatives. Clin Chim Acta 371(1-2):25–31. https://doi.org/10.1016/j.cca.2006.02.039
3. Harley CB, Futcher AB, Greider CW (1990) Telomeres shorten during ageing of human fibroblasts. Nature 345(6274):458–460. https://doi.org/10.1038/345458a0
4. Lansdorp PM, Verwoerd NP, van de Rijke FM, Dragowska V, Little MT, Dirks RW, Raap AK, Tanke HJ (1996) Heterogeneity in telomere length of human chromosomes. Hum Mol Genet 5(5):685–691
5. Canela A, Vera E, Klatt P, Blasco MA (2007) High-throughput telomere length quantification by FISH and its application to human population studies. Proc Natl Acad Sci U S A 104(13):5300–5305. https://doi.org/10.1073/pnas.0609367104
6. Baerlocher GM, Vulto I, de Jong G, Lansdorp PM (2006) Flow cytometry and FISH to measure the average length of telomeres (flow FISH). Nat Protoc 1(5):2365–2376. https://doi.org/10.1038/nprot.2006.263
7. Cawthon RM (2002) Telomere measurement by quantitative PCR. Nucleic Acids Res 30 (10):e47
8. Cawthon RM (2009) Telomere length measurement by a novel monochrome multiplex quantitative PCR method. Nucleic Acids Res 37(3):e21. https://doi.org/10.1093/nar/gkn1027
9. Baird DM, Rowson J, Wynford-Thomas D, Kipling D (2003) Extensive allelic variation and ultrashort telomeres in senescent human cells. Nat Genet 33(2):203–207. https://doi.org/10.1038/ng1084
10. Bendix L, Horn PB, Jensen UB, Rubelj I, Kolvraa S (2010) The load of short telomeres, estimated by a new method, Universal STELA, correlates with number of senescent cells. Aging Cell 9(3):383–397. https://doi.org/10.1111/j.1474-9726.2010.00568.x
11. Kim NW, Piatyszek MA, Prowse KR, Harley CB, West MD, Ho PL, Coviello GM, Wright WE, Weinrich SL, Shay JW (1994) Specific association of human telomerase activity with immortal cells and cancer. Science 266 (5193):2011–2015
12. Burger AM (2002) Standard TRAP assay. Methods Mol Biol 191:109–124. https://doi.org/10.1385/1-59259-189-2:109
13. Wege H, Chui MS, Le HT, Tran JM, Zern MA (2003) SYBR green real-time telomeric repeat amplification protocol for the rapid quantification of telomerase activity. Nucleic Acids Res 31(2):E3–E3
14. Kimura M, Stone RC, Hunt SC, Skurnick J, Lu X, Cao X, Harley CB, Aviv A (2010) Measurement of telomere length by the Southern blot analysis of terminal restriction fragment lengths. Nat Protoc 5(9):1596–1607. https://doi.org/10.1038/nprot.2010.124

Chapter 19

Generation of Hematopoietic Stem and Progenitor Cells from Human Pluripotent Stem Cells

Hideyuki Oguro

Abstract

Human pluripotent stem cells (PSCs) have the potential to provide a virtually unlimited supply of cells for transplantation therapy. When combined with recent advances in genome editing technologies, human PSCs could offer various approaches that enable gene therapy, drug discovery, disease modeling, and in vitro modeling of human development. De novo generation of hematopoietic stem cells (HSCs) from human PSCs is an important focus in the field, since it enables autologous HSC transplantation to treat many blood disorders and malignancies. Although culture conditions have been established to generate a broad spectrum of hematopoietic progenitors from human PSCs, it remains a significant challenge to generate bona fide HSCs that possess sustained self-renewal and multilineage differentiation capacities upon transplantation. In this review, recent promising advances in the efforts to generate HSCs and hematopoietic progenitors from human PSCs in vitro and in vivo or from somatic cells are discussed.

Key words Hematopoietic stem cells, Hematopoietic progenitors, Human pluripotent stem cells, Hemogenic endothelium, Hematopoietic differentiation, Hematopoietic development

1 Introduction

Hematopoietic stem cells (HSCs) reside at the top of the hierarchy in the blood system and give rise to all types of hematopoietic and immune cells throughout life [1]. Allogenic and autologous HSC transplantation is widely used in clinics to treat many blood disorders and malignancies [2]. Umbilical cord blood, mobilized peripheral blood, and bone marrow are currently available sources of HSCs for transplantation. However, in allogenic transplantation settings, availability of human leukocyte antigen (HLA)-matched donor HSCs is limited, and mismatched HSC transplantation needs intense immunosuppression regimens after transplantation with an associated risk of graft-versus-host disease. In autologous transplantation settings, substantial numbers of patients fail to reach the minimum HSC collection threshold required for transplantation with traditional HSC mobilization strategies [3].

Shin Kaneko (ed.), *In Vitro Differentiation of T-Cells: Methods and Protocols*, Methods in Molecular Biology, vol. 2048, https://doi.org/10.1007/978-1-4939-9728-2_19, © Springer Science+Business Media, LLC, part of Springer Nature 2019

Although ex vivo expansion of HSCs has made progress over the years [4], the duration that HSCs can be maintained in culture is still limited and there are detrimental changes in the characteristics of HSCs, such as restricted multipotency and homing potential, after extended culture. Genome editing of HSCs has great potential to cure many genetic blood disorders and malignancies. However, it is currently very challenging to directly engineer genomic sequences of primary HSCs because of the difficulty of selecting corrected clones.

Because of its broad potential of applications in basic research and clinical settings, in vitro differentiation of human induced pluripotent stem cells (iPSCs) into HSCs and progenitor cells has been intensively studied. Human iPSCs could provide a virtually infinite source of HSCs for autologous transplantation and iPSCs, in contrast to HSCs, can be genetically engineered for gene correction and disease modeling. Studies of embryonic hematopoietic development in model organisms have led to precise culture protocols for derivation of hematopoietic progenitors from human iPSCs. However, the derivation of long-term self-renewing HSCs from human iPSCs in culture remains elusive. Protocols that more accurately recapitulate human embryonic specification of HSCs will be the key to generate HSCs from human iPSCs. In this review, recent advances in the efforts to generate HSCs from human iPSCs in vitro and in vivo or from somatic cells are discussed.

2 Hematopoietic Development in the Embryo

Our current understanding of embryonic hematopoietic development has been instructive for the directed differentiation of hematopoietic cells from iPSCs. In mammals, embryonic blood development occurs in several waves at independent sites. Primitive hematopoiesis is the earliest wave occurring in the extraembryonic yolk sac by mouse embryonic day 7 (E7) and generates mostly macrophages, megakaryocytes, and nucleated erythrocytes to serve the immediate needs of the growing embryo [5]. Recent studies in mice revealed that embryonic hematopoietic development is more complex than previously thought—there are independent programs for the generation of other types of hematopoietic cells, such as erythro-myeloid progenitors, that form in the yolk sac as early as E8.5 and migrate and colonize the fetal liver to establish early myelopoiesis before the arrival of definitive HSCs [6, 7]. In addition, immune restricted, lymphoid-primed progenitors develop in the yolk sac as early as E9.5 prior to emergence of definitive HSCs [8]. These progenitors cannot be categorized as either primitive or definitive hematopoiesis and are considered to be the second wave of hematopoiesis.

Primitive hematopoiesis is transient and replaced by adult-type definitive hematopoiesis initiated in the embryo proper. HSCs are generated in the definitive hematopoiesis program and sustain hematopoiesis throughout life. HSCs are first detected on the ventral wall of the dorsal aorta in the aorta-gonad-mesonephros (AGM) region of the embryo proper at E10.5 and form intra-aortic hematopoietic clusters that co-express endothelial and hematopoietic markers through endothelial-to-hematopoietic transition (EHT) of hemogenic endothelium [9–11]. In addition to the AGM region, HSCs are also found in the umbilical and vitelline arteries, yolk sac, placenta, and head [11–14]. HSCs subsequently migrate to the fetal liver where HSCs are expanded until birth and, finally, migrate to the bone marrow where HSCs maintain a mostly quiescent state in this niche for the rest of life.

3 In Vitro Differentiation of Human iPSCs into Hematopoietic Cells

Several in vitro directed hematopoietic differentiation strategies from pluripotent stem cells (PSCs), which include iPSCs and embryonic stem cells (ESCs), have been developed. Kaufman et al. first reported human ESC-derived myeloid colony-forming hematopoietic progenitors by co-culturing with the murine bone marrow cell line S17 or the yolk sac endothelial cell line C166 [15]. Co-culture of mouse ESCs with OP9 murine stromal cell line derived from newborn calvaria of the macrophage colony-stimulating factor-deficient mice efficiently generated multilineage hematopoietic cells [16]. Vodyanik et al. demonstrated that OP9 co-culture system also efficiently differentiated human ESCs into myeloid colony-forming hematopoietic progenitors as well as B cells and natural killer cells in vitro [17].

In vitro aggregation of human ESCs into three-dimensional clusters of cells known as embryoid bodies (EBs) allows for the spontaneous differentiation and cell specification along the three germ layers [18]. Using the EB formation method, Chadwick et al. reported that bone morphogenetic protein-4 (BMP4), a cytokine belonging to the transforming growth factor beta superfamily and functioning throughout development in mesoderm induction and bone formation, strongly promoted differentiation of myeloid colony-forming hematopoietic progenitors [19]. Cells from dissociated EBs co-express CD34 and kinase insert domain receptor (KDR also known as VEGFR2, FLK1, and CD309), and treatment of human EBs with vascular endothelial growth factor A (VEGFA), a ligand for KDR, increased the frequency of CD34^{+}KDR^{+} hematopoietic progenitors and erythroid colony formation [20]. The combination of BMP4 and VEGF with KIT ligand (KITLG also known as SCF) and basic fibroblast growth factor (FGF2) robustly increased the number of myeloid colony-forming hematopoietic progenitors [21].

Since myeloid cells are found in both primitive and definitive hematopoiesis, it is difficult to separate these two hematopoietic programs and identify HSC-driven definitive hematopoiesis from PSC-derived myeloid cells. T cells are only found in definitive hematopoiesis; therefore T cell differentiation potential is a useful indicator for definitive hematopoiesis derived from PSCs. Galic et al. demonstrated that co-culturing human ESCs with OP9 stromal cells followed by implantation into human thymic tissues in immunodeficient mice enabled induction of the T cell program in vivo [22]. Timmermans et al. reported that co-culturing human ESCs with OP9 stromal cells followed by transfer onto OP9 cells expressing mouse delta-like canonical Notch ligand 1 (Dll1), called OP9-DL1 cells [23], in the presence of fms-related tyrosine kinase 3 ligand (FLT3LG), interleukin-7 (IL7), and KITLG generated T cells in vitro [24].

Although co-culturing PSCs with stromal cells in serum-containing medium is a robust method to generate hematopoietic cells, factors from stromal cells and serum are poorly defined and thus it is difficult to fine-tune the culture conditions. Direct hematopoietic differentiation from PSCs using defined media and factors that recapitulate embryonic hematopoietic development is a more controllable method and can be used to purify GMP-grade cells for clinical applications. Ramos-Mejia et al. reported that upregulation of Nodal/Activin signaling serves as an indicator of human PSC lines prone to differentiate toward hematopoietic lineages [25]. However, Kennedy et al. reported that development of the definitive hematopoietic progenitor population is not dependent on Nodal/Activin pathway. They demonstrated that inhibition of Nodal/Activin pathway by treating human EBs in a defined medium containing the small molecule SB-431542, a Nodal/Activin pathway inhibitor [26], blocked the primitive hematopoiesis program and induced the development of definitive hematopoietic progenitors ($CD34^{+}CD43^{-}$ population at day 9 of differentiation) with the potential to generate T cells. In contrast, treating human EBs with Activin A between days 2 and 4 of differentiation induces development of $CD34^{+}CD43^{+}$ primitive hematopoietic progenitors at day 9 of differentiation [27]. The canonical Wnt pathway is also known to promote hematopoietic differentiation from human ESCs [28, 29]. In their subsequent study, Sturgeon et al. demonstrated that treating EBs with GSK-3 inhibitor CHIR99021, which activates canonical Wnt pathway, inhibited primitive hematopoiesis and increased the number of definitive hematopoietic progenitors [30]. Furthermore, treating EBs with the Wnt-antagonist IWP2 increased the number of primitive hematopoietic progenitors. These studies showed that the bifurcation of human primitive and definitive hematopoiesis is temporally regulated by Nodal/Activin and canonical Wnt pathways.

During embryonic development, HSCs emerge from the hemogenic endothelium and can be distinguished from vascular endothelium. Choi et al. reported that hemogenic endothelium progenitors can be identified in the VE-cadherin$^+$CD73$^-$CD235a$^-$CD43$^-$ population at day 5 of co-culture of human ESCs with OP9 stromal cells [31]. This surface marker phenotype discriminated hemogenic endothelium progenitors from VE-cadherin$^+$CD73$^+$ non-hemogenic endothelium progenitors that have features of endothelial cells and VE-cadherin$^+$CD235a$^+$CD41a$^-$ angiogenic hematopoietic progenitors that possess primary hematopoietic characteristics but are also capable of generating endothelial cells. Ditadi et al. demonstrated that human PSC-derived hemogenic endothelium and vascular endothelium represent separate lineages [32]. They showed that the hemogenic endothelium undergoes EHT to generate both endothelial and definitive hematopoietic progenitors from the CD34$^+$CD73$^-$CD184$^-$ population of EBs at day 8 of differentiation. In contrast, progenitors for arterial vascular endothelium and venous vascular endothelium are segregated to the CD34$^+$CD73^{med}CD184$^+$ and CD34$^+$CD73^{hi}CD184$^-$ populations, respectively.

4 Engraftment of Human PSC-Derived Hematopoietic Cells into Immunocompromised Mice

Although established culture conditions generate a broad spectrum of hematopoietic progenitors from human PSCs [33], it is not yet possible to generate bona fide HSCs from human PSCs in culture. HSCs possess the abilities of self-renewal and multilineage differentiation. The gold standard to functionally test for HSCs is to transplant PSC-derived cells into immunocompromised recipient mice and assess long-term and multilineage reconstitution of the hematopoietic system in the recipient mice.

Tian et al. reported that hematopoietic cells derived from the co-culture of human ESCs with S17 stromal cells were able to engraft into nonobese diabetic/severe combined immunodeficient (NOD-*scid*) recipient mice, and these cells were capable of secondary hematopoietic engraftment at very low levels [34]. Ledran et al. co-cultured human ESCs with mouse primary stromal cells derived from the AGM and fetal liver [35]. They used as recipients highly immunodeficient NOD.Cg-*Prkdc*^{scid} *Il2rg*^{tm1Wjl} (NSG) mice, NOD-*scid* mice that have a complete null allele for the *Il2rg* gene and have defective mature T, B, and natural killer (NK) cells as well as innate immune system [36], and showed the human

ESC-derived hematopoietic cells were capable of primary and secondary engraftment. Instead of co-culture with stromal cells, Lu et al. pretreated human ESCs with a medium conditioned by the HepG2 human hepatocarcinoma cell line before starting the EB differentiation [37]. This pretreatment induced a mesodermal program in EBs, and cells from day 14 EBs transplanted into NOD. Cg-*Prkdc*^{scid} *Il2rg*^{tm1Sug} (NOG) mice, NOD-*scid* mice that lack the intracytoplasmic domain of *Il2rg* gene [38], engrafted in both primary and secondary recipient mice. Although these approaches that utilized stromal cell-derived factors enabled the recovery of serially transplantable cells, the chimerism of human cells in recipient mice was very low.

In another approach, overexpression of transcription factors that regulate HSC development or expansion can promote hematopoietic differentiation from human PSCs. Antennapedia-class homeobox (*HOX*) genes consist of 39 genes in four clusters that encode transcription factors. A broad range of *HOX* genes, mainly the A and the B clusters, are expressed by human hematopoietic cells [39]. Ectopic expression of mouse hematopoietic cells with *Hoxb4* induces rapid expansion of HSCs [40]. Transient ectopic expression of *Hoxb4* in mouse ESC-derived EBs co-cultured with OP9 stromal cells endowed transduced cells with the ability for robust, multilineage hematopoiesis in primary and secondary recipients [41]. However, in contrast with mouse ESCs, Wang et al. reported that *HOXB4*-transduced human ESC-derived hematopoietic cells only had limited repopulation capacity upon transplantation [42]. Transcriptome analysis identified *HOXA9* as the most upregulated gene in cord blood-derived CD34^{+}CD38^{-}CD45^{+} cells as compared with human ESC-derived CD34^{+}CD38^{-}CD45^{+} cells. Ramos-Mejia et al. reported that overexpression of *HOXA9* enhanced hematopoietic differentiation from human ESCs, but was not sufficient to confer in vivo engraftment potential [43]. Dou et al. compared transcriptome profiles of human ESC-derived hematopoietic stem and progenitors (HSPCs) and human fetal liver-derived HSPCs using the same CD34^{+}CD38$^{-/lo}$CD90^{+}CD43^{+}CD45$^{+/-}$ immunophenotype and found that medial *HOXA* genes remained suppressed in human ESC-derived HSPCs [44]. Interestingly, *Hox* genes are developmentally regulated by retinoic acid signaling in patterning vertebrate embryogenesis [45], and retinoic acid receptor-α agonist AM580 promoted HSC potential in AGM-derived hemogenic endothelium ex vivo [46]. AM580 treatment during co-culture of human EBs with OP9 stromal cells induced expression of *HOXA* genes and prolonged maintenance of hematopoietic progenitors in culture [44]. However, overexpression of multiple medial *HOXA* genes (*HOXA5*, *HOXA7*, and/or *HOXA9*) was insufficient to confer in vivo engraftment potential on human ESC-derived hematopoietic cells. Thus, additional factors are needed to generate HSCs from human PSCs.

Doulatov et al. reported a strategy to re-specify myeloid lineage-restricted progenitors derived from human PSCs into multilineage progenitors that can engraft in vivo [47]. They ectopically expressed *HOXA9*, *ERG*, *RORA*, *SOX4*, and *MYB* in $CD34^{+}CD45^{+}$ progenitors from dissociated day 14 human EBs and observed short-term engraftment with myeloid and erythroid lineages at 4 weeks after transplantation into NSG mice, but long-term multilineage hematopoiesis was not achieved. Sugimura et al. recently demonstrated a successful generation of HSCs from human PSCs [48]. They first generated day 8 EBs by a published method using SB-431542 and CHIR99021 [27, 32] and then performed lentiviral transduction of dissociated hemogenic endothelial cells to overexpress seven transcription factors (*ERG*, *HOXA5*, *HOXA9*, *HOXA10*, *LCOR*, *RUNX1*, and *SPI1*) using the tetracycline-inducible system. Transduced hemogenic endothelial cells were subsequently transplanted into NSG mice by intrafemoral injection with doxycycline treatment for 2 weeks. Human $CD34^{+}$ cells generated from transduced hemogenic endothelial cells in the bone marrow at 2 months after the primary transplantation had multilineage and long-term reconstituting potential upon transplantation into secondary recipient mice. Although it is not yet possible to generate human PSC-derived HSCs in vitro with robust repopulation capacity, this study opened up the possibility of generating HSCs from hemogenic endothelial cells in an in vivo environment.

Another approach recently reported by Tan et al. used single transcription factor, *MLL-AF4*, to generate highly engraftable HSPCs from human iPSCs [49]. Given that *MLL-AF4* is the product of the t(4;11) chromosomal translocation found in infant B cell acute lymphoid leukemia with a poor prognosis [50], they used a nonintegrating tetracycline-inducible system to transiently express *MLL-AF4* in hematopoietic cells derived from iPSCs at 12 days of differentiation and administered doxycycline in vitro for 2–3 days followed by transplantation into NSG mice with another 2 weeks of doxycycline administration in vivo. They observed more than 20% of overall human iPSC-derived cells in the bone marrow of recipient mice at 8 weeks after transplantation, which is higher than previous reports [47, 48], and these cells underwent multilineage differentiation. However, at 16 weeks after primary transplantation and 8 weeks after secondary transplantation, they observed a B cell leukemia-like phenotype and a significant decrease of cells in the myeloid lineage, even though they transiently expressed *MLL-AF4*. Nevertheless, it will be instructive to decipher the downstream pathways regulated by MLL-AF4 during the initial event of generating highly engraftable HSPCs.

5 In Vivo Generation of HSCs from Human iPSC-Derived Teratoma

PSCs possess pluripotency, the potential to differentiate into tissues from any of three germ layers, and form a tumor made up of tissues from three germ layers in vivo, called a teratoma. A teratoma formation assay is the traditional gold standard to assess the pluripotency of PSCs. Interestingly, teratocarcinoma cells possess ability to differentiate into hematopoietic cells that can give rise to cells in the multiple hematopoietic lineages, including erythrocytes, macrophages, and megakaryocytes, in culture [51–53]. More recently, two groups demonstrated successful generation of multilineage and long-term engraftable HSCs from human iPSC-derived teratoma. Amabile et al. observed the development of $CD45^{+}CD34^{+}$ HSPCs in bone marrow-like structures within teratoma, and the number of HSPCs was increased by co-transplantation of human iPSCs with OP9 stromal cells that express either DLL1 or Wnt3a [54]. These teratoma-derived HSPCs possessed multilineage repopulating ability in primary recipient mice and were engraftable in a secondary recipient, albeit at very limited levels, without a sign of leukemic transformation. Suzuki et al. also reported teratoma-derived, engraftable HSCs [55]. They co-transplanted human iPSCs with OP9 stromal cells along with implantation of an osmotic pump that releases KITLG and thrombopoietin (THPO) for 2 weeks. Interestingly, teratoma-derived human hematopoietic cells were found in the host mouse bone marrow and peripheral blood at 12 weeks of teratoma formation, suggesting migration of teratoma-derived cells into host hematopoietic tissues. Teratoma-derived human $CD45^{+}CD34^{+}$ HSPCs obtained from the host mouse bone marrow had multilineage repopulating ability upon transplantation into NOD-*scid* $JAK3^{null}$ mice. The same group subsequently demonstrated that teratoma-derived mouse HSC formation can be further improved by the forced expression of three transcription factors, *Gfi1b*, *Fos*, and *Gata2*, in mouse iPSCs [56]. The combination of these transcription factors had been reported to reprogram mouse embryonic fibroblasts into hematopoietic progenitors in vitro [57]. They also demonstrated that depopulating host HSCs using recipient mice with a hematopoietic-specific *Kit* deficiency further increased migration of teratoma-derived HSCs into the host bone marrow, and these migrated HSCs were serially transplantable. Human iPSC-derived teratomas provide an intriguing, replenishable system to study human HSC development. Future detailed molecular analyses addressing how human HSCs develop in teratomas may lead to a better understanding of human HSC development and to improving the method for HSC generation from human iPSCs.

6 Reprogramming of Somatic Cells into Hematopoietic Lineages

Direct conversion of human non-hematopoietic somatic cells into HSCs is an alternative approach for in vitro generation of HSCs. As mentioned above, forced expression of *Gfi1b*, *Fos*, and *Gata2* is sufficient to convert mouse embryonic fibroblasts into hematopoietic progenitors via an intermediate endothelial-like precursor stage [57]. Batta et al. also demonstrated direct conversion of mouse embryonic and adult fibroblasts into multilineage hematopoietic progenitors by the ectopic expression of *Erg*, *Gata2*, *Lmo2*, *Runx1c*, and *Tal1/Scl* and that loss of p53 function facilitated this conversion [58]. These converted hematopoietic progenitors possess short-term engraftment ability at 2 weeks after transplantation into NSG mice, but engraftment was not detected at 8 weeks after transplantation. More recently, Cheng et al. converted mouse embryonic fibroblasts into HSPCs with long-term engraftment ability by expressing *Tal1/Scl*, *Lmo2*, *Runx1*, and *Bmi1* [59].

Since the hemogenic endothelium give rise to both definitive HSCs and vascular endothelial cells, these two cell types are closely related and thus endothelial cells are an attractive source for reprogramming into HSCs. Sandler et al. transduced human umbilical vein endothelial cells or human adult dermal microvascular endothelial cells with *FOSB*, *GFI1*, *RUNX1*, and *SPI1* transcription factors and demonstrated that these endothelial cells were reprogrammed into HSPCs that have ability to engraft into NSG mice after primary and secondary transplantation [60]. However, reprogrammed HSPCs had limited T cell differentiation capacity both in vitro and in vivo. In their subsequent study, Lis et al. used similar approach to generate fully functional HSCs from adult mouse endothelial cells [61]. These reprogrammed HSCs were serially transplantable and had multilineage differentiation capacity including T cell differentiation capacity. In addition to the iPSC-derived HSC approach, the direct conversion approach also holds promise for generating de novo human HSCs.

7 Conclusions

Generation of fully functional HSCs from human iPSCs or somatic cells has been a long-sought goal in the field. Forced expression of a set of transcription factors in human iPSCs or somatic cells has been leading to remarkable progress to achieve this goal, but successful derivation of bona fide HSCs in vitro remains a significant challenge. Successful generation of de novo HSCs from human iPSCs can be a virtually infinite source of HSCs for transplantation in clinics and also can generate immune-humanized mice in research laboratories. Immune-humanized mice will provide in vivo models of human hematopoietic and immune systems, which can provide a genetic platform to study functions of specific genes in normal and malignant hematopoiesis by generating the hematopoietic system from genome-edited human iPSCs. Generation of patient-derived xenograft mouse models of cancer with an autologous immune system generated from human iPSCs derived from the same patient will provide a valuable tool to study cancer immunology. Several immunodeficient recipient mouse models have been recently reported to improve engraftment of HSPCs and differentiation of human hematopoietic and immune systems to efficiently generate immune-humanized mice, such as immunodeficient mice that contain *Kit* mutations to compromise endogenous HSPCs [62–64], transgenic expression of three human cytokine genes [65] or knocking in four cytokine genes [66] in immunodeficient mice to efficiently support human myeloid cells, and transplantation of human fetal liver and thymus [67, 68] or human PSC-derived thymic epithelial progenitors [69, 70] under the kidney capsule of immunodeficient mice to enhance the development of T cells.

Understanding the molecular events that regulate the development of definitive hematopoiesis will improve production of antigen-specific T cells for immunotherapy and red blood cells and platelets for blood transfusion. To generate de novo HSCs from human iPSCs or non-hematopoietic somatic cells in vitro, it will be instructive to elucidate gene regulatory networks that regulate HSC development in the embryo. It will be also important to understand the molecular signals that allow HSCs to expand in the fetal liver and neonatal bone marrow for the ex vivo expansion of HSCs derived from human iPSC or somatic cells. To generate clinically usable HSCs, however, will require the development of transgene-free methods for the derivation of de novo HSCs.

Acknowledgments

This work was supported by The Jackson Laboratory. The author thanks William Skarnes for thoughtful reading and editing of the manuscript and apologizes to authors whose papers could not be cited in this review owing to space limitations.

References

1. Orkin SH, Zon LI (2008) Hematopoiesis: an evolving paradigm for stem cell biology. Cell 132(4):631–644
2. Niederwieser D, Baldomero H, Szer J, Gratwohl M, Aljurf M, Atsuta Y et al (2016) Hematopoietic stem cell transplantation activity worldwide in 2012 and a SWOT analysis of the Worldwide Network for Blood and Marrow Transplantation Group including the global survey. Bone Marrow Transplant 51 (6):778–785
3. Giralt S, Costa L, Schriber J, Dipersio J, Maziarz R, McCarty J et al (2014) Optimizing autologous stem cell mobilization strategies to improve patient outcomes: consensus

guidelines and recommendations. Biol Blood Marrow Transplant 20(3):295–308

4. Kumar S, Geiger H (2017) HSC niche biology and hsc expansion ex vivo. Trends Mol Med 23 (9):799–819
5. Palis J, Robertson S, Kennedy M, Wall C, Keller G (1999) Development of erythroid and myeloid progenitors in the yolk sac and embryo proper of the mouse. Development 126 (22):5073–5084
6. Gomez Perdiguero E, Klapproth K, Schulz C, Busch K, Azzoni E, Crozet L et al (2015) Tissue-resident macrophages originate from yolk-sac-derived erythro-myeloid progenitors. Nature 518(7540):547–551
7. McGrath KE, Frame JM, Fegan KH, Bowen JR, Conway SJ, Catherman SC et al (2015) Distinct sources of hematopoietic progenitors emerge before HSCs and provide functional blood cells in the mammalian embryo. Cell Rep 11(12):1892–1904
8. Boiers C, Carrelha J, Lutteropp M, Luc S, Green JC, Azzoni E et al (2013) Lymphomyeloid contribution of an immune-restricted progenitor emerging prior to definitive hematopoietic stem cells. Cell Stem Cell 13 (5):535–548
9. Medvinsky A, Dzierzak E (1996) Definitive hematopoiesis is autonomously initiated by the AGM region. Cell 86(6):897–906
10. de Bruijn MF, Ma X, Robin C, Ottersbach K, Sanchez MJ, Dzierzak E (2002) Hematopoietic stem cells localize to the endothelial cell layer in the midgestation mouse aorta. Immunity 16(5):673–683
11. de Bruijn MF, Speck NA, Peeters MC, Dzierzak E (2000) Definitive hematopoietic stem cells first develop within the major arterial regions of the mouse embryo. EMBO J 19 (11):2465–2474
12. Ottersbach K, Dzierzak E (2005) The murine placenta contains hematopoietic stem cells within the vascular labyrinth region. Dev Cell 8(3):377–387
13. Li Z, Lan Y, He W, Chen D, Wang J, Zhou F et al (2012) Mouse embryonic head as a site for hematopoietic stem cell development. Cell Stem Cell 11(5):663–675
14. Gekas C, Dieterlen-Lievre F, Orkin SH, Mikkola HK (2005) The placenta is a niche for hematopoietic stem cells. Dev Cell 8 (3):365–375
15. Kaufman DS, Hanson ET, Lewis RL, Auerbach R, Thomson JA (2001) Hematopoietic colony-forming cells derived from human embryonic stem cells. Proc Natl Acad Sci U S A 98(19):10716–10721
16. Nakano T, Kodama H, Honjo T (1994) Generation of lymphohematopoietic cells from embryonic stem cells in culture. Science 265 (5175):1098–1101
17. Vodyanik MA, Bork JA, Thomson JA, Slukvin II (2005) Human embryonic stem cell-derived CD34+ cells: efficient production in the coculture with OP9 stromal cells and analysis of lymphohematopoietic potential. Blood 105 (2):617–626
18. Itskovitz-Eldor J, Schuldiner M, Karsenti D, Eden A, Yanuka O, Amit M et al (2000) Differentiation of human embryonic stem cells into embryoid bodies compromising the three embryonic germ layers. Mol Med 6(2):88–95
19. Chadwick K, Wang L, Li L, Menendez P, Murdoch B, Rouleau A et al (2003) Cytokines and BMP-4 promote hematopoietic differentiation of human embryonic stem cells. Blood 102(3):906–915
20. Cerdan C, Rouleau A, Bhatia M (2004) VEGF-A165 augments erythropoietic development from human embryonic stem cells. Blood 103 (7):2504–2512
21. Pick M, Azzola L, Mossman A, Stanley EG, Elefanty AG (2007) Differentiation of human embryonic stem cells in serum-free medium reveals distinct roles for bone morphogenetic protein 4, vascular endothelial growth factor, stem cell factor, and fibroblast growth factor 2 in hematopoiesis. Stem Cells 25 (9):2206–2214
22. Galic Z, Kitchen SG, Kacena A, Subramanian A, Burke B, Cortado R et al (2006) T lineage differentiation from human embryonic stem cells. Proc Natl Acad Sci U S A 103(31):11742–11747
23. Schmitt TM, Zuniga-Pflucker JC (2002) Induction of T cell development from hematopoietic progenitor cells by delta-like-1 in vitro. Immunity 17(6):749–756
24. Timmermans F, Velghe I, Vanwalleghem L, De Smedt M, Van Coppernolle S, Taghon T et al (2009) Generation of T cells from human embryonic stem cell-derived hematopoietic zones. J Immunol 182(11):6879–6888
25. Ramos-Mejia V, Melen GJ, Sanchez L, Gutierrez-Aranda I, Ligero G, Cortes JL et al (2010) Nodal/Activin signaling predicts human pluripotent stem cell lines prone to differentiate toward the hematopoietic lineage. Mol Ther 18(12):2173–2181
26. Inman GJ, Nicolas FJ, Callahan JF, Harling JD, Gaster LM, Reith AD et al (2002) SB-431542 is a potent and specific inhibitor of transforming growth factor-beta superfamily type I activin receptor-like kinase (ALK) receptors ALK4,

ALK5, and ALK7. Mol Pharmacol 62 (1):65–74

27. Kennedy M, Awong G, Sturgeon CM, Ditadi A, LaMotte-Mohs R, Zuniga-Pflucker JC et al (2012) T lymphocyte potential marks the emergence of definitive hematopoietic progenitors in human pluripotent stem cell differentiation cultures. Cell Rep 2(6):1722–1735
28. Woll PS, Morris JK, Painschab MS, Marcus RK, Kohn AD, Biechele TL et al (2008) Wnt signaling promotes hematoendothelial cell development from human embryonic stem cells. Blood 111(1):122–131
29. Wang Y, Nakayama N (2009) WNT and BMP signaling are both required for hematopoietic cell development from human ES cells. Stem Cell Res 3(2-3):113–125
30. Sturgeon CM, Ditadi A, Awong G, Kennedy M, Keller G (2014) Wnt signaling controls the specification of definitive and primitive hematopoiesis from human pluripotent stem cells. Nat Biotechnol 32(6):554–561
31. Choi KD, Vodyanik MA, Togarrati PP, Suknuntha K, Kumar A, Samarjeet F et al (2012) Identification of the hemogenic endothelial progenitor and its direct precursor in human pluripotent stem cell differentiation cultures. Cell Rep 2(3):553–567
32. Ditadi A, Sturgeon CM, Tober J, Awong G, Kennedy M, Yzaguirre AD et al (2015) Human definitive haemogenic endothelium and arterial vascular endothelium represent distinct lineages. Nat Cell Biol 17(5):580–591
33. Ackermann M, Liebhaber S, Klusmann JH, Lachmann N (2015) Lost in translation: pluripotent stem cell-derived hematopoiesis. EMBO Mol Med 7(11):1388–1402
34. Tian X, Woll PS, Morris JK, Linehan JL, Kaufman DS (2006) Hematopoietic engraftment of human embryonic stem cell-derived cells is regulated by recipient innate immunity. Stem Cells 24(5):1370–1380
35. Ledran MH, Krassowska A, Armstrong L, Dimmick I, Renstrom J, Lang R et al (2008) Efficient hematopoietic differentiation of human embryonic stem cells on stromal cells derived from hematopoietic niches. Cell Stem Cell 3(1):85–98
36. Shultz LD, Lyons BL, Burzenski LM, Gott B, Chen X, Chaleff S et al (2005) Human lymphoid and myeloid cell development in NOD/LtSz-scid IL2R gamma null mice engrafted with mobilized human hemopoietic stem cells. J Immunol 174(10):6477–6489
37. Lu M, Kardel MD, O'Connor MD, Eaves CJ (2009) Enhanced generation of hematopoietic cells from human hepatocarcinoma cell-stimulated human embryonic and induced pluripotent stem cells. Exp Hematol 37 (8):924–936
38. Ito M, Hiramatsu H, Kobayashi K, Suzue K, Kawahata M, Hioki K et al (2002) NOD/SCID/gamma(c)(null) mouse: an excellent recipient mouse model for engraftment of human cells. Blood 100(9):3175–3182
39. Sauvageau G, Lansdorp PM, Eaves CJ, Hogge DE, Dragowska WH, Reid DS et al (1994) Differential expression of homeobox genes in functionally distinct CD34+ subpopulations of human bone marrow cells. Proc Natl Acad Sci U S A 91(25):12223–12227
40. Antonchuk J, Sauvageau G, Humphries RK (2002) HOXB4-induced expansion of adult hematopoietic stem cells ex vivo. Cell 109 (1):39–45
41. Kyba M, Perlingeiro RC, Daley GQ (2002) HoxB4 confers definitive lymphoid-myeloid engraftment potential on embryonic stem cell and yolk sac hematopoietic progenitors. Cell 109(1):29–37
42. Wang L, Menendez P, Shojaei F, Li L, Mazurier F, Dick JE et al (2005) Generation of hematopoietic repopulating cells from human embryonic stem cells independent of ectopic HOXB4 expression. J Exp Med 201 (10):1603–1614
43. Ramos-Mejia V, Navarro-Montero O, Ayllon V, Bueno C, Romero T, Real PJ et al (2014) HOXA9 promotes hematopoietic commitment of human embryonic stem cells. Blood 124(20):3065–3075
44. Dou DR, Calvanese V, Sierra MI, Nguyen AT, Minasian A, Saarikoski P et al (2016) Medial HOXA genes demarcate haematopoietic stem cell fate during human development. Nat Cell Biol 18(6):595–606
45. Marshall H, Morrison A, Studer M, Popperl H, Krumlauf R (1996) Retinoids and hox genes. FASEB J 10(9):969–978
46. Chanda B, Ditadi A, Iscove NN, Keller G (2013) Retinoic acid signaling is essential for embryonic hematopoietic stem cell development. Cell 155(1):215–227
47. Doulatov S, Vo LT, Chou SS, Kim PG, Arora N, Li H et al (2013) Induction of multipotential hematopoietic progenitors from human pluripotent stem cells via respecification of lineage-restricted precursors. Cell Stem Cell 13(4):459–470
48. Sugimura R, Jha DK, Han A, Soria-Valles C, da Rocha EL, Lu YF et al (2017) Haematopoietic stem and progenitor cells from human pluripotent stem cells. Nature 545(7655):432–438

49. Tan YT, Ye L, Xie F, Beyer AI, Muench MO, Wang J et al (2018) Respecifying human iPSC-derived blood cells into highly engraftable hematopoietic stem and progenitor cells with a single factor. Proc Natl Acad Sci U S A 115 (9):2180–2185
50. Sanjuan-Pla A, Bueno C, Prieto C, Acha P, Stam RW, Marschalek R et al (2015) Revisiting the biology of infant t(4;11)/MLL-AF4+ B-cell acute lymphoblastic leukemia. Blood 126(25):2676–2685
51. Cudennec C, Nicolas JF (1977) Blood formation in a clonal cell line of mouse teratocarcinoma. J Embryol Exp Morphol 38:203–210
52. Cudennec CA, Johnson GR (1981) Presence of multipotential hemopoietic cells in teratocarcinoma cultures. J Embryol Exp Morphol 61:51–59
53. Cudennec CA, Salaun J (1979) Definitive red blood cell differentiation in a clonal line of mouse teratocarcinoma cultured in vivo in the chick embryo. Cell Differ 8(2):75–82
54. Amabile G, Welner RS, Nombela-Arrieta C, D'Alise AM, Di Ruscio A, Ebralidze AK et al (2013) In vivo generation of transplantable human hematopoietic cells from induced pluripotent stem cells. Blood 121(8):1255–1264
55. Suzuki N, Yamazaki S, Yamaguchi T, Okabe M, Masaki H, Takaki S et al (2013) Generation of engraftable hematopoietic stem cells from induced pluripotent stem cells by way of teratoma formation. Mol Ther 21(7):1424–1431
56. Tsukada M, Ota Y, Wilkinson AC, Becker HJ, Osato M, Nakauchi H et al (2017) In vivo generation of engraftable murine hematopoietic stem cells by Gfi1b, c-Fos, and Gata2 overexpression within teratoma. Stem Cell Reports 9(4):1024–1033
57. Pereira CF, Chang B, Qiu J, Niu X, Papatsenko D, Hendry CE et al (2013) Induction of a hemogenic program in mouse fibroblasts. Cell Stem Cell 13(2):205–218
58. Batta K, Florkowska M, Kouskoff V, Lacaud G (2014) Direct reprogramming of murine fibroblasts to hematopoietic progenitor cells. Cell Rep 9(5):1871–1884
59. Cheng H, Ang HY, AEF C, Li P, Fang HT, Liu TM et al (2016) Reprogramming mouse fibroblasts into engraftable myeloerythroid and lymphoid progenitors. Nat Commun 7:13396
60. Sandler VM, Lis R, Liu Y, Kedem A, James D, Elemento O et al (2014) Reprogramming human endothelial cells to haematopoietic cells requires vascular induction. Nature 511 (7509):312–318
61. Lis R, Karrasch CC, Poulos MG, Kunar B, Redmond D, Duran JGB et al (2017) Conversion of adult endothelium to immunocompetent haematopoietic stem cells. Nature 545 (7655):439–445
62. Cosgun KN, Rahmig S, Mende N, Reinke S, Hauber I, Schafer C et al (2014) Kit regulates HSC engraftment across the human-mouse species barrier. Cell Stem Cell 15(2):227–238
63. McIntosh BE, Brown ME, Duffin BM, Maufort JP, Vereide DT, Slukvin II et al (2015) Nonirradiated NOD,B6.SCID Il2rgamma−/− Kit(W41/W41) (NBSGW) mice support multilineage engraftment of human hematopoietic cells. Stem Cell Reports. 4(2):171–180
64. Yurino A, Takenaka K, Yamauchi T, Nunomura T, Uehara Y, Jinnouchi F et al (2016) Enhanced reconstitution of human erythropoiesis and thrombopoiesis in an immunodeficient mouse model with kit (Wv) mutations. Stem Cell Reports. 7 (3):425–438
65. Wunderlich M, Chou FS, Link KA, Mizukawa B, Perry RL, Carroll M et al (2010) AML xenograft efficiency is significantly improved in NOD/SCID-IL2RG mice constitutively expressing human SCF, GM-CSF and IL-3. Leukemia 24(10):1785–1788
66. Rongvaux A, Willinger T, Martinek J, Strowig T, Gearty SV, Teichmann LL et al (2014) Development and function of human innate immune cells in a humanized mouse model. Nat Biotechnol 32(4):364–372
67. Lan P, Tonomura N, Shimizu A, Wang S, Yang YG (2006) Reconstitution of a functional human immune system in immunodeficient mice through combined human fetal thymus/ liver and CD34+ cell transplantation. Blood 108(2):487–492
68. Melkus MW, Estes JD, Padgett-Thomas A, Gatlin J, Denton PW, Othieno FA et al (2006) Humanized mice mount specific adaptive and innate immune responses to EBV and TSST-1. Nat Med 12(11):1316–1322
69. Sun X, Xu J, Lu H, Liu W, Miao Z, Sui X et al (2013) Directed differentiation of human embryonic stem cells into thymic epithelial progenitor-like cells reconstitutes the thymic microenvironment in vivo. Cell Stem Cell 13 (2):230–236
70. Parent AV, Russ HA, Khan IS, LaFlam TN, Metzger TC, Anderson MS et al (2013) Generation of functional thymic epithelium from human embryonic stem cells that supports host T cell development. Cell Stem Cell 13 (2):219–229

Chapter 20

Using the Inducible Caspase-9 Suicide-Safeguard System with iPSC and Bioluminescent Tracking

Jonathan Villanueva, Toshinobu Nishimura, and Hiromitsu Nakauchi

Abstract

For scientists working within the field of induced pluripotent stem cells (iPSCs), this protocol will provide a thorough walk-through on how to conduct in vitro and in vivo experiments that validate the function of a particular safeguard system technology. In short, we provide instructions on how to generate inducible Caspase-9 (iC9) safeguard system with human iPSCs that act as normal or abnormal models of the cells for therapeutics to be tried after differentiation. These iC9-iPSCs should be modified prior to beginning this protocol by constitutively expressing luciferase, an enzyme capable of generating bioluminescent signals through the oxidation of the substrate luciferin. Monitoring the bioluminescent signal over time provides the information on whether a safeguard system is working or not.

Key words Immunotherapy, Cancer, Bioluminescence, Cytotoxicity, T cell

1 Introduction

Cell-based immunotherapy is slowly becoming a mainstream option for the treatment of viral infections and cancer [1]. Within the last decade, there has been a major resurgence of interest in cell-based therapies like CAR T cell therapies, with some entering the clinical setting for the treatment of certain leukemias. With this new found interest, there is a large demand for protocols describing how to validate a therapy's safety and the ability to remove the cells in the event they become cancer stem cells. In this chapter, we will describe the use of the iC9 system to remove iPSC teratomas and luciferase to track the cells' clearance.

This protocol uses iC9-iPSCs to form tumor burdens. However, any cell line constitutively expressing the iC9 system, luciferase, and known to form tumor burdens can be used to test iC9 efficacy. To visualize the luciferase signal, it is crucial that there is an in vivo imaging system capable of receiving bioluminescent signal available for these experiments. Combined, these features describe the necessities to validate the safety of a cell therapy safeguard system.

Shin Kaneko (ed.), *In Vitro Differentiation of T-Cells: Methods and Protocols*, Methods in Molecular Biology, vol. 2048,
https://doi.org/10.1007/978-1-4939-9728-2_20,

2 Materials

2.1 Cells and Animals

1. iC9-iPSCs:
   (a) For example, iC9 was introduced by retrovirus (Addgene).
   (b) For example, Firefly luciferase was introduced by lentivirus (Addgene).
2. NOD-SCID mice, 8–12 weeks of age (The Jackson Laboratory).

2.2 Equipment

1. Biosafety cabinet.
2. Hemocytometer.
3. Tissue culture incubator.
4. Imaging system (i.e., Xenogen IVIS, IVIS Spectrum, SII Lago-X).

2.3 Plasticware and Other Disposables

1. 6-Well culture plate (Thermo Fisher Scientific).
2. 96-Well clear bottom black polystyrene microplate (Thermo Fisher Scientific).
3. Cryotube (Thermo Fisher Scientific).
4. BD PrecisionGlide 27 G × ½″ hypodermic needles (Becton Dickinson).

2.4 Reagents and Media

1. Phosphate-buffered saline (PBS) (Gibco): PBS used in the chapter is free from both Ca^{2+} and Mg^{2+}.
2. StemFit AK02N medium (Ajinomoto).
3. iMatrix-511 (recombinant laminin-511) (Stemgent).
4. TrypLE Select Enzyme (1×) (Gibco).
5. ROCK Inhibitor (Y-27632) (TOCRIS).
6. Recombinant human FGF-basic (bFGF, 154aa) (PeproTech).
7. Dimethyl sulfoxide (DMSO) (Sigma-Aldrich).
8. Matrigel Growth Factor Reduced Basement Membrane Matrix, LDEV-Free (Corning).
9. CID/AP20187 (B/B Homodimerizer) (Clontech).
10. Ethanol (molecular biology grade) (Sigma-Aldrich).
11. PEG 400 (Sigma-Aldrich).
12. Tween 80 (Sigma-Aldrich).
13. D-Luciferin (Sigma-Aldrich).

2.5 Formulations

1. iPSC medium: StemFit PSC medium supplemented with 20 ng/mL bFGF.

2. Freezing solution: 90% iPSC medium, 10% DMSO, 10 μM Y-27632.
3. CID solution (for in vitro assay): Lyophilized AP20187 is reconstituted in ethanol (500 μM, 10,000×).
4. CID solution (for in vivo assay): Lyophilized AP20187 is reconstituted in ethanol to produce a 5 mg/mL stock (store at −20 °C). AP20187 is diluted in sterile water containing 10% PEG 400 and 1.4% Tween 80 freshly on the day of injection [2].
5. D-Luciferin solution: Reconstituted in distilled water (30 mg/mL).

3 Methods

3.1 Routine Culture of iC9-iPSCs

3.1.1 Thawing iC9-iPSCs

1. Thaw iC9-iPSCs quickly in a 37 °C water bath.
2. Pipette the cell suspension slowly and gently from the cryovial tube into a 15-mL centrifuge tube.
3. Add 10 mL of iPSC medium dropwise to the cell suspension.
4. Centrifuge the cells at 300 × *g* at RT for 5 min.

   During centrifugation, add 1.5 mL of iPSC medium per well of a 6-well plate with 10 μM Y-27632/well and 5 μL of iMatrix-511/well.
5. After centrifugation, aspirate the supernatant, and resuspend the cells in 1 mL of iPSC medium, creating a single cell suspension.
6. Count cells and plate 2 × 10^4 cells per well in the prepared 6-well plate.
7. Incubate at 37 °C, 5% CO_2 (*see* **Note 1**).

3.1.2 Maintaining iC9-iPSCs

1. Replace culture medium every other day with 1.5 mL of iPSC medium (*see* **Note 2**).
2. When the iC9-iPSCs are 80–90% confluent, aspirate medium and wash well to be split with 2 mL of PBS.
3. Aspirate PBS and add 1 mL of dissociation buffer (TrypLE Select Enzyme (1×)). Incubate cells at 37 °C for 4 min.
4. Remove the dissociation buffer and create a single cell suspension.
5. Count cells and plate 2 × 10^4 cells/well in a 6-well plate as done in the previous section.

3.1.3 Freezing iC9-iPSCs

1. Prepare 5 mL freezing solution for iC9-iPSCs per well. Pre-cool the solution on ice.

2. Passage the cells as described in **steps 2–6** of the previous subsection, using 5 mL of freezing medium in place of iPSC medium.
3. Disaggregate the target cells into a single cell suspension.
4. Aliquot 1 mL cell suspension per cryovial.
5. Freeze the cells at −80 °C. Transfer them to liquid nitrogen for storage.

3.2 Monitoring and Clearance of Luciferase Expressing iC9-iPSCs In Vitro

1. Seed iC9-iPSCs on 96-well black microplate as following the steps in previous section (Subheading 3.1.2).
2. Incubate at 37 °C, 5% CO_2 for 3 days.
3. Add D-luciferin (30 mg/mL) at 1:100 dilution ratio.
4. Image plate with preferred imaging system after the incubation at 37 °C for 5 min (Fig. 1a).

3.3 Monitoring and Clearance of Luciferase Expressing iC9-iPSCs In Vivo

3.3.1 Inoculation of iC9-iPSCs into NOD-SCID Mice

1. Aspirate medium from iC9-iPSC wells and wash with 2 mL of PBS.
2. Remove PBS and treat cells with 1 mL dissociation buffer for 4 min.
3. Remove dissociation buffer, add 3 mL of iPSC medium, and create a single cell suspension.
4. Count cells.
5. Resuspend cells in 2× diluted Matrigel Matrix with Y-27632 (10 μM) at a concentration of 5×10^6 cells/mL.
6. Prepare enough 27 G × ½″ needles with 200 μL (1×10^6 cells) of cell suspension from previous step for every mouse in experiment. Be sure to keep cells as a homogeneous suspension as you make aliquots for injection.
7. Inject cells into the testicular capsule of NOD-SCID mice [3].

3.3.2 Luciferase Tracking of Teratoma Formation from iC9-iPSCs

1. Two weeks after injection of iC9-iPSCs, intraperitoneally inject mice with D-luciferin (150 mg/kg) and wait for 10 min. Take bioluminescent image by imaging system of choice (*see* **Note 3**).
2. Monitor for progressive increase in luciferase activity weekly for 3 weeks (*see* **Note 4**).

3.3.3 Injection of CID for Removal of iC9-iPSC-Derived Teratoma

1. On week 5, inject 50 μg/mouse of CID intraperitoneally into each mouse for 3 successive days.
2. On week 6, intraperitoneal inject mice with D-luciferase (150 mg/kg) and image mice after 10 min with preferred imaging system (Fig. 1b).

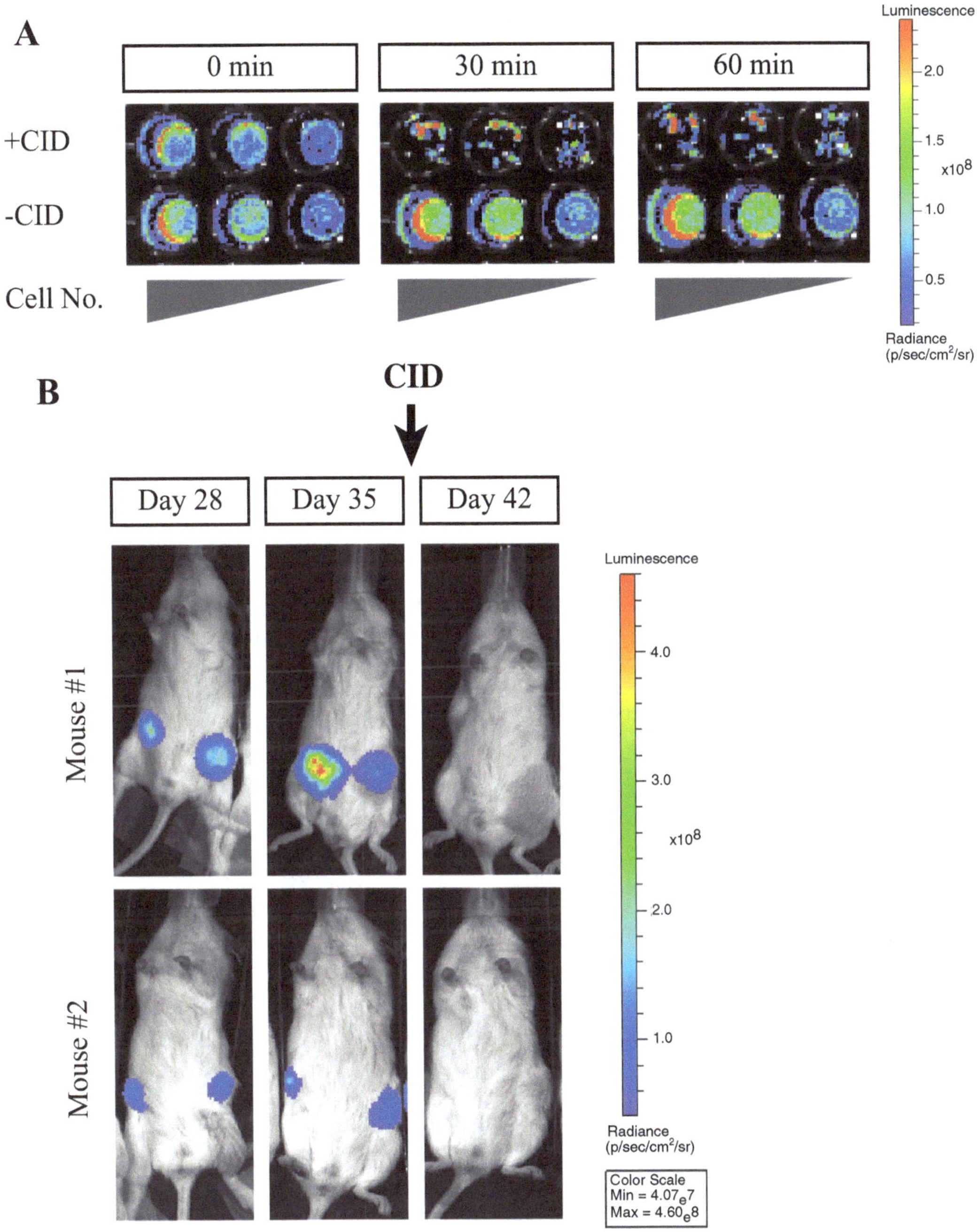

Fig. 1 iC9-iPSCs and their derivative tumors are potently killed after addition of CID in vitro and in vivo. (**a**) iC9-iPSCs are killed within 30 min of CID addition at multiple cells number dosages. (**b**) iC9-iPSC-derived teratomas drastically decrease in size after CID addition

3. To measure effectiveness of iC9 safeguard system, calculate the ratio of radiance (R) signal before and after administration of CID:

$$\%\text{Clearance} = \left[1 - \left(R^{\text{week 6}} - R^{\text{week 0}}\right)/\left(R^{\text{week 5}} - R^{\text{week 0}}\right)\right] \times 100$$

4 Notes

1. Y-27632 (10 μM) is necessary only after passaging. Unnecessary to add Y-27632 for routine feeding.
2. StemFit AK02N medium is designed to enable medium changes every other day. Other media such as mTeSR1 (STEMCELL Technologies) must be replaced daily. Be sure to check the appropriate feeding routine for your medium of choice.
3. In case of Living Image Software (Perkin Elmer), quantify bioluminescent signal using a region of interest (ROI) set to average radiance (signal measured as total photon/s/cm^2/steridian (p/s/cm^2/sr)). Sample height is normally 1.5 cm.
4. It is possible to monitor tumor growth by measuring testis size in three dimensions once a week [4].

References

1. Fesnak AD, June CH, Levine BL (2016) Engineered T cells: the promise and challenges of cancer immunotherapy. Nat Rev Cancer 16:566–581. https://doi.org/10.1038/nrc.2016.97
2. Zhao S, Weinreich MA, Ihara K, Richard RE, Blau CA (2004) In vivo selection of genetically modified erythroid cells using a jak2-based cell growth switch. Mol Ther 10:456–468. https://doi.org/10.1016/j.ymthe.2004.05.014
3. Peterson SE et al (2011) Teratoma generation in the testis capsule. J Vis Exp (57):e3177. https://doi.org/10.3791/3177
4. Ando M et al (2015) A safeguard system for induced pluripotent stem cell-derived rejuvenated T cell therapy. Stem Cell Reports 5:597–608. https://doi.org/10.1016/j.stemcr.2015.07.011

INDEX

Shin Kaneko (ed.), *In Vitro Differentiation of T-Cells: Methods and Protocols*, Methods in Molecular Biology, vol. 2048, https://doi.org/10.1007/978-1-4939-9728-2, © Springer Science+Business Media, LLC, part of Springer Nature 2019

E

F

G

H

I

L

M

N

O

FSC
www.fsc.org
MIX
Papier | Fördert gute Waldnutzung
FSC® C083411

Kolibri 360 GmbH
Ferdinand-Jühlke-Straße 7
99095 Erfurt, Deutschland
produktsicherheit@kolibri360.de